AIR FORCES

Air Forces
The Next Generation

Edited by

AMIT GUPTA

Howgate Publishing Limited

Copyright © 2020 Amit Gupta

First published in 2020 by
Howgate Publishing Limited
Station House
50 North Street
Havant
Hampshire
PO9 1QU
Email: info@howgatepublishing.com
Web: www.howgatepublishing.com

British Library Cataloguing-in-Publication Data
A catalogue record for this book is available from the British Library

ISBN 978-1-912440-08-5 (pbk)
ISBN 978-1-912440-09-2 (ebk - PDF)
ISBN 978-1-912440-14-6 (ebk - ePUB)

Amit Gupta has asserted his right under the Copyright, Designs and Patents Act, 1988, to be identified as the editor of this work.

The views expressed in this book are those of the individual authors and do not necessarily reflect official policy or position.

CONTENTS

NOTES ON CONTRIBUTORS

Mr Gregory Alegi

Gregory Alegi is a historian and journalist, with 35 years of experience in the aerospace, defense, and security fields. He teaches aerospace history at the Italian Air Force Academy (since 1998) and history of the Americas at LUISS University in Rome (since 2006). His extensive Italian and English writing includes *La storia dell'Aeronautica Militare: la nascita* (*The History of the Italian Air Force: The Birth*), winner of the ITAF Association Aerospace Book Prize in 2016. In 2015 he authored "The Italian Experience: Pivotal and Underestimated," a chapter in Karl Muller's *Precision and Purpose: Airpower in the Libyan Civil War*. He sits on the Board of Advisors of the Fondazione ICSA, for which he has authored research papers on the Italian defence budget and the F-35 program. As a journalist, he is the managing editor of the monthly *Aeronautica* and sits on the strategic committee of Airpress and the editorial board of *The Aviation Historian*.

Group Captain James R. Beldon MBE

Jim Beldon is a navigator with over 3,000 flying hours on the E-3D Sentry. He has served on operations in the Balkans, Iraq, Afghanistan, and across the globe. At home he has worked at the Ministry of Defence, Permanent Joint Headquarters, and the Joint Services Command and Staff College. He received an MBE in the 2007 New Year's Honours List, and in 2015 was appointed a Fellow of the Royal Aeronautical Society. He became the Director of Defence Studies (RAF) in March 2017.

Dr Donovan C. Chau

Donovan Chau is Director of Research Engagement in the Division of Research and Strategic Innovation at the University of West Florida. He works to connect students and faculty with industry and government—contributing directly to the needs of professions and society. Previously, Donovan Chau was associate professor of political science (tenured) at California State University, San Bernardino. Before academia, he worked in

Washington, D.C., first in the U.S. House of Representatives, then as a government contractor in homeland security and counterterrorism. He is a member of the U.S. Air Force Reserve. Donovan Chau earned a BA in literature and government from Claremont McKenna College, an MS in defense and strategic studies, and a PhD in politics and international relations from the University of Reading (U.K.).

Lt. Col. Luís E.P. Celles Cordeiro

Luís E. P. Celles Cordeiro is a graduate from the Brazilian Air Force Academy—class of 1999. Currently he is the chief of the training division at the Brazil Aeronautical Accidents Investigation and Prevention Center (CENIPA). Before assuming this position, he was an academic coordinator of two disciplines (military use of force and international humanitarian law) at the Squadron Officer College (Rio de Janeiro, Brazil) for four years and instructor of Air Force basic doctrine, law of armed conflict, and joint operations. Luís Celles Cordeiro has an MSc (political science and international relations) from the Brazilian Air University, also located in Rio de Janeiro. He is an active member of the Brazilian National Defense Studies Association and the Brazilian International Relations Association, participating in seminars and discussing the content of several articles (in Portuguese, English, and Spanish) that he has published. Areas of interest: cyber warfare, joint operations, international law of armed conflict, airpower doctrine, flight safety, and logistics.

Professor Peter W. Gray

Peter Gray retired from the Royal Air Force in June 2008, having reached the rank of Air Commodore (1*), and took up the position of Senior Research Fellow in Air Power Studies at the University of Birmingham in 2008. He retired from Birmingham in 2018 and is now Honorary Professor of Air Power Studies at the University of Wolverhampton. Peter Gray spent his early career as a navigator on the F4 Phantom aircraft and, more recently, commanded 101 Squadron flying VC10 K tanker aircraft. He has spent two staff tours in the personnel field, followed by a lengthy sojourn in the Cabinet Office, several appointments in the Ministry of Defence, and has served as Director of Defence Studies for the Royal Air Force. He is a graduate of the Higher Command and Staff Course and was assistant director on the 2001 programme. Peter Gray holds degrees from the universities of Dundee, London, Cambridge, and Birmingham (PhD). He is a Fellow of the RAeS and the Royal Historical Society.

Dr Amit Gupta

Amit Gupta is an associate professor in the USAF Air War College, Alabama. His writing has focused on arms production and weapons proliferation, South Asian and Australian security policies, diaspora politics, as well as popular culture and politics. Recently he has written on the U.S.–China rivalry and the impact of demography on U.S. foreign policy. His articles have appeared in *Orbis*, *Asian Survey*, *Security Dialogue*, *The Round Table*, *Mediterranean Quarterly*, *The International Journal of the History of Sport*, and *Sport in Society*. He is also the author or editor of seven books.

Dr Ron Gurantz

Ron Gurantz is assistant professor in the Department of Strategy at the Air War College. He joined the Air War College faculty in 2016, having previously taught political science at the University of California, Los Angeles. He holds a PhD and MA in political science from UCLA, and a BA in political economy from the University of California, Berkeley. His areas of specialization include crisis decision-making, deterrence, and air power. He is currently researching U.S. air power in the Vietnam War, with a focus on threat credibility and military effectiveness. His previous research has examined the credibility of deterrent threats and the influence of domestic politics on the course of international crises.

Associate Professor Lars Peder Haga

Lars Peder Haga studied Russian at the Norwegian Defence Intelligence College and the University of Bergen and holds a PhD in modern history from the Norwegian University of Science and Technology in Trondheim. Since 2011 he has been at the Royal Norwegian Air Force Academy as an associate professor and teaches modern history, international relations, and air power. His research interests include the development of Russian air power and Russian air operations. He is one of the two main authors of the new Norwegian air operations doctrine.

Ms Tiola Javadi

Tiola Javadi is a senior analyst at the Indonesia Programme, S. Rajaratnam School of International Studies (RSIS), Singapore. Her research mainly covers Indonesia's foreign policy, civil-military relations, and the modernization of the Indonesian military. She holds an MSc degree in strategic studies from RSIS.

Dr Sharad Joshi

Sharad Joshi is an associate professor in the Nonproliferation and Terrorism Studies Program (NPTS) in the Graduate School of International Policy and Management (GSIPM), at the Middlebury Institute of International Studies, Monterey. Sharad Joshi's research and teaching focuses on various facets of conflict, terrorism, and nonproliferation matters in South Asia. Sharad Joshi holds a PhD from the University of Pittsburgh's Graduate School of Public and International Affairs, and also served as a postdoctoral fellow at the Middlebury Institute's Center for Nonproliferation Studies. He has also been associated with the institute's Monterey Terrorism Research and Education Program (MonTREP) as a research associate and as interim director. He is also affiliated with Chatham House (U.K.) as an associate fellow of international security.

Dr Adhi Priamarizki

Adhi Priamarizki is a Visiting Fellow at the Indonesia Programme, S. Rajaratnam School of International Studies (RSIS), Singapore. His primary research interests are civil-military relations in Southeast Asia, Indonesia's defense policy, and Indonesian politics. He holds a doctoral degree in International Relations from Ritsumeikan University.

Mr Richard Shimooka

Richard Shimooka is a senior fellow at the Macdonald-Laurier Institute. He was a senior fellow at the Defence Management Studies Programme at Queen's University from 2007 to 2012, and a research fellow at the Conference of Defence Associations Institute from 2012 to 2017. His work covers a diverse array of topics, including western foreign and defence policy, modern airpower, and defence procurement.

Dr Guillaume de Syon

Guillaume de Syon teaches modern European history and the history of technology at Albright College in Reading, Pennsylvania. He is also a visiting scholar in history at Franklin & Marshall College. Guillaume de Syon is the author of *Zeppelin! Germany and the Airship 1900–1939*, and *Science & Technology in Modern European Life*, and of numerous articles and chapters on the history of technology and aviation.

Wing Commander Pete Wooding, MNZM

Pete Wooding has served with the Royal Australian Air Force since 2000 having previously served with the Royal New Zealand Air Force. He has undertaken appointments in aeroengineering, operations support, and attaché positions, as well as having deployed to Afghanistan and commanded 460 Squadron, RAAF. His educational background includes post-graduate study at the Australian Defence Force Academy, RAF College Cranwell, Royal Military College Canada, and the U.S. Air War College. WGCDR Wooding holds a Bachelor of Engineering degree and master degrees in science, defence studies, and strategic studies. He was appointed a Member of the New Zealand Order of Merit in 1997. He is currently a member of the teaching staff at the Australian Command and Staff College (Joint) in Canberra.

Dr Shang-su Wu

Shang-su Wu is a research fellow in the Regional Security Architecture Programme of the S. Rajaratnam School of International Studies (RSIS), Nanyang Technological University, Singapore. Before joining the RSIS, he taught in the National Defense University in Taiwan. He is the author of *The Defence Capabilities of Small States: Singapore and Taiwan's Responses to Strategic Desperation*. His research interests are military modernization, Taiwan issues, railways, and international relations. His articles have been published in the *Naval War College Review, Defence Studies, The Pacific Review, Asian Survey, Contemporary Southeast Asia*, among others.

Professor Xiaoming Zhang

Xiaoming Zhang is a professor in the Department of Strategy at the Air War College, Alabama, teaching strategy and subjects related China and

East Asia. He is the author of *Red Wings over the Yalu: China, the Soviet Union and the Air War in Korea* (2002) and *Deng Xiaoping's Long War: The Military Conflict between China and Vietnam, 1979–1991* (2015).

FOREWORD

Today, we are witnessing significant changes in the employment of air power as governments and militaries recognize the expanding potential of manned and unmanned aircraft to fulfill a wide array of political and military objectives. Just 20 years ago, young air cadets starting their careers were not trained to fly or work with unmanned vehicles. Now, however, air forces around the world recognize the value of integrating these vehicles into their force structures to play potentially pivotal roles in their operational strategies. At the same time, the role and importance of manned aircraft has continued to grow, fueling a technological race to achieve air mastery. Even with the rush to outfit air forces with fifth-generation aircraft, there is a simultaneous move to begin developing and deploying sixth-generation aircraft. These advanced airframes are transforming war fighting to the point that no major power can afford to be left behind in this race. Further, these new generations of aircraft will boast far more capable and lethal weaponry than their predecessors, particularly with the introduction of hypersonic missiles. The United States, China, and Russia are leading in the development of these weapons but other nations, such as India and France, are making rapid and impressive advances as well.

Not only are we witnessing significant technological advances, but we are also noting an expanded array of air power missions that reflect the changing security concerns of the international community. Air forces monitor the flow of refugees, serve as "first responders" to provide disaster relief (either singularly or in coordination with other nations' air forces as part of a coalition effort), track environmental changes, and monitor natural disasters such as volcanic eruptions, red tides, and wildfires as well as manmade disasters such as oil spills and the effects of overpopulation. The expanded capabilities of air power and the broader applications of those capabilities to security issues has propelled national air forces to a premier position as the "response force of choice" for governments dealing with a variety of current and emerging crises. As a result, this is both an extraordinarily exciting and an equally taxing time for the world's air forces. Air power advocates and practitioners must constantly reconsider their understanding as they seek to meld the tremendous array of capabilities at their disposal to an equally broad assortment of challenges.

In this context, I am particularly pleased to write this foreword, as this book explores these capabilities and challenges in both a national and regional context. It addresses the challenges and opportunities air forces face in both the western and nonwestern world. Readers may be surprised to learn how increasingly similar the challenges are, regardless of regional location or the size of the air power force structure. Clearly, budgetary constraints, the expanding array of missions, and the rapidly changing technological environment are complicating the decision-making calculus for air power advocates and practitioners. Air force leaders must make difficult choices. This book reveals how various air forces are seeking to overcome the challenges and embrace the opportunities they face. The breadth of the discussion and accompanying analysis makes this volume valuable to air power practitioners, advocates, and those charged with developing tomorrow's air power leaders.

Steven L. Kwast
Lieutenant General, USAF
Commander

ACKNOWLEDGMENTS

Like any work, this one could not have been done without the help and encouragement of a group of people. Every one of the contributors was low maintenance because they delivered their chapters on time and promptly made editorial changes. The contributors want to thank our publisher Kirstin Howgate for all the hard work that she has put in to make this volume into a first-rate product.

Special thanks to Lt. General Steven L. Kwast who at short notice read the manuscript and wrote a foreword. Also to Janet MacMillan and Madeline Koch of Editing Globally who provided such excellent copyediting and typesetting.

As always, I want to thank my mother, Ambika Gupta, for her encouragement and support of my academic efforts.

LIST OF ABBREVIATIONS

A2/AD	anti-access / area denial
AAA	anti-aircraft artillery
AAM	air-to-air missiles
AAR	air-to-air refueling
ACM	Air Chief Marshall)
ADF	Australian Defence Force
AESA	active electronic scanned array
AEW	airborne early warning
AEW&C	airborne early warning and control
AFC	Australian Flying Corps
AFDP	Air Force Development Plan
AGS	Alliance Ground Surveillance
AI	artificial intelligence
ALAT	aviation légère de l'armée de terre (land army light aviation, France)
AMISOM	African Union Mission in Somalia
AMRAAM	advanced medium-range air-to-air missiles
ANSV	Agenzia Nazionale Sicurezza Volo (National Flight Safety Agency, Italy)
ANZUS	Australia, New Zealand, United States Treaty
ASEAN	Association of Southeast Asian Nations
ASI	Agenzia Spaziale Italiana (Space Agency, Italy)
ASM	air-to-surface missile
ASRAAM	advanced short-range air-to-air missile
ASW	anti-submarine warfare
ATG	Air Task Group
ATO	Attack Option
ATS	Air Traffic Service
AWACS	airborne warning and control system

BAF	Brazilian Air Force (Força Aérea Brasileira)
BAFO	best and final offer
BAS	battlefield air strike
BDA	battle damage assessment
BKR	Badan Keamanan Rakyat (Indonesian People's Security Agency)
BLA	Baluchistan Liberation Army
BMD	ballistic missile defense
BMD	Ministério da Defesa (Brazilian Ministry of Defense)
BRI	Belt and Road Initiative
BRICS	Brazil, Russia, India, China, and South Africa
BVR	beyond visual range
C2	command and control
C3	communications, command, and control
CAMM-ER	Common Anti-air Modular Missile Extended Range
CAOC	Combined Air Operations Centre
CAS	Chief of Air Staff
CAST	Centre for Analysis of Strategies and Technologies
CFC	Comando Forze di Combattimento (Combat Forces Command, Italy)
CFSS	Comando Forze di Supporto e Speciali (Support and Special Forces Command, Italy)
CHOD	Capo di SMD (Chief of Defense Staff, Italy)
CIA	Central Intelligence Agency (United States)
CID	Ministry of Defense Innovation Center (Italy)

CIOC	Comando Interforze Operazioni Cibernetiche (Joint Cyber Operations Command, Italy)
CLD	Comando Logistico della Difesa (Defense Logistics Command, Italy)
CMC	Central Military Commission
CNES	Centre national d'études spatiales (National Center for Space Studies, France)
COA	Comando Operazioni Aeree (Air Operations Center, Italy)
COAC	Combined Air Operations Centre
COFS	Comando Operativo delle Forze Speciali (Special Forces Operational Command, Italy)
COI	Operativo di vertice Interforze (Joint Operations Command, Italy)
COM	Codice Ordinamento Militare (Code of Military Organization, Italy)
COMAER	Comando da Aeronáutica (Aeronautical Command, Brazil)
COIN	counterinsurgency
COMLOG	Comando Logistico (Logistics Command, Italy)
CPEC	China-Pakistan Economic Corridor
CPGS	conventional prompt global strike
CSA	Comando della Squadra Aerea (Force Command, Italy)
CSF	Coalition Support Funds
CSO	composante spatiale optique (optical space component, France)
CSV	Centro Sperimentale Volo (Flight Test Center, Italy)
DACCC	Deployable Air Command and Control Centre

DARPA	Defense Advanced Research Program Agency
DCDC	Development, Doctrine, and Concepts Centre
DEW	directed energy weapon
D Def S	Director of Defence Studies
DNA	Direttore Nazionale degli Armamenti (National Director for Armaments, Italy)
DNAL	Direttore Nazionale degli Armamenti e della Logistica (National Director for Armaments and Logistics, Italy)
DPP	Democratic Progressive Party (Taiwan)
DWP	Defence White Papers
EATC	European Air Transport Command
ECM	electronic countermeasures
ECR	electronic combat and reconnaissance
EDA	European Defence Agency
EEZ	exclusive economic zone
EDF	European Defense Fund
EH	employment hypotheses
ELINT	Electronic Signals Intelligence
ENAC	Ente Nazionale Aviazione Civile (National Civil Aviation Agency, Italy)
END	Estratégia Nacional de Defesa (National Defense Strategy, Brazil)
EODAS	Earth Observation Data Archiving Service
EU	European Union
EURAC	European Air Chiefs
EuroMALE	European medium altitude long-endurance
EW	electronic warfare
FAB	Força Aérea Brasileira (Brazilian Air Force)
FACO	final assembly and check-out
FATA	Federally Administered Tribal Areas

FCAS	future combat air system
FMF	foreign military financing
FPDA	Five Power Defence Arrangement
FPWP	Foreign Policy White Paper (Australia)
GCI	Ground Control Intercept
GES	geostationary satellites
GDP	gross domestic product
GMF	Global Maritime Fulcrum
GPS	global positioning network
GST	Global Strategic Trends
HAL	Hindustan Aeronautics Limited
HMD	head-mounted display
IAF	Indian Air Force
IAI	Istituto Affari Internazionali
ICEFCOM	Information, Communications, and Electronic Force Command (Taiwan)
IDF	Indigenous Defensive Fighters
IDS	interdictor/strike
IED	improvised explosive device
IMET	International Military Education Training
IMF	International Monetary Fund
IOC	initial operational capacity
IS	Islamic State
ISIL	Islamic State of Iraq and the Levant
ISIS	Islamic State of Iraq and Syria
ISR	intelligence, surveillance, and reconnaissance
ISTAR	Intelligence Surveillance Target Acquisition and Reconnaissance
IT	information technology
ITAF	Italian Air Force (Aeronautica Militare, AM)
JAMMAS	Joint Airborne Multimission Multisensor Systems
JAPCC	Joint Air Power Competence Center
JDCC	Joint Doctrine and Concepts Centre

JDP	Joint Doctrine Publication
JeM	Jaish-e-Mohammed
JFAC	Joint Force Air Command
KAF	Kenya Air Force
KDF	Kenya Defense Force
KFX/IFX	Korean Fighter Xperiment/Indonesia Fighter Xperiment
KMT	Kuomintang (Chinese Nationalist Party, Taiwan)
KOHANUDNAS	Komando Pertahanan Udara Nasional (National Air Force Command, Indonesia)
KOOPSAU	Komando Operasi Angkatan Udara (Air Force Operational Command, Indonesia)
LAAD	Latin American Aero and Defense conference
LAC	line of actual control
LCA	light combat aircraft
LIFT	lead-in fighter trainer
LOC	line of control
LPM	Loi de programmation militaire (Military Programming Act, France)
MALE	medium altitude long endurance
MANPADS	man-portable air defense systems
MEADS	medium extended air defense system
MEF	minimum essential force
MICA	Missile d'interception de Combat et d'Autodéfense (Interception Missile Combat and Self Defense, France)
MIRV	multiple independent re-entry vehicles
MISE	Ministry for Industry and Economic Development
MLV	microsatellite launch vehicles
MMRCA	medium multi-role combat aircraft
MND	Ministry of National Defence (Taiwan and ROC)

MOD	Ministry of Defence (U.K.)
MOOTW	military operations other than war
MPA	maritime patrol aircraft
MRBM	medium-range ballistic missiles
MRTT	multi-role tanker transport
MTOW	maximum take-off weight
MUIR	Ministry of Education, University and Research
NATO	North Atlantic Treaty Organization
NCA	National Command Authority (Pakistan)
NGS	non-geostationary satellite
NGWS	Next Generation Weapon System
NHS	National Health Service (U.K.)
NSG	Nuclear Suppliers Group
OLUO	Air Offensive Fight Operation
Opsdukud	Air Support Operation
Opshanud	Air Defense Operation
Opsinfo	Information Operation
OSUS	Strategic Air Attack Operation
OCCAR	Organisation for Joint Armament Cooperation
PAC	Patriot Advanced Capacity
PAF	Pakistan Air Force
PDN	National Policy of Defense (Brazil)
PESA	passive electronically scanned array
PESCO	Permanent Structured Cooperation
PGM	precision guided munitions
PIRA	Provisional Irish Republican Army
PAK-DA	Perspektivnyi aviatsionnyi kompleks dalnei aviatsii (strategic bomber, Russia)
PKI	Partai Komunis Indonesia (Indonesian Communist Party)
PLA	People's Liberation Army
PLAAF	People's Liberation Army Air Force
PLAN	People's Liberation Army Navy
PLANAF	People's Liberation Army Navy Air Force

PND	Política Nacional de Defesa (National Defense Policy, Brazil)
POD	Preliminary Operational Demand
POL	petrol, oil, and lubricant
PRC	People's Republic of China
PT DI	PT Dirgantara Indonesia (Indonesian Aerospace)
PTLIR	Preliminary Technical, Logistic and Industrial Requirements
RA	Regia Aeronautica (Royal Air Force, Italy)
RAF	Royal Air Force
RAAF	Royal Australian Air Force
R&D	research and development
RDS	Rapid Deployment Squadron (Kenya)
RMA	Revolution in Military Affairs
RN	Royal Navy
ROC	Republic of China
ROCA	Republic of China Army
ROCAF	Republic of China Air Force
ROCN	Republic of China Navy
ROP	request of proposal
RSAF	Republic of Singapore Air Force
RSIS	S. Rajaratnam School of International Studies
SAM	surface-to-air missile
SAMP/T	Sol-Air Moyenne Portée/Terrestre (ground-based, road-mobile tactical ballistic missile defense system)
SAR	sea-air-rescue or search and rescue
SATCOM	satellite communication
SDI	strategic defence interests
SDO	strategic defence objectives
SDR	Strategic Defence Review (U.K.)
SDSR	Strategic Defence and Security Review (U.K.)

SEAD   suppression of enemy air defense
SEATO   South East Asian Treaty Organization
SEF   standard essential force
SFC   Strategic Forces Command
SG   Segretariato Nazionale della Difesa
                    (National Secretariat for Defense, Italy)
SIPRI   Stockholm International Peace Research
                    Institute
SIT   industry and technology strategy
SIVAM   Amazon Surveillance System (Brazil)
SLBM   submarine-launched ballistic missile
SMA   Stato Maggiore dell'Aeronautica (Air
                    Force General Staff, Italy)
SMD   Stato Maggiore della Difesa (Defense
                    Staff, Italy)
SOC   Space Operations Center
SPD   Strategic Plans Division
SRBM   short- range ballistic missiles
SSF   Strategic Support Force
SSSP   Space Systems Strategic Program (Brazil)
STOL   short takeoff and landing
TK   Tien-Kung
TNI   Tentara Nasional Indonesia (Indonesian
                    Armed Forces)
TNI AD   Tentara Nasional Indonesia Angkatan
                    Darat (Indonesian Army)
TNI AU   Tentara Nasional Indonesia Angkatan
                    Udara (Indonesian Air Force)
TVC   thrust vector nozzle
UAS   unmanned aircraft system
UAV   unmanned aerial vehicle
UCAV   unmanned combat aerial vehicle
UGS   Ufficio Generale Spazio (Space General
                    Office, Italy)
UN   United Nations
UOR   urgent operational requirements

USAF	U.S. Air Force
VKS	Vozdushno-kosmicheskie sily (Air and Space Forces, Russia)
VTOL	vertical takeoff and landing
V/STOL	vertical or short takeoff and landing
VVKO	Voiska vozdushno-kosmicheskoi oborony (Aerospace Defense Forces, Russia)
VVS	Voenno-vozdushnye sily (Air Force, Russia)
WGS	wideband global satcom

INTRODUCTION
Amit Gupta

We live in an era where significant changes are happening in the use of air power even while financial considerations are constraining the development of future air assets. On the one hand, we are witnessing air forces being used in an increasing number of military and nonmilitary operations to ensure national and international security. On the other, nations are increasingly seeking to reduce their air forces because they need to commit scarce governmental resources to social welfare programs like health care, pensions, and education. As a consequence, we are in a classic guns versus butter dilemma that is shaping the development of force structures, missions, and the employment of national air power. This volume discusses this central tension, as well as other issues in a cross-national perspective. It does so by examining the missions, technologies, force structures, and modernization plans of both western and nonwestern countries, and what emerges from the various chapters is that regardless of geographical location, form of government, and threat scenarios, the challenges faced by air forces are remarkably similar.

Air Forces and Air Power in the Modern Age

Today, there are several trends that influence, and will continue to influence the development of air forces and air power. First, we are witnessing the blurring of civilian and military lines regarding the use of aircraft to ensure domestic and international security. In the United States, as an unintended consequence of the war on terror, the Central Intelligence Agency created its own air force and it has been one of the most effective in the war on terror. Increasingly, for border security, civilian and military assets, often including unmanned aircraft, are employed as countries recognize the utility of such platforms. Border protection forces are likely to employ more such assets as they face the challenges of illegal immigration, narco-trafficking, and human trafficking.

Second, is the continued demand for the modernization of forces and we see the clash of technologies and technological choices among rival nations. China, as Zhang's chapter shows, is building an air force with an advanced missile capability to make it difficult for the United States Air Force and the United States Navy to project power in the western Pacific. Instead, because of the threat posed by Chinese anti-access/area-denial systems, the United States has been pushed back to the second island chain in the Pacific and forced to make the northern Australian city of Darwin the place through which to rotate Marine detachments. Indeed, one of the major elements of the air power arms race is the quest by several nations—particularly the United States, China, and Russia—to develop hypersonic missiles that will take on the current mission of combat aircraft to penetrate defenses at long distances.

There is also the more conventional race between Russia and China, as described in the Haga and Zhang chapters, to develop fifth-generation combat aircraft that can match the American F-22 and F-35 fighters. France and Germany are going one step further to build a sixth-generation fighter that will take air power into the latter half of the twenty-first century.

Third, the era of techno-nationalism that marked the development of combat aircraft is coming to an end as nations, beset with ballooning developmental costs, seek partnerships and to integrate foreign components into their systems. In the past, fighter aircraft were viewed as symbols of national technological prowess as well as symbols of autonomy, but the costs of such independent developments are now prohibitive. It is also difficult to develop every part of an aircraft domestically since some of these technologies are both complex and demanding. Therefore the Chinese have imported engines from Russia and would love the European Union to rescind its arms embargo against Beijing so that China could itself avail of advanced European technologies. France, which believed in strategic autonomy and thus, at a time when most European countries were collaborating on aircraft production, decided to expend significant resources to build the Rafale fighter, now sees the advantages of collaboration.

Country	Army 1972	Army 2017	Aircraft 1972	Aircraft 2017
Britain	180,458	85,600	500	254
France	328,000	111,650	500	281
Germany	327,000	111,650	459	235
Italy	306,600	102,200	320	244

Table 1   Manpower and Aircraft Cuts in the Major NATO Nations 1972–2017
Source: The Military Balance [IISS London], 1972 and 2017 editions.

Fourth, is the larger question of what will be the nature of air power and air forces in the future—will the emphasis be on manned or unmanned platforms? The versatility of unmanned aircraft has been proven in a series of conflicts around the world but air forces remain the domain of those who would like a pilot in the cockpit. This struggle may ultimately be decided by the budgets of nations who seek to deter threats at an affordable cost.

One of the key challenges for air power around the world is not the existence of threats that need to be deterred but, rather, the shrinking budgets for military expenditure as countries invest in human capital rather than weapons systems.

In the aftermath of the Cold War, with the collapse of the Warsaw Pact, the North Atlantic Treaty Organization (NATO) countries, as well as neutral nations like Sweden, decided to take advantage of the peace dividend and rapidly reduce their armed forces. A united Germany, which in the Cold War could have fielded thousands of Leopard tanks, shrunk its armored force down to 311 tanks (although how many of these remain operational is a matter of conjecture). The Netherlands, in 2011, went one step further and actually disbanded their armored regiments, only to later lease German Leopards and integrate their forces with the Bundeswehr. Sweden, which at the height of the Cold War could put nearly 800,000 active and reserve troops in the field, was by the 2015 only able to sustain 14,000 active troops and 26,000 reserves. The table above shows how deep the cuts have been in the four major European nations.

Admittedly, one has to be cautious in making comparisons between aircraft of the 1970s and contemporary aircraft because the latter are far more capable, and in some cases carry greater payloads, but the shrink-ing of squadrons is a matter of concern for any air force as it means loss of manpower and the ability to plan and sustain large-scale operations. More significantly, these air forces do not have an unlimited supply of munitions—especially precision guided ones—and this was brought home in the Libya crisis when the NATO air forces ran out of precision guided munitions and the United States had to supply weapons to the Belgians, Norwegians, and Danes who flew F-16s and, therefore, could use U.S. munitions. After the successful Libyan regime change, the NATO nations met in Chicago and promised to build up their inventories, but the ability to sustain another Libya-like campaign remains questionable.

It is unlikely that western democracies will ever be able to bring their force levels back to those of the glory days of the Cold War. Most NATO nations fall below the two percent of gross domestic product (GDP) expenditure on defense that they are supposed to maintain. What quite a few have done is meet the NATO requirement for 20 percent of defense expenditure to be on acquiring new equipment and on research and devel-opment. This is, however, far too little. It is likely that these countries will face greater cuts to their forces as politicians reallocate money to press-ing social expenditures like pensions, health care, and education. Canada walked out of the F-35 program, preferring to buy mothballed F-18s from Australia, because it balked at the escalating costs associated with the F-35. Italy, has complained about the cost of the F-35 and its defense minister, Elizabetta Trenta, stated that the country was unlikely to buy more planes beyond those already committed to.

The gun versus butter dilemma is no longer restricted to western liberal and social democracies. In the nonwestern world, politicians have begun to make similar choices as they too realize that the path to reelection is investment in human capital through improved education and health care. Brazil, as discussed in the chapter by Celles, had to cut the number of Gripen it initially intended to buy. In India, successive governments have kept the defense budget around 1.5–1.6 percent of GDP and focused on social programs. Consequently, the centrist Congress party starved the military and brought in a series of welfare schemes for the impoverished. The recently reelected prime minister, Narendra Modi, heads a right-wing coalition that wants a more muscular foreign policy, but at the same time has sought to improve national welfare programs and recently decided to provide health coverage of 500,000 rupees (approximately $7,200) for each citizen. Not surprisingly, defense expenditure has taken a hit as the Modi government reduced the order for 126 Rafales to 36 for cost reasons. India has now asked for tenders for another 114 combat aircraft but, again, it is far more likely that it will make a smaller purchase.

The other factor that is likely to shape future generations of air forces is the renewed emergence of internal threats to a nation's security, ranging from terrorism to insurgencies to nontraditional security challenges. Taking the last issue first, environmental degradation and environmental disasters increasingly concern nations as they are likely to have devastating effects on a national economy and the standing of a nation with their neighbors. For Brazil, the environmental degradation of the Amazon poses long-term economic- and climate-related challenges, and the effective way to monitor these is by using aerial vehicles. In fact, Brazil's Amazon surveillance system (SIVAM) uses both land- and air-based sensors to track the degradation of the Amazon. In Norway, security concerns now range from needing to monitor potential oil spills to controlling the flow of illegal human trafficking.

Terrorism and insurgencies remain major problems for nations around the world and we have seen countries increasingly turn to appropriate platforms to counter such threats. Iraq, Afghanistan, and Nigeria have, therefore, turned to turboprop aircraft rather than fast jets to counter these threats. Other countries are likely to follow suit, the most notable being the United States, which is now considering the purchase of either the Super Tucano or the AT-6 Texan to provide this type of capability to the Air Force's special operations.

Lastly, there is the growing debate on the role of unmanned vehicles in the air forces of the future. When unmanned vehicles or drones entered service they were met with institutional resistance because they were seen as being slow-moving aircraft, unable to meet the requirements of the modern battlefield, and most importantly, they threatened the role of the pilot in cockpit of the future. Yet given the role drones are playing in surveillance, intelligence gathering, and recognizance, as well as the role they will play with naval air forces, the United States is already working with unmanned aerial vehicles (UAVs) that are launched from its aircraft carriers. What will happen in the future is the emergence of unmanned combat aerial vehicles (UCAVs). As the costs of fighter aircraft escalate, the UCAVs become a cheaper option, especially when it comes to the launch of standoff weapons like the long-range standoff cruise missile that the United States is seeking to develop to penetrate China's increasingly formidable air defenses.

The other problem that UCAVs resolve is that of losing pilots. We live in an age where public opinion is increasingly casualty averse and the downing of aircraft and the capture of pilots leads to a national media frenzy. We saw this most recently in February 2019 when an Indian MiG-21 pilot was shot down by the Pakistani Air Force, whipping up nationalist passions in India. When Donald Trump recalled aircraft that were launched to conduct a strike on Iran, he stated that the fact that Iranian forces had shot down an unmanned drone meant that the response had to be one that did not lead to a disproportionate loss of life on the Iranian side. However, as the American president said, had a manned American aircraft been shot down with a loss of lives, it would have been a very different matter. What role UCAVs and UAVs will play in the future will be one of the intriguing issues that divide air forces.

The various countries covered in this book face all of the compulsions and problems mentioned above and they are seeking solutions that are similar as well as unique given their particular national threat environments, their coalitional partnerships, and their perception of their role in world affairs. Nations seeking a greater role in world affairs are not only building up their combat capabilities, but also their ability to project power over long distances using heavy-lift transport capabilities. Others are seeking smaller but more potent forces to meet their security challenges. This book examines these efforts and we hope it is valuable to the reader seeking to make a comparative assessment of air power.

1 UNITED STATES Air Power Interventionism in a New Era

Ron Gurantz

In the wake of the 1991 Gulf War, many observers declared that the United States was at the leading edge of a revolution in military affairs. Utilizing a combination of precision-guided munitions, stealth, and advanced navigation and surveillance, the United States had acquired the ability to strike targets with extraordinary precision from long range.[1] The United States later added unmanned aerial vehicles (UAVs) to this arsenal. From the Kosovo War and the no-fly zones over Iraq in the 1990s to the war against the Islamic State of Iraq and Syria (ISIS) and the continuing drone strikes in Pakistan today, the United States has used these capabilities to destroy enemy armed forces, punish recalcitrant regimes, and assassinate individual terrorists without having to commit ground troops. They became the backbone of a pattern of behavior that I label "air power interventionism": repeatedly conducting military operations against weaker adversaries by relying primarily or entirely on air power.

Air power interventionism has become one of the defining features of American foreign policy in the post–Cold War era. However, the missions that most typically involve air power interventions, such as counterterrorism and humanitarian intervention, have been deemphasized in recent national security documents. Instead, officials have placed "great power competition" with China and Russia at the top of the U.S.'s list of national

---

1   Benjamin S. Lambeth, *The Transformation of American Airpower* (Ithaca, NY: Cornell University Press, 2000); Thomas G. Mahnken, *Technology and the American Way of War Since 1945* (New York: Columbia University Press, 2008). These capabilities are also enabled by other factors, such as the U.S.'s superiority in materiel production, logistics, and training.

security priorities.[2] This has resulted in a reallocation of resources toward preparing for warfare with technologically advanced adversaries and away from warfare with minor powers and nonstate actors.[3]

Will this new era of great power competition result in the decline of air power interventionism? In this chapter, I attempt to shed light on this question by evaluating the effects of great power competition on the U.S.'s foreign policy calculations regarding military interventions. I take two separate approaches to this inquiry. First, I examine the effects of great power competition on interventionism in the U.S.'s last great power rivalry, the Cold War. I show that great power competition, while placing some restraints on U.S. foreign policy, did not eliminate U.S. interests in other parts of the world and sometimes created new reasons to intervene. Second, I evaluate how new technologies being developed for great power warfare are likely to affect the U.S.'s propensity to intervene against less-advanced adversaries. I show that there are many reasons to believe that they will increase the willingness of the United States to conduct military interventions using air power.

My findings lead me to conclude that the competition with China and Russia will not lead to the end of air power interventionism. Despite a reallocation of resources toward great power competition, American leaders will continue to face strong incentives to conduct military operations against weaker powers and to rely on air power as a relatively cheap and effective way to do so. While the future of U.S. foreign policy is impossible to predict with any certainty, any claim that the United States is getting out of the business of air power interventions should be met with a healthy dose of skepticism. The United States defense establishment, despite its current focus on "high-end" combat against China and Russia, should not ignore the need to be prepared for "low-end" interventions against less-advanced adversaries. Nor should those adversaries be tempted by the belief that antagonistic policies may become less likely to provoke a U.S. military response.

This chapter proceeds as follows. First, I review the history of air power interventionism in the post–Cold War era. Then, I review changes in

---

2   Idrees Ali, "U.S. Military Puts 'Great Power Competition' at Heart of Strategy: Mattis." *Reuters*, Jan. 19, 2018.

3   *National Security Strategy of the United States* (Washington, DC: President of the United States, 2017); *National Defense Strategy of the United States* (Washington, DC: Department of Defense, 2018).

the international environment and in U.S. defense policy that suggest that air power interventionism may decline in coming years. Next, I evaluate how great power rivalry influenced U.S. interventionism in the Cold War. Last, I evaluate how experimental military technologies may influence the willingness of the United States to conduct air power interventions. I conclude with thoughts about the implications of this argument for the future of U.S. military and foreign policy.

Air Power Interventionism and the Return of Great Power Competition

Since the fall of the Berlin Wall, the United States has fought in three major wars and repeatedly conducted smaller military interventions and operations. Art and Greenhill identify 11 instances of conventional military intervention or military threats since 1991: "Iraq (1990–1991), Somalia (1992–1993), Haiti (1993), North Korea (1994), Bosnia (1995), Kosovo and Serbia (1998–1999), Afghanistan (2001), Iraq again (2003–2011), Libya (2011), Syria (2014–), and Iraq yet again (2015–)."[4] Their list doesn't identify less conventional military actions such as the enforcement of no-fly zones against Iraq (1991–2003), air strikes against al-Qaeda in Sudan and Afghanistan (1998), and continuing drone operations in Pakistan, Yemen, and Somalia (2004–).

In most of these operations, air power has played the primary role. In Libya and recent operations in Iraq, U.S. air forces supported allied ground forces by striking fielded forces and other targets such as command and control and economic infrastructure.[5] In Serbia and Syria, the United States used air strikes to destroy military and civilian targets to punish the governments for their military actions.[6] The enforcement of no-fly zones in Iraq

---

4   Robert J. Art and Kelly M. Greenhill, "Coercion: An Analytical Overview," in Kelly M. Greenhill and Peter Krause, eds., *Coercion* (New York: Oxford University Press, 2018), 3. All involved combat actions, except for Haiti and North Korea.

5   Christopher S. Chivvis, *Toppling Qaddafi* (Cambridge, U.K.: Cambridge University Press, 2014); "Special Report: Inherent Resolve" (Washington, DC: U.S. Department of Defense, September 2018), <https://dod.defense.gov/OIR/>.

6   Dag Henrikson, *NATO's Gamble* (Annapolis, MD: Naval Institute Press, 2007); Michael R. Gordon, Helene Cooper, and Michael D. Shear, "U.S. Launches Missiles into Syria," *New York Times*, A1, Apr. 6, 2017.

combined all of these types of actions with a long-term effort to keep Iraqi air forces grounded.[7] In Afghanistan, Pakistan, Yemen, and elsewhere, the United States has used UAVs to track and kill terrorist suspects.[8] Indeed, in the majority of post–Cold War military operations, the United States has relied on air power to the exclusion of other types of military force.

Why has the United States so consistently turned to air-only interventions? The U.S.'s technological advantage against these adversaries enables it to operate in foreign air space almost without restriction.[9] As a result, air power has offered the promise of accomplishing foreign policy objectives at a low risk to American lives. Air-only operations resulted in a Serbian withdrawal from Kosovo and helped Libyan rebels overthrow that country's government, and the United States lost no lives in combat in either operation.[10] The precision with which the United States can deliver its munitions also limits civilian casualties. Of course not everyone agrees that air power has delivered on its promise. Critics argue it encourages excessive interventionism and rarely delivers a cheap and easy victory.[11] The most recent *National Security Strategy* echoes these criticisms, rejecting the notion that wars can "be fought and won quickly, from stand-off distances and with minimal casualties."[12] Still, presidents have so far been unable to resist the opportunity that air power provides to conduct relatively safe and inexpensive military operations.

As great power competition reemerges, the changing national security environment may demand that the United States refocus its efforts. Air power interventionism has focused almost entirely on what Betts labels

---

7   Jon B. Alterman, "Coercive Diplomacy Against Iraq, 1990–98," in Robert J. Art and Patrick M. Cronin, eds., *The United States and Coercive Diplomacy* (Washington, DC: United States Institute of Peace, 2003).

8   Brian Glyn Williams, *Predators: The CIA's Drone War on al Qaeda* (Washington, DC: Potomac Books, 2013).

9   Barry Posen, "Command of the Commons," *International Security*, Vol. 28, No. 1 (Summer 2003).

10 Daniel L. Byman and Matthew C. Waxman, "Kosovo and the Great Airpower Debate," *International Security*, Vol. 24, No. 4 (Spring 2000); Chivvis, *Toppling Qaddafi*.

11 For a prominent critique, see Eliot Cohen, "The Mystique of U.S. Air Power," *Foreign Affairs*, Vol. 73, No. 1 (Jan.–Feb., 1994). Also see Stephen Biddle, "The Libyan Dilemma: The Limits of Air Power," *Washington Post*, Mar. 25, 2011.

12 *National Security Strategy*, p. 27. These criticisms are by the former National Security Advisor in H.R. McMaster, "On War: Lessons to be Learned," *Survival*, Vol. 50, No. 1 (Feb.–Mar. 2008).

as "second- and third-order challenges: rogue states, medium-sized wars, terrorists, peacekeeping operations, and humanitarian relief."[13] However, both China and Russia have emerged as what Betts labels "first-order dangers." China and Russia can pose serious military threats to U.S. allies and interests and can inflict catastrophic damage on the United States itself. A combination of long-standing tensions and militarily assertive policies by both countries have created the possibility in the minds of American officials of having to eventually confront them with military force.[14]

In late 2017 and early 2018, the Trump administration released a series of documents that explicitly labeled the competition with China and Russia as the "central challenge to US prosperity and security" and the country's top national security priority.[15] This has not involved a wholesale rejection of previous priorities, as North Korea, Iran, and terrorism are also listed as major threats.[16] However, official statements make clear that these threats are being downgraded in importance. According to former Secretary of Defense Jim Mattis, "Great power competition, not terrorism, is now the primary focus of US national security."[17] Secretary of the Air Force Heather Wilson has been more explicit about relegating these missions, saying that "the only way that we're going to restore enough time to prepare for a high-end fight" is "by reducing our commitment to the fight against violent extremism."[18] Furthermore, missions such as deterring local aggression, nation-building, peacekeeping, and humanitarian intervention are largely absent from the documents.

Defense officials appear to assume that resource constraints will force the United States to turn away from peripheral interventions as it refocuses on great power competition. Military acquisitions, planning, and training are already shifting in this direction. Both China and Russia are developing

---

13 Richard K. Betts, "Pick Your Battles," *Foreign Affairs*, Vol. 93, No. 6 (Nov./Dec. 2014), <https://www.foreignaffairs.com/articles/united-states/2014-10-20/pick-your-battles>.

14 Ronald O'Rourke, "A Shift in the International Security Environment: Potential Implications for Defense Issues for Congress," *Congressional Research Service* Report R43838, October 24, 2018.

15 *National Security Strategy* 2017; *National Defense Strategy* 2018; *Nuclear Posture Review* (Washington DC: Department of Defense, 2018).

16 *National Security Strategy* 2017.

17 Ali, "U.S. Military Puts 'Great Power competition' at Heart of Strategy: Mattis."

18 Stephen Losey, "Secretary Wilson to Lay Out 'Air Force We Need' at AFA," *Air Force Times*, Sep. 10, 2018.

long-range, precision-strike weapons to create an anti-access (A2)/area denial (AD) capability, which could deny the United States the ability to easily project force into regions near these countries by destroying bases, surface ships, and aircraft.[19] The U.S. military is acquiring capabilities to counter these systems, and this focus may squeeze out cheaper capabilities that are meant for low-intensity conflicts. Since as early as 2015, the U.S. Air Force has been explicit about wanting a "high-end focused force" to counter these threats, declaring that "we will not posture for extended stabilization operations, nor will low-intensity operations be the primary focus of our capabilities development."[20]

This is clear from Air Force planning and acquisitions. Two of its top three near-term procurement priorities, the F-35 Joint Strike Fighter and the B-21 Raider Long-Range Strike Bomber, were designed to operate in the A2/AD environment.[21] The F-35A is meant to replace the A-10 and F-16, which offer relatively affordable options for striking ground targets but are vulnerable to modern air defenses.[22] The replacement of these jets with one that is useful for high-end warfare but far more costly to use in low-end warfare suggests that the Air Force is not anticipating many low-end missions. The Air Force has been planning for a "light attack aircraft" that would offer an affordable alternative to operate in permissive environments, but the number would surely be small and development is currently on hiatus.[23] The Air Force is also focusing heavily on developing its cyber and space capabilities, anticipating that those domains will become part of high-end warfare.[24] In the longer term, it is planning for

---

19 *Joint Operational Access Concept* (Washington, DC: Department of Defense, 2012).

20 *Air Force Strategic Master Plan* (Washington, DC: Department of Defense, May 2015).

21 Jeremiah Gertler, "Air Force B-21 Raider Long-Range Strike Bomber," *Congressional Research Service* Report R44463, Oct. 12, 2018; David A. Deptula, "Beyond the Bomber: The New Long-Range Sensor-Shooter Aircraft and United States National Security," *Mitchell Institute for Aerospace Studies* Report, 2015. The third priority is the KC-46A Tanker, which is meant to replace an aging tanker fleet.

22 Jeremiah Gertler, "F-35 Joint Strike Fighter (JSF) Program," *Congressional Research Service* Report RL30563, Apr. 23, 2018. Similarly, the F-35B will replace the Marine Corps' F/A-18 and AV-8B and the F-35C will replace the Navy's F/A-18.

23 Brian Everstine, "Light Attack Basing, Deployment Model, and Command Assignment Determined by Size of Buy," *Air Force Magazine*, Dec. 7, 2018; Valerie Insinna, "US Air Force's Plan to Launch Light-Attack Aircraft Competition Is Now deferred Indefinitely," *Defense News*, Jan. 18, 2019.

24 *America's Air Force* (Washington, DC: Department of Defense, Jul. 2014).

other high-technology platforms and researching a suite of advanced technologies, such as hypersonic missiles, nanotechnology, lasers, unmanned aerial systems, and autonomous systems, in the hope of reclaiming a significant technological edge over China and Russia.[25]

On the other hand, shifts in defense policy are only one element in the calculations that go into the development of foreign policy, and they have not yet seemed to have influenced presidential calculations. Presidents Obama and Trump promised to focus on great powers and exercise restraint in using force, but have repeatedly initiated or escalated the use of air power in peripheral conflicts. Obama escalated the program of targeted killings in Pakistan using UAVs, participated in the North Atlantic Treaty Organization (NATO) intervention in Libya, and initiated the air campaign against ISIS. Trump continued the war against ISIS and launched airstrikes against Syria in response to chemical weapons use. More broadly, the United States still maintains many global commitments. Despite President Trump's rhetoric and apparent willingness to exit international agreements, even many worried observers have acknowledged that the basic features of U.S. foreign policy—global engagement, a commitment to alliances, and a willingness to use military force—have shown little change so far.[26]

Interventionism in an Era of Great Power Competition

There does appear to be a contradiction between the announced reprioritization toward great power competition and continued interventions in peripheral areas. To better understand how this contradiction may be

---

25 Ibid; *Air Superiority Flight Plan 2030* (Washington, DC: Department of Defense, May 2016). Other high technology platforms include a new advanced jet fighter and a "stand-off arsenal plane" meant to fire large amounts of munitions from long distances. Beyond the Air Force, development is focusing on advanced warships such as the Virginia-class attack submarine and the DDG-51 Aegis destroyer, and advanced weapons such as railguns and hypervelocity projectiles. Training and organization is also shifting in every service. See O'Rourke, "A Shift in the International Security Environment: Potential Implications for Defense Issues for Congress"; Otto Kreisher, "Marines Shifting Focus to Readying for the 'High End' Fight," *Seapower Magazine*, Jun. 4, 2018.

26 Richard N. Haass, "Where to Go From Here: Rebooting American Foreign Policy," *Foreign Affairs*, Vol. 96, No. 4 (Jul./Aug. 2017); Eliot A. Cohen, "Trump's Lucky Year: Why the Chaos Can't Last," *Foreign Affairs*, Vol. 97, No. 2 (Mar./Apr. 2018); Emma Ashford and Joshua R. Itzkowitz Shifrinson, "Trump's National Security Strategy: A Critic's Dream," *Texas National Security Review*, Vol. 1, No. 2 (Feb. 2018).

resolved, I seek to understand how great power competition is likely to influence future foreign policy calculations around interventionism. I do this by first examining how great power competition influenced U.S. interventionism in the Cold War. I show that great power competition both restrained and encouraged interventionism in different ways, as the United States sought to avoid provoking major war but also to contain Communist expansionism. Neither of these effects dominated the other, and in many cases, they had little effect either way. These findings suggest that air power interventionism is unlikely to disappear with the return of great power conflict.

U.S. foreign policy during the Cold War was dominated by great power competition with the Soviet Union and, to a lesser extent, China. While the Soviet Union could never muster more than 45 percent of the United States' gross domestic product (GDP) and was regularly behind in the technological race, it developed a military force that could conquer large swaths of Eurasia and a nuclear arsenal that could devastate American society.[27] Preparation for direct conflict with these powers consumed substantial attention and resources, and the United States found itself in repeated crises and confrontations with both countries.

Nevertheless, the United States also found itself conducting many peripheral interventions during the Cold War. It landed troops in the Dominican Republic in 1965 and Grenada in 1983, conducting the same number of direct military interventions in that region as it did after the Cold War.[28] It conducted several military operations in the Middle East, landing troops in Lebanon in 1958 and 1982, clashing repeatedly with Libyan air and naval forces and launching punitive air strikes on Libya in 1986, launching a failed hostage rescue attempt in Iran in 1980, and fighting Iranian forces during the Tanker War of 1987–1988.[29] It conducted its largest

---

27 The size of the Soviet economy as a percentage of the U.S. economy peaked in 1975 at over 44% according to data in Angus Maddison, *The World Economy: A Millennial Perspective* (Paris: OECD, 2001). For a summary of Soviet military capabilities relative to the United States, see Mahnken, *Technology and the American Way of War Since 1945*.

28 Odd Arne Westad, *The Global Cold War* (Cambridge: Cambridge University Press, 2007); Peter Huchthausen, *America's Splendid Little Wars* (New York: Viking, 2003). The United States invaded Panama in 1989 and prepared to invade Haiti in 1994, though the Haitian leadership capitulated before combat operations became necessary.

29 On the 1958 landing in Lebanon, see Alexander L. George and Richard Smoke, *Deterrence in American Foreign Policy: Theory and Practice* (New York: Columbia University Press, 1974).

interventions in East Asia, fighting major wars in Korea and Vietnam that resulted in tens of thousands of American casualties.

No one could reasonably call the United States "non-interventionist" during this period, but some scholars argue that great power rivalry resulted in less intervention compared to the post–Cold War era. Art and Greenhill write that post–Cold War era interventions became more frequent because "the United States found itself freed from the restraints on action imposed by another superpower."[30] Monteiro shows that the United States spent less time at war during the Cold War than afterward.[31] Direct U.S. military interventions during the Cold War tended to be small in scale, Korea and Vietnam aside. However, counting the number or scale of interventions is an unreliable way to determine differences in the propensity to intervene, given the many other factors that influence interventionism. It is more instructive to examine specifically how great power rivalry impacted American calculations.

Consistent with the views of Art and Greenhill, the United States was certainly deterred from interventions that could lead to a direct confrontation with the Soviet Union. For example, the United States respected the Soviet sphere of influence in Eastern Europe and tolerated the Soviet invasion of Hungary in 1956.[32] In the Vietnam War, the United States restricted operations against North Vietnam because of the fear of provoking Chinese intervention.[33] The United States also limited its military actions to avoid having its strength sapped in peripheral "brushfire" wars so it would be capable of directly resisting the Soviet Union. This motivated Eisenhower to limit the American commitment of troops in the Korean War and motivated his decision to stay out of Indochina in 1954.[34]

At the same time, the United States was highly motivated to intervene when communist influence was felt within its own sphere of influence, as in

---

30 Art and Greenhill, "Coercion: An Analytical Overview," p. 3. For elaboration on Art's views, see Robert J. Art, "Introduction," in Robert J. Art and Patrick M. Cronin, eds., *The United States and Coercive Diplomacy* (Washington, DC: United States Institute of Peace Press, 2003).
31 Nuno P. Monteiro, "Unrest Assured," *International Security*, Vol. 36, No. 3 (Winter 2011/2012).
32 George and Smoke, *Deterrence in American Foreign Policy*.
33 George Donelson Moss, *Vietnam: An American Ordeal* (Upper Saddle River, NJ: Pearson, 2006).
34 Mark Clodfelter, *The Limits of Air Power* (New York: The Free Press, 1989); "Discussion at the 194th Meeting of the National Security Council, Thursday, April 29, 1954," *Eisenhower: Papers, 1953–1961 (Ann Whitman File)*, Dwight D. Eisenhower Library.

the Caribbean.[35] Even outside its traditional sphere of influence, the United States often perceived high stakes for fear that failing to stop the spread of communism in one place would lead to "falling dominoes" elsewhere. This motivated its most serious direct interventions in Korea and Vietnam.[36] Fears of falling dominoes also motivated many covert interventions. The United States organized and supplied military operations, supported military coups, and regularly attempted to influence elections around the world.[37] Covert interventions offered a relatively inexpensive means for exercising influence, and their frequency demonstrates that the United States had an interest in cheap interventions long before air power interventionism.[38] It is worth noting that the United States launched precision air strikes during the Cold War against Libya once the technology was developed.

Great power rivalry appears to have both prevented and encouraged interventionism, with no effect clearly dominating. What may be more striking than these effects, however, are the ways in which great power rivalry seemed to make little difference. The goals of many Cold War interventions—defending allies, maintaining stability, preventing terrorism, and protecting access to resources—were not so different from the goals of more recent interventions.[39] They often took the form of weighing in on internal conflicts, much like today, even if the purposes were somewhat different. They were sometimes only peripherally related to the Cold War, as in the hostage rescue attempt in Iran or the air strikes to punish Libyan terrorism.[40]

Going beyond the specific interventions, many of the commitments acquired during the Cold War remain relevant to this day despite the lack

---

35 Westad, *The Global Cold War*; Huchthausen, *America's Splendid Little Wars*.

36 William Stueck, "The Korean War," in Melvyn Leffler and Odd Arne Westad, eds., *The Cambridge History of the Cold War, Vol. 1* (Cambridge: Cambridge University Press, 2010); Fredrik Logevall, "The Indochina Wars and the Cold War, 1945–1975," in Melvyn Leffler and Odd Arne Westad, eds., *The Cambridge History of the Cold War, Vol. 2* (Cambridge: Cambridge University Press, 2010).

37 Westad, *The Global Cold War*; Dov H. Levin, "When the Great Power Gets a Vote: The Effects of Great Power Electoral Interventions on Election Results," *International Studies Quarterly*, Vol. 60, No. 2 (June 2016). Support for military operations included those in Indonesia in 1957, Cuba in 1961, and Congo in 1964. Coups included those in Iran in 1953, Guatemala in 1954, and Chile in 1973.

38 Huchthausen, *America's Splendid Little Wars*.

39 These refer to the Korean War, the interventions in Lebanon, air strikes against Libya, and naval operations in the Persian Gulf, respectively.

40 Huchthausen, *America's Splendid Little Wars*.

of great power competition. The United States remains committed to the defense of NATO, the free flow of oil from the Persian Gulf, and the security of smaller allies like Israel and South Korea. Though these were initially justified by Cold War imperatives, they clearly have other compelling motivations like maintaining international stability or satisfying domestic constituencies.[41] Many other security concerns that had their origins in the Cold War also continue to this day. Nuclear proliferation, terrorism, and even human rights eventually acquired great salience during the Cold War.[42] While the United States has conducted more humanitarian interventions in the post–Cold War era, the non-interventions in the genocides in Rwanda and Darfur suggest that its tolerance for human rights abuses may not have changed much from the Cold War. This all suggests that the return of great power competition is unlikely to dramatically change U.S. interests in peripheral areas or its propensity to intervene.

Great power conflict may have influenced not only U.S. decision-making but also the opportunities for interventionism. Monteiro and Art suggest that the Cold War suppressed conflicts in peripheral areas and that the interventionism of the post–Cold War era can be partly explained by an increase in global instability.[43] If true, then perhaps the United States will face fewer demands for interventionism as great power competition returns. However, this effect is disputed. Wohlforth argues that the existence of a single superpower has imposed order on the international system.[44] Fearon finds that the world has seen a reduction in civil war following a brief increase following the end of the Cold War.[45] Pinker

---

41 John J. Mearsheimer and Stephen M. Walt, *The Israel Lobby and U.S. Foreign Policy* (New York: Farrar, Straus and Giroux, 2007). For a strong argument about the Cold War logic of the America alliance with Israel, see A.F.K. Organski, *The $36 Billion Bargain* (New York: Columbia University Press, 1991). For a study of domestic politics influencing U.S. policy toward Israel during the Cold War, see Galen Jackson, "The Showdown That Wasn't: U.S.-Israeli Relations and American Domestic Politics, 1973–1975," *International Security*, Vol. 39, No. 4 (Spring 2015).
42 Francis J. Gavis, *Nuclear Statecraft* (Ithaca, NY: Cornell University Press, 2012); David P. Forsythe, "Human Rights in U.S. Foreign Policy: Retrospect and Prospect," *Political Science Quarterly*, Vol. 105, No. 3 (Autumn 1990).
43 Monteiro, "Unrest Assured"; Art, "Introduction." For predictions of increasing post–Cold War strife, see Robert D. Kaplan, "The Coming Anarchy," *Atlantic Monthly*, Vol. 212, No. 2 (Feb. 1994).
44 William C. Wohlforth, "The Stability of a Unipolar World," *International Security*, Vol. 24, No. 1 (Summer 1999).
45 James D. Fearon, "Why Do Some Civil Wars Last So Much Longer than Others?" *Journal of Peace Research*, Vol. 41, No. 3 (May 2004).

finds that deaths related to violent conflict have fallen in the post–Cold War world.[46] Snyder suggests that great power conflict can have a contradictory effect in the "stability-instability paradox," with nuclear weapons ensuring peace between great powers but giving them the protection to engage in more low-intensity operations against each other's interests in the periphery.[47]

An important final point is that the U.S. military's consistent focus on great power competition seemed to have had little effect in preventing peripheral interventions. After the Korean War, the U.S. Air Force committed itself primarily to strategic bombing to the detriment of tactical missions like interdiction and close air support.[48] After the Vietnam War, the U.S. military again refocused on fighting the Soviet Union and ignored problems of low-intensity conflict and counterinsurgency.[49] In neither case did this prevent further low-intensity conflicts and limited wars. Politicians may have faced domestic pressures that led them to ignore military advice, and the military may have had fiscal interests in high technology and other biases that led them to ignore national strategy.[50] The military may have also simply faced an inescapable strategic dilemma, where preparing for major war demands the bulk of resources but doesn't eliminate the need to fight minor wars. Whatever the explanation, the pattern of the U.S. military being repeatedly drawn into peripheral interventions that it vowed never to fight again is apparent in the Cold War.

The experience of the Cold War shows that great power competition can have contrasting effects on the calculations that lead to peripheral

---

46 Steven Pinker, *The Better Angels of Our Nature* (New York: Penguin Books, 2012).

47 Glenn Snyder, "The Balance of Power and the Balance of Terror," in Paul Seabury, ed., *The Balance of Power* (San Francisco, CA: Chandler, 1965).

48 Clodfelter, pp. 28, 36–37; Dennis M. Drew, "Air Theory, Air Force, and Low Intensity Conflict: A Short Journey to Confusion," in Phillip S. Meilinger, ed., *The Paths of Heaven* (Maxwell AFB, AL: Air University Press, 1997).

49 Drew, "Air Theory, Air Force, and Low Intensity Conflict." On the U.S. Army, see Conrad Crane, "Avoiding Vietnam: The U.S. Army's Response to Defeat in Southeast Asia," *Strategic Studies Institute* Monograph (Sep. 2002).

50 On domestic pressures forcing interventionism, see Jack S. Levy, "Domestic Politics and War," *Journal of Interdisciplinary History*, Vol. 18, No. 4 (Spring 1988). On domestic pressures forcing suboptimal strategies when intervention occurs, see Jonathan D. Caverley, "The Myth of Military Myopia: Democracy, Small Wars, and Vietnam," *International Security*, Vol. 34, No. 3 (Winter 2009/2010). On parochial military interests, see Stephen M. Walt, "The Search for a Science of Strategy: A Review Essay," *International Security*, Vol. 12, No. 1 (Summer 1987).

interventions, or no effect at all. Nothing in the Cold War experience indicates that in future any one of these effects is likely to dominate great power rivalries. Perhaps the more important lessons are that the United States has long attempted to conduct limited interventions, and the contrary priorities of the defense establishment have not prevented them. These findings suggest that air power interventionism is unlikely to disappear with the return of great power conflict.

Technology and the Future of Air Power Interventionism

The previous section showed that great power competition can influence foreign policy calculations and global stability. Another major effect is encouraging technological arms races. The Cold War saw major arms races in weapons systems such as nuclear weapons and ballistic missiles. Even the technologies that made air power interventionism possible in the 1990s and afterward were initially developed in the 1970s and 1980s to counter Soviet technological advances.[51]

The United States again faces an eroding technological advantage against potential great power competitors and is developing new technologies to stay ahead.[52] If history is any guide, these new technologies could provide capabilities that will enable and even encourage air power interventionism against less technologically advanced adversaries. In this section, I examine how future military technologies are likely to influence foreign policy calculations regarding air power interventions. I show that new technologies may magnify the incentives for air power interventionism. They promise to improve upon the current U.S. ability to strike any ground target within a rapid timeframe, with a low risk of American casualties, and with enough precision to avoid collateral damage.[53] This will only increase the temptation to rely on air power when facing difficult international challenges.

---

51 Mahnken, *Technology and the American Way of War Since 1945*; Barry D. Watts, "The Evolution of Precision Strike," (Washington, DC: Center for Strategic and Budgetary Assessments, 2013).
52 Watts, "The Evolution of Precision Strike."
53 Current Air Force doctrine stresses three concepts: global vigilance, "the ability to gain and maintain awareness – to keep an unblinking eye on any entity"; global reach, "the ability to project military power ... to any point on or above the earth"; and global power, "the ability to hold at risk or strike any target in the world." See *Air Force Basic Doctrine* 1-1 (Washington, DC: Department of Defense).

The challenges posed by the emerging A2/AD threat have led the United States to begin developing and testing a series of technologies including unmanned and autonomous aerial vehicles, directed energy weapons, and hypersonic missiles.[54] Hypersonic missiles can fly at extremely high speeds for long distances and at low trajectories and accurately strike targets.[55] While their main use will probably be to deliver nuclear weapons, they are also being developed for the purpose of the conventional prompt global strike (CPGS). CPGS envisions the United States having the ability to rapidly launch conventional strikes at any target on the globe. It is being primarily considered for the rapid destruction of enemy nuclear forces and A2/AD capabilities deep inside an adversary's territory. Directed energy weapons are also in development and may eventually be introduced to the force. While these weapons have CPGS potential, for now they are most promising as air and missile defense weapons mounted on airplanes or ships.[56] The United States is also developing a suite of longer-range intelligence, reconnaissance, and surveillance capabilities from aircraft to satellites.[57]

What difference will they make against less advanced adversaries? First, they promise to extend into the foreseeable future the current U.S. ability to defeat less sophisticated air defenses. While other countries are unlikely to field integrated air defense systems as sophisticated as those of China and Russia, they are likely to obtain some advanced capabilities from the great powers.[58] Improving the U.S.'s ability to hit enemy air defenses at long range and shoot down enemy missiles should help maintain U.S. air superiority over minor powers. Along with the U.S.'s continued development of electronic countermeasures and cyber capabilities meant to disable

---

54 "Reagan Defense Forum: The Third Offset Strategy," Speech by Deputy Secretary of Defense Bob Work, Nov. 7, 2015, accessed <https://dod.defense.gov/News/Speeches/Speech-View/Article/628246/reagan-defense-forum-the-third-offset-strategy/>.

55 Amy F. Woolf, "Conventional Prompt Global Strike and Long-Range Ballistic Missiles: Background and Issues," *Congressional Research Service* Report R41464, Apr. 6, 2018; John A. Tirpak, "The Great Hypersonic Race," *Air Force Magazine*, Vol. 101, No. 8 (Aug. 2018). They are likely to enter the arsenal in the mid-2020s

56 Jason D. Ellis, "Directed-Energy Weapons: Promise and Prospects" (Washington, DC: Center for a New American Security, Apr. 2015). For more ambitious possibilities about lasers being used for prompt global strike, see Doug Beason, *The E-Bomb* (Cambridge, MA: Da Capo Press, 2005).

57 *America's Air Force*, 2014.

58 Thomas Grove, "The New Iron Curtain: Russian Missile Defense Challenges U.S. Air Power," *Wall Street Journal*, Jan. 23, 2019.

enemy surveillance and targeting, the United States should have the option of conducting air power interventions for the foreseeable future.

Second, CPGS promises to give the President a greater ability to quickly and confidently destroy targets around the world. In theory, CPGS provides the capability for the President of the United States to put their finger on a map anywhere in the world and destroy it within the hour. Outside of high-end warfare, this capability has mostly been considered for the assassination of terrorists.[59] Other uses could include striking unconventional weapons production and storage facilities and striking fielded forces while being mobilized or in transit. The ability to strike these targets rapidly could encourage preemptive action when threatening movement is identified. While such attacks may not be militarily useful without follow-on strikes, they may be considered psychologically useful for dissuasion and deterrence.

Perhaps most revolutionarily, the United States is developing a new generation of UAVs. The United States already operates a range of drones, from small surveillance vehicles to remotely controlled strike aircraft like the Predator or Reaper.[60] Future developments will include miniaturizing and autonomy.[61] This technology is in its early stages, with developers still facing challenges such as placing usable weapons on small drones, identifying targets autonomously, and developing long battery life.[62] If these issues can be solved, the future could see swarms of UAVs operating with relative autonomy, with great speed, coordination, reaction time, and firepower. They may even develop to the point where they are better at discriminating between civilians and enemy fighters than humans are.[63]

Drone swarms could be used for the same missions currently handled by existing air forces such as air superiority, close air support, interdiction or coercion, but at less risk to American lives.[64] They could also provide

---

59 James M. Acton, *Silver Bullet: Asking the Right Questions about Conventional Prompt Global Strike* (Washington, DC: Carnegie Endowment for International Peace, 2013) <https://carnegieendowment.org/2013/09/03/silver-bullet-asking-right-questions-about-conventional-prompt-global-strike-pub-52778>.
60 Sarah E. Kreps, *Drones* (Oxford: Oxford University Press, 2016).
61 Ibid.
62 Paul Scharre, *Army of None* (New York: W.W. Norton & Co., 2018).
63 Ibid. This remains a major technological challenge.
64 Paul Calhoun, "DARPA Emerging Technologies," *Strategic Studies Quarterly*, Vol. 10, No. 3 (Fall 2016); Kreps, *Drones*.

a greatly improved capacity for patrolling urban and rural environments and hunting enemy fighters. Currently, air power can provide surveillance, transport, and air strikes in counterinsurgency warfare. However, the need to avoid excessive civilian casualties, discriminate between civilians and guerrillas, fight in dense urban combat environments, and hold and police territory limits air power's usefulness.[65] Small, fast, light, long-lasting drones could solve these issues. They could monitor and track enemies in remote locations, conduct urban surveillance and reconnaissance, and even target enemies within buildings.[66] With less-than-lethal options involved, drones can also take on a policing role rather than a warfighting one.

Much of this section is necessarily speculative. How these technologies will continue to develop and how they will be incorporated into the armed forces is currently unclear. Nevertheless, each has features that suggest that the United States could see a repeat of the "revolution in military affairs": systems developed for high-technology war enabling interventions against low-technology adversaries. CPGS promises to extend U.S. air superiority and improve its ability to hold at risk any target on the earth. UAVs promise to reduce the risk to American service members in military operations, and could offer the President new, lower-risk options in defeating insurgency and controlling territory. All of these could increase the effectiveness and decrease the cost of military missions from conventional warfare to peacekeeping, increasing the incentives facing the President to intervene in peripheral conflicts.

Conclusion

Since the end of the Cold War, air power interventionism has been a central feature of U.S. foreign policy. Empowered by a favorable geopolitical environment and technological superiority, the United States has regularly employed air power to conduct relatively low-cost military interventions

---

65 James S. Corum and Wray R. Johnson, *Airpower in Small Wars* (Lawrence, KS: University of Kansas Press, 2003), 430.
66 Scharre, *Army of None*, 7; USAF *Unmanned Aircraft Systems Flight Plan 2009–2047* (Washington, DC: Department of Defense, 2009). Dr. Stuart Russell, an artificial intelligence researcher and critic of autonomous weapons systems, has described the nightmare possibility of millions of small anti-personnel drones wiping out entire urban populations. A more discriminating set of drones could be targeted on enemy fighters to rid cities of hostile forces.

against non-peer competitors. I have shown that the reorientation of U.S. foreign policy toward great power competition is unlikely to change this. The United States will continue to encounter situations where it wants to use force against minor powers while limiting risk to its soldiers. In fact, the investments in new weapons for great power competition may actually motivate the United States to rely more heavily on air power in its military operations.

This is a discouraging conclusion for many in the United States military. The military's experience with interventionism has been frustrating. The desire to avoid small wars and focus only on the "high-end fight" is a common sentiment. However, it is unlikely that U.S. foreign policy will be so cooperative. It has become common to note that the United States has had to relearn counterinsurgency multiple times.[67] It should not make the same mistake with regard to air power interventionism. This is not to imply that the decisions of how to balance between high-end and low-end investments will be easy, but that they should not be made on the assumption that low-end fights are a thing of the past.

Similarly, U.S. adversaries should not make the same mistake. It is a common observation that the United States has trouble convincingly signaling its willingness to conduct military interventions to smaller adversaries.[68] If those states believe the United States is shifting away from air power interventionism, they may become more inclined to challenge U.S. interests and to provoke confrontation. It is in everyone's interest, therefore, to recognize that the United States is unlikely to scale back its commitments or reduce its reliance on air power interventionism.

Finally, these findings suggest that opponents of air power interventionism have yet to deal with some of the systematic factors making it such a common policy. Many writers in the "realist" tradition argue that the United States should use military force only to protect its most vital interests.[69] They view most wars on the periphery as motivated by naïve

---

67 Robert R. Tomes, "Relearning Counterinsurgency Warfare," *Parameters*, Vol. 34, No. 1 (Spring 2004).

68 Art and Cronin, eds., *The United States and Coercive Diplomacy*; Diane Pfundstein Chamberlain, *Cheap Threats* (Washington, DC: Georgetown University Press, 2016).

69 Betts, "Pick Your Battles"; Barry R. Posen, "Pull Back," *Foreign Affairs*, Vol. 92, No. 1 (Jan./Feb. 2013); Sebastian Rosato and John Schuessler, "A Realist Foreign Policy for the United States," *Perspectives on Politics*, Vol. 9, No. 4 (Dec. 2011).

ideological impulses, costly in resources and reputation, and prone to failure. Nevertheless, the United States continues to engage in these kinds of interventions. It is possible that American leaders simply never learn these lessons. But it is far more likely that they face compelling reasons to keep returning to air power interventions. Regional aggression, internal conflicts, and transnational terrorism all threaten American interests and values to varying degrees and demand the attention of U.S. leaders. Further, for all its shortcomings, air power has proven that it has some ability to defeat enemy armies, coerce hostile regimes, and eliminate discrete threats. The demand to address foreign policy challenges at an acceptable cost puts leaders in a position where air power is often seen as the best option. These issues are unlikely to disappear despite the rise of new, more powerful threats. Despite calls to exercise restraint and eschew these interventions, it will be very difficult for American leaders to abandon air power interventionism.

References

Acton, J.M. "Silver Bullet: Asking the Right Questions about Conventional Prompt Global Strike." *Carnegie Endowment for International Peace*, 2013.

Ali, I. "U.S. Military Puts 'Great Power Competition' at Heart of Strategy: Mattis." *Reuters*, Jan. 19, 2018.

Air Force Basic Doctrine 1-1. Washington, DC: Department of Defense.

Air Force Strategic Master Plan. Washington, DC: Department of Defense, May 2015.

Alterman, J.B. "Coercive Diplomacy Against Iraq, 1990-98," in Robert J. Art and Patrick M. Cronin, eds., *The United States and Coercive Diplomacy*. Washington, DC: United States Institute of Peace Press, 2003.

America's Air Force. Washington, DC: Department of Defense, July 2014.

Art, R.J. "Introduction," in Robert J. Art and Patrick M. Cronin, eds., *The United States and Coercive Diplomacy*. Washington, DC: United States Institute of Peace Press, 2003.

Art, R.A. and J.S. Cronin, eds., *The United States and Coercive Diplomacy*. Washington, DC: United States Institute of Peace Press, 2003.

Art, R.A. and Kelly M. Greenhill. "Coercion: An Analytical Overview," in Kelly M. Greenhill and Peter Krause, eds., *Coercion*. New York: Oxford University Press, 2018, p. 3–32.

Ashford, E. and J. R. Itzkowitz Shifrinson. "Trump's National Security Strategy: A Critic's Dream," *Texas National Security Review*, Vol. 1, No. 2 (Feb. 2018).

Beason, D. *The E-Bomb*. Cambridge, MA: Da Capo Press, 2005.

Betts, R.K. "Pick Your Battles," *Foreign Affairs*, Vol. 93, No. 6 (Nov./Dec. 2014). <https://www.foreignaffairs.com/articles/united-states/2014-10-20/pick-your-battles>.

Biddle, S. "The Libyan Dilemma: The Limits of Air Power." *Washington Post*, Mar. 25, 2011.

Byman D.L. and M.C. Waxman. "Kosovo and the Great Airpower Debate," *International Security*, Vol. 24, No. 4, Spring 2000.

Calhoun, P. "DARPA Emerging Technologies." *Strategic Studies Quarterly*, Vol. 10, No. 3, Fall 2016.

Caverley, J.D. "The Myth of Military Myopia: Democracy, Small Wars, and Vietnam." *International Security*, Vol. 34, No. 3, Winter 2009/2010.

Chamberlain, D.P. *Cheap Threats*. Washington, DC: Georgetown University Press, 2016.

Chivvis, C.S. *Toppling Qaddafi*. Cambridge, U.K.: Cambridge University Press, 2014.

Clodfelter, M. *The Limits of Air Power*. New York: The Free Press, 1989.

Cohen, E.A. "The Mystique of U.S. Air Power." *Foreign Affairs*, Vol. 73, No. 1, Jan.–Feb., 1994.

Cohen, E.A. "Trump's Lucky Year: Why the Chaos Can't Last." *Foreign Affairs*, Vol. 97, No. 2, Mar./Apr. 2018.

Corum, J.S. and W. R. Johnson. *Airpower in Small Wars*. Lawrence, KS: University of Kansas Press, 2003.

Crane, C. "Avoiding Vietnam: The U.S. Army's Response to Defeat in Southeast Asia." *Strategic Studies Institute* Monograph, Sep. 2002.

Deptula, D.A. "Beyond the Bomber: The New Long-Range Sensor-Shooter Aircraft and United States National Security." *Mitchell Institute for Aerospace Studies* Report, 2015.

Drew, D.M. "Air Theory, Air Force, and Low Intensity Conflict: A Short Journey to Confusion," in Phillip S. Meilinger, ed., *The Paths of Heaven*. Maxwell AFB, AL: Air University Press, 1997.

Ellis, J.D. "Directed-Energy Weapons: Promise and Prospects." Washington, DC: Center for a New American Security, April 2015.

Everstine, B. "Light Attack Basing, Deployment Model, and Command Assignment Determined by Size of Buy." *Air Force Magazine*, Dec. 7, 2018.

Fearon, J.D. "Why Do Some Civil Wars Last So Much Longer than Others?" *Journal of Peace Research*, Vol. 41, No. 3, May 2004.

Forsythe, D.P. "Human Rights in U.S. Foreign Policy: Retrospect and Prospect." *Political Science Quarterly*, Vol. 105, No. 3, Autumn 1990.

Gavis, F.J. *Nuclear Statecraft*. Ithaca, NY: Cornell University Press, 2012.

George, A.L. and R. Smoke. *Deterrence in American Foreign Policy: Theory and Practice*. New York: Columbia University Press, 1974.

Gertler, J. "F-35 Joint Strike Fighter (JSF) Program." *Congressional Research Service* Report RL30563, Apr. 23, 2018.

Gertler, J. "Air Force B-21 Raider Long-Range Strike Bomber." *Congressional Research Service* Report R44463, Oct. 12, 2018

Gordon, M.R., H. Cooper, and M. D. Shear. "U.S. Launches Missiles into Syria." *New York Times*, A1, Apr. 6, 2017.

Grove, T. "The New Iron Curtain: Russian Missile Defense Challenges U.S. Air Power." *Wall Street Journal*, Jan. 23, 2019.

Haass, R.N. "Where to Go From Here: Rebooting American Foreign Policy." *Foreign Affairs*, Vol. 96, No. 4, Jul./Aug. 2017.

Henrikson, D. *NATO's Gamble*. Annapolis, MD: Naval Institute Press, 2007.

Huchthausen, P. *America's Splendid Little Wars*. New York: Viking, 2003.

Insinna, V. "US Air Force's Plan to Launch Light-Attack Aircraft Competition is Now Deferred Indefinitely." *Defense News*, Jan. 18, 2019.

Jackson, G. "The Showdown That Wasn't: U.S.-Israeli Relations and American Domestic Politics, 1973–1975." *International Security*, Vol. 39, No. 4, Spring 2015.

Joint Operational Access Concept. Washington, DC: Department of Defense, 2012.

Kaplan, R.D. "The Coming Anarchy." *Atlantic Monthly*, Vol. 212, No. 2, Feb. 1994.

Kreisher, O. "Marines Shifting Focus to Readying for the 'High End' Fight." *Seapower Magazine*, June 4, 2018.

Kreps, S.E. *Drones*. Oxford: Oxford University Press, 2016.

Lambeth, B.S. *The Transformation of American Airpower*. Ithaca, NY: Cornell University Press, 2000.

Levin, D.H. "When the Great Power Gets a Vote: The Effects of Great Power Electoral Interventions on Election Results." *International Studies Quarterly*, Vol. 60, No. 2, June 2016.

Levy, J.S. "Domestic Politics and War." *Journal of Interdisciplinary History*, Vol. 18, No. 4, Spring 1988.

Logevall, F. "The Indochina Wars and the Cold War, 1945–1975," in Melvyn Leffler and Odd Arne Westad, eds., *The Cambridge History of the Cold War, Vol. 2*. Cambridge: Cambridge University Press, 2010.

Losey, S. "Secretary Wilson to Lay Out 'Air Force We Need' at AFA." *Air Force Times*, Sep. 10, 2018.

Maddison, A. *The World Economy: A Millennial Perspective*. Paris: OECD, 2001.

Mahnken, T. G. *Technology and the American Way of War Since 1945*. New York: Columbia University Press, 2008.

McMaster, H.R. "On War: Lessons to be Learned." *Survival*, Vol. 50, No. 1, Feb.–Mar. 2008.

Mearsheimer J.J. and S.M. Walt. *The Israel Lobby and U.S. Foreign Policy*. New York: Farrar, Straus and Giroux, 2007.

Monteiro, N.P. "Unrest Assured." *International Security*, Vol. 36, No. 3, Winter 2011/2012.

Moss, G.D. *Vietnam: An American Ordeal*. Upper Saddle River, NJ: Pearson, 2006.

National Security Strategy of the United States. Washington, DC: President of the United States, 2017.

National Defense Strategy of the United States. Washington, DC: Department of Defense, 2018.

Organski, A.F.K. *The $36 Billion Bargain.* New York: Columbia University Press, 1991.

O'Rourke, R. "A Shift in the International Security Environment: Potential Implications for Defense Issues for Congress." *Congressional Research Service* Report R43838, Oct. 24, 2018.

Pinker, S. *The Better Angels of Our Nature.* New York: Penguin Books, 2012.

Posen, B.R. "Command of the Commons." *International Security*, Vol. 28, No. 1, Summer 2003.

Posen, B.R. "Pull Back." *Foreign Affairs*, Vol. 92, No. 1, Jan./Feb. 2013.

"Reagan Defense Forum: The Third Offset Strategy." Speech by Deputy Secretary of Defense Bob Work, Nov. 7, 2015. <https://dod.defense.gov/News/Speeches/Speech-View/Article/628246/reagan-defense-forum-the-third-offset-strategy/>.

Rosato, S. and J. Schuessler. "A Realist Foreign Policy for the United States." *Perspectives on Politics*, Vol. 9, No. 4, Dec. 2011.

Scharre, P. *Army of None.* New York: W.W. Norton & Co., 2018.

Snyder, G. "The Balance of Power and the Balance of Terror," in Paul Seabury, ed., *The Balance of Power.* San Francisco, CA: Chandler, 1965.

"Special Report: Inherent Resolve." Washington, DC: U.S. Department of Defense, September 2018. <https://dod.defense.gov/OIR/>.

Stueck, W. "The Korean War," in Melvyn Leffler and Odd Arne Westad, eds., *The Cambridge History of the Cold War, Vol. 1.* Cambridge: Cambridge University Press, 2010.

Tirpak, J.A. "The Great Hypersonic Race." *Air Force Magazine*, Vol. 101, No. 8 (Aug. 2018).

Tomes, R.R. "Relearning Counterinsurgency Warfare." *Parameters*, Vol. 34, No. 1, Spring 2004.

USAF, *Unmanned Aircraft Systems Flight Plan 2009–2047.* Washington, DC: Department of Defense, 2009.

Walt, S.M. "The Search for a Science of Strategy: A Review Essay." *International Security*, Vol. 12, No. 1, Summer 1987.

Watts, B.D. "The Evolution of Precision Strike." Washington, DC: Center for Strategic and Budgetary Assessments, 2013.

Westad, O.A. *The Global Cold War.* Cambridge: Cambridge University Press, 2007.

Williams, B.G. *Predators: The CIA's Drone War on al Qaeda.* Washington, DC: Potomac Books, 2013.

Wohlforth, W.C. "The Stability of a Unipolar World." *International Security*, Vol. 24, No. 1, Summer 1999.

Woolf, A.F. "Conventional Prompt Global Strike and Long-Range Ballistic Missiles: Background and Issues." *Congressional Research Service* Report R41464, Apr. 6, 2018.

2 CHINA
The Quest for a Modern Air Force

Xiaoming Zhang

On November 11, 2018, the People's Liberation Army Air Force (PLAAF) unveiled a three-step roadmap for becoming a modern air force. First, by 2020, the PLAAF will be developing into a strategic service that integrates air and space power, and offensive and defensive capabilities, in which the fourth- and fifth-generation weapon systems play a major role and the systematic combat capabilities are significantly enhanced. Second, the PLAAF will improve its strategic capabilities and modernize its theory, organizational structure, service personnel, and weaponry. Finally, the building of a modern and strategic air force will be completed by 2035.[1] This roadmap demonstrates China's ongoing efforts to build a modern air force.

For years, many policy makers, strategists, and scholars have had concerns about a Chinese military modernization that might lead to a new perspective on global security for years to come. At the center of this concern is that the PLAAF has gained offensive capability by equipping itself with an increasing number of fourth- and fifth-generation aircraft and long-range antimissile systems. But what matters most is not the growth of Chinese air power capability per se, but rather, how China might use its new military strength, especially its air and space power, in future warfare.

Since its inception in 1949, the development of the PLAAF has faced many challenges. China had long been an agricultural country with little industry. The People's Liberation Army (PLA) was an army-centric and -dominated military. Analyzing the current and future state of the PLAAF's

---

1 "Chinese Air Force Announces Its Roadmap for Building a Powerful and Modernized Air Force," November 11, 2018, *Xinhuanet*, <http://www.xinhuanet.com/politics/2018-11/11/c_129991031.htm>.

modernization necessitates examining the historical development of the Chinese air force and its experiences (during the Korean War, the 1950s Taiwan Strait crises, and the air defense engagements against Nationalist Party and American intrusions); it also requires an examination of the evolution of China's air doctrine over the years, including its influence on force structure and employment. This assessment of the current development of the PLAAF is made against this historical background. The conclusion is that the PLAAF's combat experience, Chinese defensive thinking, and China's lack of industrial infrastructure and technological knowhow served as roadblocks against the Chinese efforts to develop modern air power. China's growing economic power since 2009, along with changes in security interests and technology, have dramatically transformed the PLAAF into a new air force. The chapter examines the successes and challenges in China's quest to become a strategic service.

Early Combat Experiences

The present understanding of the PLAAF's development cannot be disassociated from an overview of its early experiences. Since its establishment on November 11, 1949, the doctrinal guidance for the PLAAF's development was Chinese defensive thinking. From the outset, the PLAAF leadership preferred to build an air force that possessed more fighters than bombers. Its theory was that the role of fighters dovetailed well with the defensive cast of Chinese military thought. Bombers attacked enemy countries and territories—an aggressive act—but fighters were defensive in nature and if successful in fending off attacks would ensure air superiority.[2] The PLAAF's immediate mission, therefore, was to attain air superiority over the Nationalists, provide support to the planned amphibious assault on Taiwan, and then develop into a force capable of defending China's airspace and waters.

The Korean War provided the impetus for the rapid expansion of the air force in both aviation personnel and equipment. A large number of officers and troops were transferred from ground forces to form 26 aviation divisions and four independent regiments, and eight aviation and three

---

2   Lü Liping, *Tongtian zhi lu* [*The Road to the Sky*] (Beijing: Liberation Army Press, 1989), 144.

mechanical schools operating throughout the conflict on the Korean Peninsula.[3] The existing ground force structure was simply grafted onto the air force, and army officers were chosen to command the air force. The PLAAF leadership was accustomed to believing that building an air force on the foundation of the ground forces was necessary for its future success. Thus, the air force's primary mission was to provide support for ground troops, and it would take the victories of ground operations as its own. The air force was created as an independent service of the PLA under the direct control of the Central Military Commission (CMC), the highest military authority of the People's Republic of China (PRC).[4] This ground-centric army gave biased accounts for the PLAAF leadership's inclination at the time to perceive the air force as a support unit of the PLA. It did not consider airpower essential in a strategic sense, valuing it only for the tactical support it could provide to the ground forces during operations.[5] Such thinking, moreover, justified an army-centric and -dominated PLA system that subsequently prevented the air force from operating as an independent service.

The Korean War experience was a driving force for the PLAAF to further emphasize air defense and procurement of fighters as the largest and most important element of the Chinese air force. Throughout the war, the PLAAF limited its air operations to protecting key transportation lines and military targets in an area of the so-called MiG Alley south of the Yalu River. While recognizing the U.S.'s air superiority, Chinese leaders discounted the role China's air power had played. They found it particularly interesting that air bombardment inflicted fewer casualties upon communist forces than did ground fire. Given their confidence in the human factor—that men could overcome weapons—and their own guerrilla war experience, they remained convinced that PLA ground forces could overwhelm stronger opponents and win any future war, with air power used to supplement the power of the army.[6]

---

3   Liu Yalou, "The Initial Seven Years of Establishing the Air Force," in *Lantian zhi lu* [The Road to Blue Sky], 2 (Beijing: The Political Department of Air Force, 1992), 4–5.

4   He Tingyi, "The Establishment of Air Force Headquarters," in *Lantian zhi lu,* 1.

5   This also results in western misunderstanding of the PLAAF. See Kenneth W. Allen, Glenn Krumel, and Jonathan D. Pollack, *China's Air Force Enters the 21st Century* (Santa Monica, CA: RAND, 1995), 37.

6   Liu Pushao et al., "Biography of Zhu Guang," in *Zhongguo renmin zhiyuanjun renwu zhi* [*Biographical Notes of the Chinese Volunteer Army*] (Nanjing: Jiangsu People's Press, 1992), 309.

After the Korean War, the PLAAF constantly engaged in air combat against the Nationalist air force for the control of airspace over the coastal areas of Zhejiang and Fujian provinces.[7] As during the Korean War, the PLAAF took a passive stance and waited to respond to intrusions by the Nationalist air force, which was much smaller, but was free to choose the time and method of aerial combat. The PLAAF, by contrast, had to depend on the ground control intercept to scramble its fighters. Furthermore, the capability of the air force was restricted by political considerations and the limited range of the MiG-17 fighters. Nevertheless, operations against the Nationalists over the southeast coastal areas in the 1950s gave the Chinese valuable experience in employing air power in air defense.

This trend continued into the late 1950s and then the 1960s. A major focus of the PLAAF's day-to-day activity was on constantly scrambling its fighters to intercept intruding Nationalist and American aircraft, many of them spy planes, and unmanned drones that routinely flew over Chinese airspace, but also engaging and shooting down aircraft that accidently approached or overflew its borders, including several American aircraft shot down during the Vietnam War.[8] This air defense experience resulted in the PLAAF continuing to emphasize an air defense strategy and the development of fighter planes, radar, and ground anti-aircraft systems, while devoting only a small portion of the overall force structure to delivering limited air-to-surface ordnance.

Air defense combat experience led the PLAAF to define a set of operational principles that stressed the use of overwhelming force to: achieve the protection of friendly forces and the destruction of enemy forces; subordination of military objectives to political ones through strict adherence to the central authority's operational policy; and study the application of the PLAAF experience and tactics drawn from the Korean War and combat against the Nationalist air force. Preparing for the enemy's bombing attacks, the air force developed some countermeasures, including mental

7   Xiaoming Zhang, "Air Combat for the People's Republic: The People's Liberation Army Air Force in Action, 1949–1969," in Chinese Warfighting: The PLA Experience Since 1949, ed. Mark A. Ryan, David M. Finkelstein, and Michael A. McDevitt (Armonk, NY: M. E. Sharpe, 2003), 279–82.
8   Lin Hu, Baowei zuguo lingkong de zhandou [Fight to Protect the Motherland's Airspace] (Beijing: Liberation Army Press, 2002), 96.

preparation, camouflage, quick repair, and deployment of Chinese bombers to the region at risk in order to strike back. This policy eventually led the PLAAF leadership to believe that the use of air power was more for deterrence than for offense.[9]

Evolving Thinking on Air Power

The PLAAF's earliest effort to develop its thinking on air power can be traced back to the late 1950s. In May 1959, it organized a team of more than 100 officers to compile *kongjun zhandou tiaoling* (the Field Manual of Air Force Operations). It was not until 1963 that the air force officially accepted this field manual, along with several other regulations and handbooks, as its official guidance.[10] Documentation about the PLAAF manuals and regulations is fragmentary. The PLAAF literature provides illuminating insights into the formation and development of the Chinese air force strategy and basic models and principles for air force operations.[11]

The PLAAF, like most of the world's air power nations, regarded command of the air as the air force's primary mission, stating that during its engagement in a struggle for command of the air, the air force should keep a lookout over major airspace, and seize and maintain command of the air at critical moments over certain areas. The major difference to other nations was that for China such command of the air was exclusively over China's own airspace. The air force's critical role was to protect vital domestic objectives and the main military forces so that they did not suffer from systematic air attacks by the enemy.[12] Perceiving that China still faced the danger of large-scale invasion by powerful enemies, the PLA strategy required the air force to, first, participate in resistance against the enemy's strategic surprise attacks to shield the nation's transition from peacetime to wartime; second, provide air cover for the strategic deployment of mass

---

9   Pei, Zhigeng, and Xie, Bin, "Rushed into Fujian to Combat for Control of the Air," in *Lantian zhi lu*, 1, 193.

10 Cao Lihuai, "Recollection of the Work on Compilation of Regulations and Teaching Materials," *Lantian zhi lu*, 2, 158, 165.

11 Min Zengfu, chief ed., *Kongjun junshi shixiang gailun* [An Introduction to Air Force Military Thinking] (Beijing: Liberation Army Press, 2006); and Hua Renjie, et al., *Kongjun xueshu sixiang shi* [*The History of the Academic Thinking of the Air Force*] (Beijing: Liberation Army Press, 2008).

12 Hua Renjie, et al., *Kongjun xueshu sixiang shi*, 359.

troops; and third, to enhance domestic air defense with strategic center points, while providing support in land and naval battles.[13]

Based on these guidelines, for the first time the PLAAF determined the principles for specific utilization of the air force and its development. The PLAAF's two primary roles were to support the ground force, taking its victories as its own victories, and maintain local air dominance through air defense. Given the fact that the PLAAF remained the weaker force, preservation of its strength became another important principle that emphasized adherence to protracted operations and waiting for opportunities to destroy the enemy. As a result, the PLAAF concluded that the development of the air force must see developing air defense troops on a fairly large scale as important.[14] No matter how the PLAAF attempted to define its strategy and doctrine, it never saw the air force as an independent arm, only as a subsidiary service for both the army and navy.[15]

Moreover, the manual emphasized that the air force's primary mission was to take protracted defensive operations to wear down the adversary while the enemy was strategically on the offensive and attacking. This defense strategy also specified operational principles, emphasizing that:

- the air force should try to gain the initiative with counteroffensive attacks so as to be able to seize air superiority
- the entire air force should be used in a concentrated fashion instead of dividing the forces
- the air force should operate in a surprising and concealed manner with flexible strategy and tactics
- the air force should carefully plan and fully prepare before each operation
- the air force should implement positive measures to enhance its protection in order to sustain combat capability and
- the air force should maintain a combat style of bravery and tenacity, and a perfect mastery of combat skill so as to be able to defeat the enemy.[16]

---

13 He Weirong, "Military Thought on Air Force," in Yao Wei, chief ed., *Zhongguo kongjun baike quanshu* [*Encyclopedia of Chinese Air Force*], Vol. 1 (Beijing: Aviation Industry Press, 2005), 3.
14 Ibid.
15 Min Zengfu, chief ed., *Kongjun junshi shixiang gailun*, 394.
16 Zhao Zhongxin, "Air Force Tactics," in *Zhongguo kongjun baike quanshu*, 1, 108.

The PLAAF officially adopted this strategy and doctrine in 1963, but never had the opportunity to implement it. While continuing to perceive threat from the United States, South Korea, Taiwan, and Indochina, no large-scale invasion ever took place against China throughout the remaining years of the 1960s. The 1963 field manual slowly but surely became irrelevant. One major focus of the PLAAF's daily job was to find ways to bring down those manned reconnaissance aircraft (initially PV2 and then U-2) and unmanned drones that routinely flew over China air space. Failure to intercept and shoot them down would indicate the PLAAF was incapable of defending China. The Chinese leadership would have been humiliated because any such incident would create "a bad impression" of China's air defense capability.[17]

Engagements against intruding reconnaissance aircraft were considered as rare combat opportunities for Chinese pilots to sharpen their combat skill. Special air combat units (including surface-to-air missile units) were organized and deployed to several key locations around the country. Despite frustrations in the many attempts to shoot down the spy planes, the air force claimed that it brought down more than two dozen manned and unmanned spy planes throughout the 1960s.[18]

Combat action against reconnaissance overflights took place in the midst of an upsurge in political radicalism emphasizing the command and promotion of Chairman Mao's cult of personality. Each time an intruding flight was brought down it was viewed more like a political victory than a military one. Celebrations were held and those who were involved in the combat actions received awards and were always met by senior party and state leaders, including Chairman Mao and Premier Zhou Enlai, making headline news across the country. Senior military leaders also made use of the celebratory events to promote the air force, proclaiming "all military services must learn from the air force."[19] The study of Mao Zedong's thought and his people war theory became dominant in the PLA's daily training and education. Field manuals along with other air force teaching materials compiled in the early 1960s were refuted and many copies were even burned.[20]

---

17 Lin Hu, *Baowei zuguo lingkong de zhandou*, 96.
18 Wang Dinglie, chief ed., *Dangdai Zhongguo kongjun* [*Contemporary China's Air Force*] (Beijing: Social Science Press, 1989), 385–86.
19 Lin Hu, *Baowei zuguo lingkong de zhandou*, 272.
20 Cao Lihuai, "Recollection of the Work on Compilation," 166–67.

The zealous promotion of Mao's theory of people's war further stirred unrealistic thinking on the development of the air force in order to satisfy the radical political sentiment. In the late 1960s, the PLAAF came up with a new vision that urged the air force to "fit in with the requirements of the people's war and carry out guerrilla warfare in the air."[21] It then requested the aviation industry develop and then build a supersonic fighter that was small, light, and agile with a short-takeoff-and-landing capability.[22] The first prototype made a debut flight at the end of 1970, and later during a demo flight, senior PLA military leaders named the plane "Li Xiangyang in the air," after a famous guerrilla fighter in an anti-Japanese war movie popular at the time. The aircraft, however, never entered into service because of its inadequate thrust and firepower.[23] This failed project unmistakably suggested that it is impossible to develop any healthy and objective theories on airpower in a political system dominated by absolute authority and arbitrary decisions by key individuals.

Worship of the top leader during the 10 years of political upheaval during the Cultural Revolution (1966–1976) not only impeded the PLAAF's further definition and perfection of its strategy and doctrine, but also handed down a legacy that continued to convince the PLAAF scholars to regard the instructions of the senior leaders as an important component part of Chinese military thought on air force. That is why Mao's thinking on war and warfare, especially his people's war strategy, remains relevant for the PLAAF when developing its own strategy and doctrine. Although the PLAAF failed to fully implement its strategic vision for air force, by the early 1970s it had grown into a force with 50 air divisions, of which two thirds flew short-range fighters.[24] This trend continued in following decades.

---

21 Ling Qiang, "Revealing the Secret of China's Jian-12 Fighter," *Xiandai bingqi* [*Modern Arsenal*], No. 2 (2001), 2–4.

22 Duan Zijun, chief ed., *Dangdai Zhongguo de hangkong gongye* [*Contemporary China's Aviation Industry*] (Beijing: China Social Science Press, 1998), 163–65.

23 Ling Qiang, "Revealing the Secret of China's Jian-12 Fighter," 3.

24 Yao Jun, chief ed., *Zhongguo hangkong shi* [*China Aviation History*] (Zhengzhou: Daxiang Press, 1997), 655–65.

New Strategy for Simultaneous Offense and Defense

In the early 1990s, the PLAAF realized that China had fallen far behind the West in both technology and doctrinal thinking about airpower. Time and space were no longer the allies of those who were once so confident that China's existing air defense systems could prevent any attacks deep into the nation's heartland. Serious doubts were raised about the traditional interpretation of China's defense capabilities, including the common belief that an inferior force could overcome a superior enemy. Drawing on lessons learned from Iraq's defeat in the 1991 Gulf War, the Chinese central military leadership pointed out that "a weaker force relying solely on defensive operations would place itself in the position of having to receive blows," and that only by "taking active offensive operations" could the weaker now seize the initiative.[25]

China's evolving security interests, including the long-standing prospect of a decisive confrontation with Taiwan, also favored consideration of augmenting the PLAAF's offensive capabilities. Since 1993, Beijing has adopted a new military strategy, placing an emphasis on fighting and winning a future regional war using high technology along China's periphery. The momentum of the independence movement in Taiwan was also viewed as an increasingly serious challenge to China's sovereignty and security.[26] The central military leadership appropriately readjusted the air force's strategic missions, requiring it to maintain strong capabilities, not only for defensive operations but also for offensive ones.

Throughout the 1990s, the PLAAF pondered how China's air space could be defended by the implementation of an offensive air strategy. It became very interested in Douhet's argument that "a weaker air force could also defeat a stronger enemy provided it can compensate for the difference in strength by showing more intelligence, more intensity, and more violence in its offensive actions."[27] Subsequently, the PLAAF researchers advanced a new theory for the development of the Chinese air force that would give it capability in both offensive and defensive operations. Taking

---

25 Hua Renjie et al., *Kongjun xueshu sixiang shi*, 368.
26 Liu Huaqing, *Liu Huaqing huiyilu* [*Memoirs of Liu Huaqing*] (Beijing: Liberation Army Press, 2004), 581–82; Ye Huinan, "Four Major Changes of Our Country's National Defense Strategy Since Its Founding," *Dangdai Zhongguo shi yanjiu* [*Studies of Modern Chinese History*], No. 3 (1999), 8.
27 Hua Renjie et al., *Kongjun xueshu sixiang shi*, 369.

into account the defensive nature of China's national defense strategy, they believed that while maintaining a necessary number of ground air defense force, the PLAAF should possess an equally large number of offensive and defensive aircraft that could carry out not only air defense missions but also offensive operations, including striking deep into enemy territory.[28]

Moreover, drawing lessons from both the 1991 Gulf War and Operation Allied Force in Kosovo, the Chinese military leadership realized that future warfare would be all dimensional involving land, sea, air, space, and electronics.[29] This understanding of modern warfare facilitated the PLAAF strategists' discussions on how to develop its capability to engage in informatized warfare. For them, air offensive operations in a war to unify Taiwan with the mainland would depend on the PLAAF's electronic warfare capability. They therefore declared that the development of the Chinese air force's offensive capability must include the development of integrated firepower and information systems.[30]

In March 1999, the Chinese leadership endorsed the PLAAF's new strategic vision, directing it to transform gradually from a homeland air defense force to one that was capable of both offensive and defensive operations with Chinese characteristics. Because China is a large country with many important targets to be protected, they believed that the only effective air defense was to destroy enemy attacking forces on the ground and at sea. They charged the air force to "bear the brunt of, and be employed throughout the entire course" of the conflict and "to complete certain strategic missions independently."[31] Late that year, for the first time, the PLAAF set an ambitious goal to develop itself into a strategic air force. The PLAAF adopted a three-step implementation strategy for air force development over the next several decades, with a focus on building three operational systems: air offense, air and space defense, and airlift and airborne.[32]

According to its 2008 Defense White Paper, China expected to lay a solid foundation for the development of the PLA into a more high-tech and more balanced network-centric joint force by 2010, to accomplish

---

28 Dong Wenxian, *Xiandai kongjun lun* [*On the Modern Air Force* (continuation)] (Beijing: Blue Sky Press, 2005), 247–48.

29 Liu Huaqing, *Liu Huaqing huiyilu*, 610.

30 Dong Wenxian, *Xiandai kongjun lun*, 57.

31 He Weirong, "Military Thought on the Air Force," 15.

32 Dong Wenxian, *Xiandai kongjun lun*, 47.

mechanization and make major progress in informatization by 2020, and to reach the goal of modernizing national defense and the armed forces by the middle of the twenty-first century.[33] This was a logical follow on to the strategic vision the PLAAF introduced in 2004. That year, the PLAAF enunciated a new strategic vision calling for the development of a long-range strategic air force and the active involvement of integrated air and space (*kongtian yiti*) operations with integrated information and firepower systems (*xinxi huoli yiti*).[34]

Under the guidance of this developmental strategy, the PLAAF embarked on a two-stage transformation. The first stage was to lay a framework for a force capable of both offensive and defensive operations, by increasing the number of high-performance offensive aircraft, combat support aircraft, and advanced surface-to-air missile systems. The second stage was to wield fighter aircraft, surface-based defense, and command, control, communication, and intelligence elements into an integrated operational system able to conduct both air offensive and defensive operations under informatized conditions.

The development of China's air force capabilities focuses upon four areas:

1. Offensive capability that is capable of protecting national security and national interests from the air and space.
2. Integrated air defensive and anti-missile capability that is capable of monitoring both air and space flying objects and attacking them.
3. Superior capability over its main opponent (presumably Taiwan) and certain counter-information capability against its strategic opponent (presumably the United States).
4. Strategic airlift capability that is capable of conducting both airlift and airdrop operations.[35]

---

33 Information Office of the State Council of the People's Republic of China, China's National Defense Paper in 2008, <http://www.gov.cn/english/official/2009-01/20/content_1210227.htm>.

34 Dai Xu, "Goodbye, Old J-6 Fighters: A Complete Examination of the Service History of the Last Meritorious Fighter in the Chinese Air Force with Combat Victory Record," *Guoji zhanwang* [*World Outlook*], No. 19 (2005), 21.

35 Shang Jinsuo, chief ed., *Kongjun jianshe xue* [*Science of Air Force Construction*] (Beijing: Liberation Army Press, 2009), 557–58.

The Quest for Advanced Air Power

Since the inauguration of the PRC, Beijing's quest for a modern air force had experienced hardship and tortuousness due to the country's backward economy, lack of industrial infrastructure, and limited access to western technology. It had pursued a "walking on two legs" policy to modernize the air force through purchases of foreign systems and the development of domestic technology. China had historically sought to be self-reliant in military production through either reverse engineering or incorporating foreign technology. By the end of the twentieth century, China had manufactured more than 6,000 J-6 (Chinese-made MiG-19) and J-7 (Chinese-made MiG-21) aircraft.[36] These aircraft were not only short range but also shortsighted. China's indigenous defense industry offered no immediate solution to achieve the PLAAF's strategic vision.[37]

In the wake of the June 1989 event at Tiananmen Square, the West imposed an arms embargo on China. As a result, the Beijing government had to turn to Russia for assistance. Beginning in 1992, China bought three batches of Su-27s—a total of 74 aircraft—along with their accessories. The Su-27 was one of the world's most modern aircraft, with a state-of-the-art weapon system but with limited offensive capabilities as a fighter. In late 1999, China fulfilled a $1.85 billion contract with Russia to purchase 38 Su-30MKK fighter-bombers with upgraded avionics, larger weapon payloads, and air-refueling capabilities for the PLAAF. Thereafter, China placed two additional orders of 38 aircraft for the air force and 24 Su-30 MK2 for the PLA naval aviation forces.[38] These purchases were perceived as a stopgap measure for the PLAAF to create a sizable fleet of fourth-generation aircraft. More importantly, the Chinese acquired first-hand exposure to the latest technology of modern fighters. In 1996, China entered into another agreement with Russia on the licensed production of 200 Su-27s at the Chinese aircraft factory in Shenyang. This inaugurated China's production of Su-27 aircraft under the name of J-11 and later J-11A.[39]

---

36 Tang Jianguang and Yang Xinggeng, "From 'Jian-7' to 'Thunder Dragon,'" *Zhongguo xin-wen zhoukan* [*China Newsweek*], No. 149 (2000), 37. Dai Xu, "Goodbye, Old Jian-6 Fighters," 21.
37 Song Yichang, "The Startup of China's Modern Aviation Industry and Reflections on It," *Zhanglue yu guanli* [*Strategy and Management*], No. 4 (1996), 104.
38 Si Gu, "Sukhoi Fighters in China," *Bingqi zhishi* [*Armament Knowledge*], No. 8 (2007), 27–28.
39 Ibid., 26–27.

Beginning in 2004, building on Sukhoi technology, the Chinese commenced the production of J-11B with their own technology, including the domestically produced WS-10A engines, new radar, avionics systems, and air-to-air missiles. These improvements enabled the J-11B to outperform the Russian-made Su-27s. A further improved variant—J-11D—with a new active electronic scanned array (AESA) radar and upgraded WS-10 engines was handed over to the air force for operational testing and evaluation. What the PLAAF values the most is the domestic version of Su-30MKK under the name of J-16. This heavy multi-role fighter/fighter-bomber has a more powerful AESA fire control radar with a greater weapon load, and a longer range, and is capable of carrying a variety of domestic guided weapons for air-to-air, air-to-ground, and air-to-sea missions. SIt entered PLAAF service in 2015, and will be the main strike capability of the Chinese air force in the coming decades. Meanwhile, an electronic warfare (EW) Wild Weasel variant of the J-16, equivalent to the American EA-18D, is under development.[40]

While engaged in stopgap efforts, China has also concentrated on the development of domestically made aircraft. In the early 2000s, after more than a decade of effort, the Chinese aviation industry successfully developed J-10s with the characteristics of fourth-generation jet fighters, powered by a Russian AL-31F-type engine. Since then, the J-10 has been continuously upgraded with enhanced fourth-generation electronics, including both passive and active electronically scanned array (PESA/AESA) radar and a Chinese-made WS-10 turbofan engine, currently known as the J-10B/C. Moreover, another upgraded version powered by a new turbofan engine with a stealth swath thrust vector nozzle (TVC) is upcoming.[41] The PLAAF is reportedly very enthusiastic about this new multi-role fighter that, as part of a rapid replacement program, is enabling the PLAAF to compete against American F-15, F-16, and F-18 fighters flown by its potential adversaries.

The latest achievement by the Chinese aviation industry has been the development of a series of "20s" military aircraft, including the J-20 stealth fighter, the heavy-load transport aircraft Y-20, the medium helicopter Z-20, and the long-range strategic bomber H-20. Except for bomber H-20, which

---

40 Hui Tong, "Chinese Military Aviation," <http://chinese-military-aviation.blogspot.com/>.
41 Ibid.

is a four-engine stealth flying wing design similar to the American B-2 with an expectation to take a maiden flight by 2020, the Y-20, and the Z-20, and the J-20 entered service with the PLAAF and the PLA Army in 2016, 2018, and 2019, respectively. The J-20 is a truly meaningful fifth-generation fighter equipped with AESA, EODAS, Intel Rapid Storage Technology, advanced cockpit design and head-mounted display (HMD) that improves pilot's situation awareness by providing the highest degree of "information fusion," allegedly competitive with the American F-22 and F-35.[42] Its overall performance will continue to improve in the coming years as more powerful WS-15 engines with TVC capability become available.[43] More significantly, the appearance of the J-20 has already reshaped the air power landscape in the East Asian and western Pacific regions, prompting Japan and South Korea to pursue the development of their own fifth-generation stealth fighters or to purchase the U.S. F-35s.

For the past 10 years, increasing focus has been also placed on informationalization as a leapfrog measure to close the PLAAF's cyber and EW gap with the United States and western Europe. The development of sophisticated communications, command, and control (C3), and intelligence, surveillance, and reconnaissance (ISR) capabilities has been the PLAAF's most urgent priority.[44] Following earlier experimental trials using an obsolete Soviet-legacy Tupolev Tu-4 modified with turboprop engines and rudimentary search radar in a saucer dome, China initially developed two "high-low" versions of a domestic AWACS (Airborne Warning and Control System): the high-end KJ-2000 based on the Russian IL-76MD airframe; and the low-end KJ-200 based on the Y-8F-200 transport platform. These platforms were handed over to the PLAAF in 2005 and 2006, respectively, to coordinate fighters and bombers via secure datalinks.

Simultaneously, China launched the "High New" (Gaoxin) project to develop a series of AWACS and EW aircraft first based on Y-8 and then on Y-9 turboprop transport platforms. There are 12 types of such aircraft, one of which is a medium-sized AWACS—KJ-500—carrying a new digital

---

42 Ibid.
43 Rick Joe, "China's Stealth Fighter: It's Time to Discuss J-20's Agility," Dec. 7, 2018, *The Diplomat*,<https://thediplomat.com/2018/12/chinas-stealth-fighter-its-time-to-discuss-j-20s-agility/>.
44 Cai Fengzhen and Tian Anping, *Kongtian yiti zuozhan xue* [*Study of Integrated Aerospace Operations*] (Beijing: Liberation Army Press, 2006), 554–56.

radar with the same capability as a KJ-2000, a satellite communication system (SATCOM) and additional electronic signals intelligence (ELINT) antennas. The PLAAF and the PLA Navy Air Force (PLANAF) received this aircraft in 2014 and 2015 respectively. Currently, mass production is under way. The PLAAF has more AWACS and EW capabilities than ever before.[45]

In retrospect, although the U.S. government successfully pressurized Israel to cancel the sale of the Phalcon AWACS system to the PLA in 1999, China appears to have pulled together sufficient talent and resources to build its own system despite this seeming setback. The chief engineer and designer of the Chinese AWACS project recently claimed that China's radar technology has reached the same level as leading foreign countries and in some areas it is even better.[46] Efforts by the United States and European countries to prevent China from obtaining high-tech weapons similarly do not seem to have succeeded. China's current achievement in developing airborne radars, sensors, and avionics has prompted the PLAAF to explore system-of-system operations.

Continuing Constraints and Challenges

The PLAAF still faces many challenges in reaching its development goals. To understand China's claims about the development of Chinese aerospace power for the present and the near future, it is necessary to consider several issues, including strategy, force structure, the officer corps, the enlisted force, unit training, logistics, and maintenance. The PLAAF's strategic vision, calling for the development of a strategic air force with long-range capabilities, as well as integrated air and space operations employing firepower systems that incorporate advanced information technology, is modeled on U.S. practices as the Chinese have perceived them.

However, the Chinese military system had long consisted predominantly of ground forces; accordingly, army-centric thinking and leadership have played a dominant influence in shaping the PLA's command and

---

45 Hui Tong, "Chinese Military Aviation," <http://chinese-military-aviation.blogspot.com/>.
46 He Yi, "Special Interview of Wang Xiaomo, Member of Chinese Academy of Engineering, and Chief Designer of Our Country's Early Warning Aircraft," *Bingqi zhishi* [*Ordnance Knowledge*] No. 11A (2009), 12–16.

control, and force structure. The PLA's four general departments—the General Staff Department, General Political Department, General Logistics Department, and General Armament Department—served as the headquarters for all services, namely the ground force, navy, air force, and the second artillery force. Military regions served as the command organization for ground troops, and only played a concurrent leadership role for the personnel of other services located within their regions during wartime. Ground force officers commanded the military regions, and the commanders of the other services could only serve as their deputies. Since there was no permanent joint organization at the military region level, when a joint command organization was formed, air force officers could only assume assistant (hence subordinate) positions.

The PLAAF has also organized along these lines with four departments in its headquarters at the top as well as in each military region. It has combined aviation with ground-to-air defense forces, consisting of aviation, surface-to-air missiles (SAM), and anti-aircraft artillery (AAA), and airborne units, as well as communications, radar, electronic countermeasures (ECM), chemical defense, technical reconnaissance, and other specialized elements working together. These units followed the hierarchy of division, regiment/field station, group/battalion, and squadron/company. The PLAAF's organizational structure had multiple components and layers, many of which overlapped, spawning redundancies. Despite reorganization and restructuring because of force reduction over the years, the traditional organizational structure of the PLAAF remained intact.[47]

The new round of reforms since 2015 downgraded the four general departments and reorganized them into 15 functional departments and offices of the CMC. A Strategic Support Force was created and is responsible for developing and employing most of the PLA's space, cyber, and EW capabilities. The PLAAF similarly reorganized. Five new theater commands that replaced military regions were created to command ground, naval, air, and rocket forces assigned to their theater to undertake operations. The military services are responsible only for organizing, training, and equipping units.[48] The PLAAF's command hierarchy has shrunk from five levels

47 Shang Jinsuo, chief ed., *Kongjun jianshe xue*, 398.
48 "Central Military Commission Opinion on Deepening the Reform of National Defense and the Armed Forces," <http://news.xinhuanet.com/mil/2016-01/01c_117646695.htm>.

(air force, military region air force, air corps, division, and regiment) to three levels (air force, theater air force, and air units). The PLAAF also reorganized its aviation, SAM, radar, and airborne forces into brigade-sized units. Supposedly, the reform streamlined the PLA's command and control and the PLAAF's force structure.

A conventional academic consensus is that instituting change in military organization is at best difficult. The PLA remains a Party-controlled armed service with senior PLA leaders socialized by the unique Party–army relationship that has also rewarded them with promotion to the higher ranks. As a result, the PLA senior officers are unlikely to seek greater autonomy, and its junior officers and enlisted personnel are probably less interested in professional creativity. Currying favor with the leadership is a cultural phenomenon in any political system dominated by absolute authority and arbitrary decisions by key individuals. It represents not only military service subordination to the Party (strongly entrenched in Chinese military culture) but also demonstrates the political reliability and loyalty of military service to individual senior Party leaders. The PLAAF is no exception to this. Its leadership is confident that, when they bring their requests to the Party leadership's personal attention, they will receive favorable approval. Nothing should upset the continuity of this entwining Party–military bond of mutual support.

Despite the PLAAF's current efforts to streamline and optimize its force structure, the Chinese air force remains a mixed force of aviation, ground-based air defense, radar, and airborne units, while space assets remain separate from it. This mixed-force structure will continue to complicate China's air and space decisions, particularly with regard to training and allocating roles and missions among the services and branches, and will influence resource allocation for Chinese air force modernization. Currently, the PLAAF still has quite a number of the older J-7 and J-8 fighters in service, and as do its air defense forces, which remain equipped with obsolete SAM systems. The development of advanced aircraft engines remains a holdup, slowing the ongoing productivity of J-20s and Y-20s. As a result, the PLAAF's offensive and long-range airlift capabilities will remain limited over the next few years. Moreover, it is worth noting that the PLAAF does not possess its own space assets or strategic missiles, which are controlled by the PLA Rocket and Strategic Support Forces, respectively. It is still unclear how the PLAAF will become a service with integrated air and space capabilities.

One ultimate objective of the ongoing military reform is to enhance the PLA's ability to carry out joint operations in modern warfare. The PLA has not fought a war for more than 30 years, becoming increasingly lacking in warfighting awareness and capabilities. Meanwhile, China has faced growing security challenges, both conventional and non-conventional, especially along its maritime periphery and the areas separated by the first island chain between the East China and South China Seas. Domination of the skies over these two areas would give China a decisive advantage in defending its sovereign territories, such as Taiwan, and disputed islands, such as Diaoyu/Senkaku, Paracel, and the Spratly Islands. Since 2015, the PLAAF has increasingly conducted high seas training over the western Pacific and the South China Sea waters, including flying near and around Taiwan with H-6K bombers, Su-30 and J-11 fighters, and Y-9 EW aircraft.

By transitioning from older generation aircraft to new aircraft with significantly improved capabilities, the PLAAF is also enhancing its training by incorporating new systems and methods. It has placed emphasis increasingly on technical and tactical training in complex environments, combined arms and aircraft-type training, and joint training under mission-oriented and confrontational conditions. To develop similar fighter weapons and tactics training programs like those of the U.S. Air Force (USAF), the PLAAF has reorganized its testing and training base at Dingxin, turning it into a modern combat training base with a new "blue force" (brigade-sized unit) equipped with J-20s, J-10s, Su-30s, and J-16s, playing the role of the "opposition force," and combat electromagnetic environment simulators. Air units from the PLAAF and PLANAF are required to go to Dingxin to participate in the annual "Red Sword" exercises that conduct uninterrupted and unrestricted confrontations around the clock. These exercises are the equivalent of the USAF's Red Flag series, which involve various troops from the air force, such as aviation, ground-to-air missile, radar, and electronic countermeasure units, and aim to beef up the PLAAF's systematic combat capabilities.[49]

The PLAAF established the "Golden Helmet" competition for fighter pilots in 2011 and the "Golden Dart" competition for attack aircraft pilots

---

49 "The Red Sword–2018 Exercise of the Air Force Is Devoted to Improve Systematic Capabilities for Winning," <http://www.xinhuanet.com/politics/2018-05/24/c_129879430.htm>.

in 2013, using a multi-information airborne digital information recording system to record the air combat information. Most winners of these competitions are young pilots who are more likely to accept the less-scripted training than older pilots who grew up with a follow-the-plan mindset. The program has stimulated the PLAAF units to conduct their training more aggressively than ever.[50] Nevertheless, training gaps remain as exposed in the PLAAF's joint exercises with the Russian air force, whose pilots are accustomed to taking a narrow path to launch air-ground attacks, engaging in a series of aerobic maneuvers and hard turns in bombing runs, and landing by a high angle of attack approach.[51]

The PLA has also recognized systems confrontation (contest between opposing operational systems) as the basic mode of warfare in the twenty-first century. Chinese military researchers have developed their concepts and ideas about a system of systems and systems warfare by drawing lessons from the war experience of the United States and its allied forces since the early 1990s, and by learning from western military literature. Systems thinking and systems concepts appear to be a guiding logic behind the recent PLA reorganization and force restructure.[52] For years, the PLA carried out the cross-military-region and scenario-based joint exercises to explore and study system of system operations. The PLAAF participates in these joint multiservice exercises, conducting precision strikes, intelligence gathering, command and control, and anti-missile and air defense. A domestic-made datalink similar to the American Link-16 has been deployed to exchange combat information in near real time among the PLA's air, sea, and ground assets in order to improve the efficiency of its envisioned operational systems. Many of the PLA's concepts and theories about how to fight system confrontational warfare remain on paper, and its envisioned operational system has yet to be fielded. The ongoing efforts and commitment by the Chinese will turn such an aspiration into

---

50 Fan Jianghuai and Wang Tianyi, "How Much Do You Know About the Evolutionary History of the Air Force's 'Golden Helmet' Competitions," <http://jz.chinamil.com.cn/n2014/tp/content_8097644_2.htm>.

51 Huang Ming, Zhang Yuqing, and Zhang Mimi, "War Eagle, Flying High for the Motherland—An Account of Air Force Special Grade Pilot, Brigade Commander Hao Jinwen," <http://china.chinadaily.com.cn/a/201901/03/WS5c2e07b9a3100a343d6f1ccf.html>.

52 For systems confrontation and operations, see Jeffrey Engstrom, *System Confrontation and System Destruction Warfare: How the Chinese People's Liberation Army Seeks to War Modern Warfare* (Santa Monica, CA: the RAND Corporation, 2018).

the reality, as the recent PLAAF's roadmap for a modern air force has illustrated.

Conclusion

China's quest for a modern air force has been arduous and tortuous. The PLAAF was created on the basis of a war-tested army. As a result, several factors, such as the Chinese defensive mindset, the PLAAF's combat experience, and lack of domestic industrial capability, shaped China's initial efforts to build a modern air force. The critical role played by air power in modern warfare, since the early 1990s, prompted the PLAAF to adopt a new strategic vision emphasizing integrated air and space, simultaneous offensive and defensive, and integrated firepower and information. However, it has been the rise of China's economic power in the past two decades that has allowed the PLAAF to slowly but steadily turn itself into a modern air force. With increasingly advanced weapon systems entering into the service, the PLAAF has reorganized and restructured itself, modeling itself on U.S. practices as the Chinese have perceived them with an emphasis placed on system-of-system operational capability.

Despite the progress the PLAAF has made in recent years, the Chinese air force faces challenges in becoming a strategic service with integrated air and space and offensive and defensive capabilities to fight system-of-system operations. No doubt, change is underway within the ranks of the PLAAF, which continues to modernize its theory, organizational structure, service personnel, and weaponry. The most severe constraint is perhaps the PLA's political and organizational culture—a relic from its founding—demanding absolute loyalty to the Party. The PLAAF contiguously maintains this tradition. With the current military reforms, change is underway within the ranks of the PLAAF, which embraced a new concept of a modern air force and operations that emphasized developing an air force with an offensive mindset and a new joint force culture. Nevertheless, it will take a long time for the PLAAF to cultivate new ways of doing things and to develop a new organizational system. One thing is for sure, China will not quit until the PLAAF becomes a real, modernized, strategic service.

References

Allen, K.W., G. Krumel, and J. D. Pollack. *China's Air Force Enters the 21st Century*. Santa Monica, CA: RAND, 1995.

Cai, F. and A. Tain. *Kongtian yiti zuozhan xue* [*Study of Integrated Aerospace Operations*]. Beijing: Liberation Army Press, 2006.

Cao, L. "Recollection of the Work on Compilation of Regulations and Teaching Materials." In *Lantian zhi lu* [*The Road to Blue Sky*], 2. Beijing: The Political Department of Air Force, 1992, 158, 165.

Dai, X. "Goodbye, Old J-6 Fighters: A Complete Examination of the Service History of the Last Meritorious Fighter in the Chinese Air Force with Combat Victory Record." *Guoji zhanwang* [*World Outlook*], No. 19, 2005, 21.

Dong, W. *Xiandai kongjun lun* [*On the Modern Air Force* (continuation)]. Beijing: Blue Sky Press, 2005.

Duan, Z., chief ed. *Dangdai Zhongguo de hangkong gongye* [*Contemporary China's Aviation Industry*]. Beijing: China Social Science Press, 1998.

Engstrom, J. *System Confrontation and System Destruction Warfare: How the Chinese People's Liberation Army Seeks to Wage Modern Warfare*. Santa Monica, CA: RAND Corporation, 2018.

Fan, J and T. Wang. "How Much Do You Know About the Evolutionary History of the Air Force's "Golden Helmet" Competitions." <http://jz.chinamil.com.cn/n2014/tp/content_8097644_2.htm>.

He, T. "The Establishment of Air Force Headquarters." In *Lantian zhi lu* [The Road to Blue Sky], 1. Beijing: The Political Department of Air Force, 1992.

He, W. "Military Thought on Air Force." In Yao Wei, chief ed., *Zhongguo kongjun baike quanshu* [*Encyclopedia of Chinese Air Force*], 1. Beijing: Aviation Industry Press, 2005.

He, Y. "Special Interview of Wang Xiaomo, Member of Chinese Academy of Engineering, and Chief Designer of Our Country's Early Warning Aircraft." *Bingqi zhishi* [*Ordnance Knowledge*], No. 11A, 2009.

Hua, R. et al. *Kongjun xueshu sixiang shi* [*The History of the Academic Thinking of the Air Force*]. Beijing: Liberation Army Press, 2008.

Huang, M., Y. Zhang, and M. Zhang. "War Eagle, Flying High for the Motherland— An Account of Air Force Special Grade Pilot, Brigade Commander Hao Jinwen." <http://china.chinadaily.com.cn/a/201901/03/WS5c2e07b9a3100a343d6f1ccf.html>.

Hui, T. "Chinese Military Aviation." <http://chinese-military-aviation.blogspot.com/>.

Information Office of the State Council of the People's Republic of China. *China's National Defense Paper in 2008*. <http://www.gov.cn/english/official/2009-01/20/content_1210227.htm>.

Joe, R. "China's Stealth Fighter: It's Time to Discuss J-20's Agility." Dec. 7, 2018, *The Diplomat*.<https://thediplomat.com/2018/12/chinas-stealth-fighter-its-time-to-discuss-j-20s-agility/>.

Lin, H. *Baowei zuguo lingkong de zhandou* [*Fight to Protect the Motherland's Airspace*]. Beijing: Liberation Army Press, 2002.

Ling, Q. "Revealing the Secret of China's Jian-12 Fighter." *Xiandai bingqi* [*Modern Arsenal*], No. 2, 2001, 2-4.

Liu, H. *Liu Huaqing huiyilu* [*Memoirs of Liu Huaqing*]. Beijing: Liberation Army Press, 2004.

Liu, P. et al. "Biography of Zhu Guang." In *Zhongguo renmin zhiyuanjun renwu zhi* [*Biographical Notes of the Chinese People's Volunteer Army*]. Nanjing: Jiangsu People's Press, 1992.

Liu, Y. "The Initial Seven Years of Establishing the Air Force." In *Lantian zhi lu* [*The Road to Blue Sky*], Vol. 2. Beijing: The Political Department of Air Force, 1992.

Lü, L. *Tongtian zhi lu* [*The Road to the Sky*]. Beijing: Liberation Army Press, 1989.

Min, Z., chief ed. *Kongjun junshi shixiang gailun* [*An Introduction of Air Force Military Thinking*]. Beijing: Liberation Army Press, 2006.

Pei, Z. and B. Xie. "Rushed into Fujian to Combat for Control of the Air." In *Lantian zhi lu* [*The Road to Blue Sky*], 1. Beijing: The Political Department of Air Force, 1992.

Shang, J., chief ed. *Kongjun jianshe xue* [Science of Air Force Construction]. Beijing: Liberation Army Press, 2009.

Si, G. "Sukhoi Fighters in China." *Bingqi zhishi* [*Armament Knowledge*], No. 8, 2007.

Song Y. "The Startup of China's Modern Aviation Industry and Reflections on It." *Zhanglue yu guanli* [*Strategy and Management*], No. 4, 1996.

Tang, J. and X. Yang. "From 'Jian-7' to 'Thunder Dragon'." *Zhongguo xinwen zhoukan* [*China Newsweek*], No. 149, 2000.

Wang, D., chief ed. *Dangdai Zhongguo kongjun* [*Contemporary China's Air Force*]. Beijing: Social Science Press, 1989.

Xinhuanet. "Central Military Commission Opinion on Deepening the Reform of National Defense and the Armed Forces." <http://news.xinhuanet.com/mil/2016-01/01c_117646695.htm.>

Xinhuanet. "Chinese Air Force Announces Its Roadmap for Building a Powerful and Modernized Air Force." <http://www.xinhuanet.com/politics/2018-11/11/c_129991031.htm>.

Xinhuanet. "The Red Sword–2018 Exercise of the Air Force Is Devoted to Improve Systematic Capabilities for Winning." <http://www.xinhuanet.com/politics/2018-05/24/c_129879430.htm>.

Yao, J., chief ed. *Zhongguo hangkong shi* [*China Aviation History*]. Zhengzhou: Daxiang Press, 1997.

Ye, H. "Four Major Changes of Our Country's National Defense Strategy since Its Founding." *Dangdai Zhongguo shi yanjiu* [*Studies of Modern Chinese History*], No. 3, 1999.

Zhang, X. "Air Combat for the People's Republic: The People's Liberation Army Air Force in Action, 1949–1969." In Mark A. Ryan, David M. Finkelstein, and Michael A. McDevitt, eds., *Chinese Warfighting: The PLA Experience since 1949.* Armonk, NY: M.E. Sharpe, 2003.

Zhao, Z. "Air Force Tactics." In *Zhongguo kongjun baike quanshu* [*Zhongguo Air Force Encyclopedia*], 1.

3 RUSSIA
Modernizing Air and Space
Lars Peder Haga

Introduction

Russia remains among the world's great military powers, and a member of the exclusive club of nuclear states. For more than a decade it has rebuilt and reformed its armed forces with the intent of being a significant actor both regionally and globally. Since 2015, the modernized and modernizing Russian air and space forces has projected Russian power in Syria, demonstrating the will and ability to use air power in the "western way," comparable to the operations of the North Atlantic Treaty Organization (NATO) and various ad hoc U.S.-led alliances in the post–Cold War decades.

In the following chapter, Russia's threat perception, economic prospects, and other factors that appear to be shaping the form of Russian air power will be discussed. Russia does not publish domain-specific doctrine so it follows that an assessment of Russian air power doctrine has to be made from observations of exercise patterns and operations, statements to the media by political and military leaders, and inferences from fleet and equipment composition. In addition, there is a rich debate on air power as well as military power in general in professional journals, formal websites, and in the dynamic Russian blogosphere. While these kinds of sources present some methodological challenges, it is quite possible to form a comprehensive image of current debates and developments of Russian air power.

Organizational Structure

Organizationally, Russian air and space force units have taken a nominal step back from the reforms named after then Minister of Defense, Anatolii Serdiukov, that started in 2008. Serdiukov disbanded several aviation units and consolidated them into air bases, which often encompassed several air stations and airfields. The new air bases were subordinate to Air Force and Air Defense commands, numbered one to four, and in turn subordinate to the four new military districts that were created by amalgamating the pre-existing seven.[1] Strategic aviation, transport aviation, and special aviation, which comprises airborne early warning and control, specialized reconnaissance, and the like, remained under centralized control.

Beginning in 2015, while the substance of the structural reform endured, the Air Force and Air Defense commands were renamed as Air Force and Air Defense armies, and air bases were renamed using the familiar divisions, regiments, brigades, and squadrons. Now each military district and the Northern Fleet (the Northern Fleet has, since December 2014, in all but name, been Russia's fifth military district[2]) has an air defense and aviation army. Centralized units are also in the process of being reorganized, or rather renamed, as armies. The numbering of the units as well as honorary titles are inherited from Soviet formations, in line with the prominent role that Soviet military heritage has in the Russian military's public image.

In 2015 the Air Force (Voenno-vozdushnye sily (VVS)) and the Aerospace Defense Forces (Voiska vozdushno-kosmicheskoi oborony (VVKO)) were amalgamated into the Air and Space Forces (Vozdushno-kosmicheskie sily (VKS)). The cited reasons were to streamline command and control and further integrate all ground-based air- and missile-defense forces, which formerly were split between the two branches of the armed forces.[3]

---

1   Anton Lavrov, "Reform of the Russian Air Force," in *Russia's New Army*, ed. Mikhail Barabanov, Moscow: Centre for Analysis of Strategies and Technologies, 2011, 57–59.
2   Iurii Gavrilov, "'Sever Arktiki'. Sozdano novoe strategicheskoe komandovanie," *Rossiyskaya Gazeta*, No. 6546 (274), Dec. 1, 2014 <https://rg.ru/2014/12/01/komandovanie-site.html>.
3   Aleksei Nikolskii, "Vozdushno-kosmicheskie sily Rossii obedinili VVS i vojska Vozdushno-kosmicheskoy oborony," *Vedomosti*, Aug. 3, 2015, <https://www.vedomosti.ru/politics/articles/2015/08/04/603322-vozdushno-kosmicheskie-sili-rossii-obedinili-vvs-i-voiska-vozdushno-kosmicheskoi-oboroni>.

Threat Perception

Russia's threat perception is complex and stresses multiple threats and likely vectors of attack. The military doctrine of 2014 notes that while the risk of a full-scale war is reduced, other threats are increasing. Among these is a tendency for military threats to "move into the information domain and the internal sphere of the Russian Federation."[4]

A more fine-grained threat perception can be found in the Russian General Staff's Chief of Staff Valerii Gerasimov's series of articles in *Voenno-promyshlennyi kur'er*, starting with the "Value of science in prediction"[5] from 2013. This article gained notoriety as the so-called Gerasimov doctrine after the events in Ukraine during the winter and spring of 2014, as it was widely interpreted as a blueprint for Russian "hybrid operations." While Gerasimov's articles are of course intended to stimulate Russia's appropriation and development of similar ways of warfare, they should primarily be read as a threat analysis. Gerasimov outlines the Russian military establishment's contemporary military threats, and also how they are expected to materialize in the near future.

The existential threat to Russia's sovereignty stems from what the Russians view as a U.S.-dominated NATO, which is expansionist by default, and aims to either maintain or reestablish a unipolar world order.[6] While the West is portrayed as preferring to reach its goals by engineering "color revolutions," it will also eventually back up a faltering or failed revolt with military power, with air power as its first choice. The wars against Iraq, Yugoslavia, Libya and attacks on Syria in the 1990s and 2000s are considered blueprints for how this modern style of warfare looks, all with a heavy emphasis on air power.

Finally, the threat from NATO is perceived to be comprehensive, or hybrid. Gerasimov in 2016 defined hybrid methods as "the complex application of political, economic, informational, and other non-military means, realized with the support of military force."[7]

---

4   *Voennaya doktrina Rossiiskoy Federatsii*, 4–5.
5   Valerii Gerasimov, "Tsennost nauki v predvidenii," *Voenno-promyshlennyi kurer*, No. 8 (476) Feb. 27, 2013.
6   See e.g., Sergei Shoigu, "Organizatsiia protiv khaosa," *Voenno-promyshlennyi kurer*, No. 14 (727), Apr. 10, 2018, <https://vpk-news.ru/articles/42095>.
7   Valerii Gerasimov, "Po opytu Sirii," *Voenno-promyshlennyi kurer*, No. 9 (624), Mar. 9 2016.

This threat picture has resulted in a continuation of the Soviet Union's emphasis on aerospace control and defense, and on strategic deterrence. The long-standing air power truth, shared also by Russia, that control of the air is the prerequisite for carrying out any kind of air operations, has led to the prioritization of combat aircraft in the procurement of new platforms.[8] The addition of so-called hybrid threats, in particular information warfare, is likely to be a driving force behind Russia's impressive media promotion of the armed forces, including air power.

Economic Prospects

Now the Russian economy is growing again after the economic setbacks following the financial crisis in 2008, the collapse of oil prices in 2013, and the onset of sanctions after the annexation of Crimea in 2014. The growth is, however, sluggish and with poor prospects for serious improvement unless Russia carries out substantial structural reforms.[9]

Russia in 2017 was the fourth largest spender on the military in the world after the United States and China with a wide margin and Saudi Arabia with a narrower margin.[10] A new state program for modernization and rearmament of Russia's military forces until 2027 (GPV-2027) was launched in February 2018. The post-2014 economic setbacks, likely infighting between the branches of the military over priorities, as well as conflict with the department of the treasury over the amount of funding, caused the adoption of the plan to be postponed several times,[11] but it was finally approved in February 2018. The program is supposed to spend slightly less than the nominal amount of money that was set aside for the preceding program (GPV-2020), but the real amount will be lower due to the mounting inflation. On the other hand, as the import substitution

---

8   Anton Lavrov, "Reform of the Russian Air Force," in *Russia's New Army*, ed. Mikhail Barabanov, Moscow: Centre for Analysis of Strategies and Technologies, 2011, 62.
9   OECD, "Russia," *OECD Economic Outlook*, Vol. 2018, Issue 2, Preliminary Version, 178–80.
10   Nan Tien et al., "Trends in World Military Expenditure," *SIPRI Fact Sheet*, May 2018, 2–3, <https://www.sipri.org/sites/default/files/2018-04/sipri_fs_1805_milex_2017.pdf>.
11   Aleksandra Dzhordzhevich and Ivan Safronov, "U trillionov est dva soiuznika – armiia i flot," *Kommersant*, No. 235 (6229), Dec. 12, 2018, <https://www.kommersant.ru/doc/3500710>.

campaign necessitated by western sanctions creates results and most defense purchases will be made domestically, the rouble will go a longer way than before, and predictions that Russia's military modernization is facing a collapse are exaggerated. The current Russian regime has also for a time declared an ambition to the military's share of gross domestic product (GDP). Russia is, thus, most likely going to continue modernization of the armed forces at a relatively steady pace in the close future.[12]

Training

After the collapse of the Soviet Union and well into the 2000s, the training standards of Russian pilots were low because of the abysmally poor availability of flight hours.[13] To carry out the most complicated missions during the Five-Day War against Georgia (Russo-Georgian War), Russia mobilized test pilots from the training centers, as pilots in line units simply lacked the necessary training and experience with the relevant weapon systems.[14]

The Serdiukov reforms slashed the number of military schools and education centers, and there was a genuine fear of underproduction of pilots as a result.[15] However, increasing the number of flight hours has been a priority, and in 2018, the Ministry of Defense (MoD) claimed that Russian pilots had an average of 100 flight hours, and 120 for "young pilots."[16] The real significance of the numbers are hard to assess, as it is impossible to know what quality and kinds of training the 100–120 hours contain. One hundred hours also seems low by western standards, where close to

---

12 Michael Kofman, "The Collapsing Russian Defence Budget and Other Fairy Tales," *Russia Military Analysis Blog*, May 23, 2018, <https://russianmilitaryanalysis.wordpress.com/2018/05/23/the-collapsing-russian-defence-budget-and-other-fairy-tales/>; Michael Peck, "Why Russia Won't Go Bankrupt Paying for Its Military Buildup," *National Interest*, Aug. 8, 2018, <https://nationalinterest.org/blog/buzz/why-russia-wont-go-bankrupt-paying-its-military-buildup-28092>.
13 Lavrov, 2011, 53.
14 "Shturmany Shpitonkovy," *Krasnaia Zvezda*, Aug. 6, 2015, <http://redstar.ru/index.php/newspaper/item/25216-shturmany-shpitonkovy>.
15 Aleksandr Babakin, "Kadrovyi shtopor aviatsii," *Nezavisimoe voennoe obozrenie*, No. 25, Jul., 25, 2014.
16 Russian MoD, "V komandovanii Voenno-vozdushnykh sil VKS podveli itogi za 2018 god," Dec. 4, 2018, <https://function.mil.ru/news_page/country/more.htm?id=12206756@egNews>.

200 flight hours or more seems to be considered the minimum for maintaining, as well as building, advanced skill sets for combat pilots.[17]

When the operations in Syria began however, the training situation had improved sufficiently to allow the deployment of pilots from regular units, and Russia has taken advantage of the Syrian campaign to rotate most combat aircrews through the theater in order to gain combat experience.[18] Even though this is an environment devoid of a serious and persistent air threat against Russian operations, the combat experience and operational flight hours are likely to have given a substantial boost to the skill levels of Russian pilots.

Priorities, Roles, and Modernization Plans

Russia is one of a few military powers with a realistic ambition to build a complete air power system on its own, without relying on the capabilities of allies. After the end of the "acquisitions break" that lasted from the early 1990s until 2012, Russia's aerospace forces and naval aviation have received consistent deliveries of both new-built and upgraded existing aircraft. However, it is necessary to take into account a Russian tendency to announce hyper-ambitious plans and projections that later are quietly downplayed when assessing Russian fleet modernization. For example, in 2013, Viktor Bondarev, Commander of the Russian Air Force, claimed that the T-50 (now known as the Su-57) fifth-generation fighter should begin to enter service from 2017, and that deliveries of the PAK-DA future bomber would begin in 2020.[19]

---

17 Steven Losey, "Pilots Are Flying 17 Hours Per Month, But It's Still Not Enough, Air Force Secretary Says," *Airforce Times*, Mar. 1, 2018, <https://www.airforcetimes.com/news/your-air-force/2018/03/01/secaf-air-force-pilots-are-flying-17-hours-per-month-but-its-still-not-enough/>.

18 Pavel Nastin, "Shojgu: dve treti letnego sostava VKS poluchili boevoyu praktiku v Sirii," *TV Zvezda*, Aug. 12, 2018, <https://tvzvezda.ru/news/forces/content/201808120023-nfy6.htm>.

19 Aleksandr Pinchuk, "VKS: Na novoi tekhnike k novym vozmozhnostiam," *Krasnaia Zvezda*, No. 159, Sep. 4, 2013, 1.

Strategic Aerospace Defense

Russia's geography, with an immensely long border, large stretches of which are in very sparsely populated areas, has made Russia vulnerable to incursions from the air. Border-crossing spy flights were a more or less regular occurrence in the post-WW II Soviet Union, until they were finally ended by the downing of Gary Powers and his Central Intelligence Agency-operated U-2 by a ground-to-air missile in 1960. A post on identification of foreign spy flights along Russia's borders has been permanently and prominently placed on the Russian MoD's weekly reports on their activities since the beginning of the reports' publication in 2017[20] and is a reminder of this sense of constantly being watched and challenged in the airspace along the border.

This historical vulnerability and the fact that the above-mentioned regime-change operations that are central to Russia's threat perception were carried out by a massive use of air power explain why Russia strongly emphasizes a functioning and efficient defensive air protection system. The radars and command-and-control infrastructure of aerospace control units, known in Russia as radio-technical troops, are being renewed and overhauled.[21] Russia is a world leader in the development of surface-to-air missile systems (SAM), building, deploying, and exporting systems across the full spectrum, from man-portable air defense systems to extreme long-range systems. The qualities of Russian SAM systems have attracted a great number of international customers, including NATO member Turkey. Since 2011, Russia has acquired 19 or 20 regimental sets of the S-400 long-range air defense missile system[22] out of 28 planned in the state armament plan for 2011–2020.[23] Together with other Russian defensive and offensive missile systems, the density and quality of Russian air defense systems have brought the concept of anti-access and area denial (A2/AD) into the strategic debate in Europe.[24]

---

20 Russian Ministry of Defence (@mod_russia), Aug. 10, 2017, <https://twitter.com/mod_russia/status/895897917755666434>.

21 Anastasia Evdokimova, "Vozdushno-kosmicheskie sily Rossii poluchili bolee 70 noveishikh RLS," *TV Zvezda*, Jan. 6, 2018, <https://tvzvezda.ru/news/forces/content/201801060824-5ies.htm>.

22 BMPD, "Ministerstvo oborony poluchilo tretii v 2017 godu polkovoi komplekt zenitnoi raketnoi sistemy S-400," Oct. 27, 2017, https://bmpd.livejournal.com/2914426.html>.

23 Lavrov, 2011, 75.

24 Luis Simon, "Demystifying the A2/AD Buzz," *War on the Rocks*, Jan. 4, 2017, <https://warontherocks.com/2017/01/demystifying-the-a2ad-buzz/>.

Strategic Deterrence and Long-Range Bombers

Just like Russia sees strategic air defenses as vital for the preservation of its sovereignty, the ultimate guarantor is its nuclear deterrent force. While then deputy prime minister for armaments Dmitri Rogozin questioned the continued utility of an air-based component back in 2012,[25] the current Russian regime appears firmly determined to retain a nuclear triad—with air-based weapons as well as submarine and ground-launched missiles.

The program to develop a new strategic bomber (PAK-DA) is still ongoing. However, the first, wildly optimistic plans for delivery of the first aircraft in the early 2020s have been revised to delivery in 2028.[26] That it will be years before a new bomber is operational is quite clear from the stop-gap plan to renew production of the Tu-160 Blackjack strategic bomber,[27] and the modernization of existing as well as development of new air-launched cruise missiles for the Tu-22 Backfire.[28] The Tu-95 fleet is also being upgraded, giving the venerable Bear a lifespan comparable to the B-52.

In addition to nuclear deterrence, Russia is also developing airborne conventional deterrence capabilities. The need to create conventional deterrence is one of the few points in the current Russian military doctrine that likely is pointing specifically to air power.[29] The long-range strikes by strategic aviation assets in Syria, are clearly intended to not only test the capabilities of new conventional air-launched cruise missiles, but also to signal to the world that Russia is developing a long-range, high-precision strike capability. Also, the ability to carry out intermediate and long-distance strikes drives demand for long-range reconnaissance systems. Russia has few of these, at least with the capability to operate in disputed airspace, and this is likely the driving force behind the many programmes to develop medium- and high-altitude, long endurance unmanned aerial vehicles (UAVs).

---

25 Iegor Sozaev-Guriev, "Rogozin i Makarov posporili iz-za novogo bombardirovshchika," *Izvestiia*, June 5, 2012, <https://iz.ru/news/526580>.

26 Luiza Ignateva, "Модернизация 2.0: Казанский авиазавод замахнулся на 20 миллиардов," *Realnoevremia,* Aug. 16, 2018, <https://realnoevremya.ru/articles/109582-19-mlrd-na-modernizaciyu-kazanskogo-aviazavod-im-gorbunova≥>.

27 Svetlana Bocharova and Aleksei Nikolskii, "Dalniaia aviatsiia poluchit 10 Tu-160M2 za 160 mlrd rublei," *Vedomosti*, Jan. 25, 2018, <https://www.vedomosti.ru/technology/articles/2018/01/25/748964-tu-160m2>.

28 Aleksei Ramm, "Ubijtsam avianostcev popolniat arsenal," *Izvestiia,* Nov. 17, 2017, <https://iz.ru/663570/aleksei-ramm/ubiitcam-avianostcev-popolniat-arsenal>.

29 *Voennaya Doktrina Rossijskoj Federatsii,* 2014, 14.

Combat Aircraft

The think tank the Centre for Analysis of Strategies and Technologies (CAST) assessed in 2016 that Russia had slightly less than 700 fighter aircraft of various types, with little more than a third of these being newer than 10 years old.[30] The rest of the airframes were built in the 1980s and early 1990s. The long hiatus in acquisitions since the early nineties, poor maintenance, and less than optimal storage has likely taken its toll on the aircraft and they are gradually being overhauled or replaced.

The first contracts for deliveries of Russia's Su-57 fifth-generation fighters were signed in 2018 after several delays. The announcement in 2019 of an order for 76 aircraft to be delivered by 2028 means that the fifth-generation fighter will make up roughly a tenth of Russian combat aircraft by then. The Russian MoD has on the other hand so far ordered 116 Su-30SMs and 95 Su-35Ss,[31] and received at least 129 Su-34s.[32] It is possible that an order for as many as 60 Su-30SMs with new engines and domestic avionics will be concluded in the near future, with deliveries in the period 2021–2017.[33]

Heavy Transport

The sheer size of Russia and its armed forces creates a great demand for military air transport. The Russian aerospace forces have a fleet of more than a hundred Il-76 aircraft that remain the workhorse of Russian military transport aviation. The conflict with Ukraine has put an end to deliveries and support for the heavier Antonov transports—delivery of new Antonov aircraft is highly unlikely, unless there is a dramatic shift in Russo-Ukrainian relations. When Ilyushin struggled to deliver new-built modernized Il-76s in cargo and tanker versions within agreed costs, it caused Russia's MoD to

---

30 BMPD, "VKS Rossii planiruet imet 700 istrebitelei," Dec. 16, 2016, <http://bmpd.livejournal.com/2326878.html>.
31 BMPD, "Vse, chto vy khoteli znat ob istrebitele Su-35, no stesnialis sprosit," Dec. 31, 2017, <https://bmpd.livejournal.com/3029961.html>.
32 BMPD, "VKS Rossii poluchili tri pervykh v 2019 godu frontovykh bombardirovshchika Su-34," June 12, 2019, <https://bmpd.livejournal.com/3672219.html>.
33 Aleksandra Dzhordzhevich and Andrei Safronov, "Voennoi aviatsii obeshchano skoreishee ispravlenie," *Kommersant*, No. 17, Jan. 31, 2019, <https://www.kommersant.ru/doc/3868348?from=main_6>; BMPD, "Planiruetsia novyi kontrakt na istrebiteli Su-30SM s dvigateliami AL-41F-1S," Aug. 3, 2019, <https://bmpd.livejournal.com/3726885.html>.

acquire a smaller number of aircraft.[34] However, the existing fleet of Il-76s is likely to be serviceable for another 15 to 20 years.[35] While there have been statements that Russia will build its own heavy transports, it remains to be seen whether these plans will materialize, and in any event it will be a process that will take many years. Serial production of new light and medium transports by Ilyushin also seem to be several years away. In the longer run, Russian transport aviation may face a shortage of aircraft.[36]

An obvious answer would be to look into tapping the potential of civilian airlines and cargo companies for strategic mobility. Russian air cargo companies are partaking in the international growth in aerial transport and increasing their capacities.[37] Land and naval forces are currently developing cooperation with civilian shipping, but the author has found no evidence of similar systematic cooperation with civilian air transport companies. When the Volga-Dnepr company in the spring of 2018 refused to bid on a new tender for air transport for NATO, in retaliation for western sanctions, there was no mention of providing air transport services to the Russian military, but rather an emphasis on providing civilian services only.[38]

Air-to-Air Refueling

Russia likely has, as at mid 2019, fewer than 20 operational Il-78M and Il-78M air-to-air refueling (AAR) aircraft. The numbers are sufficiently small that they are considered a centralized resource, parceled out to the military districts on demand, and with priority to strategic aviation. The Russian MoD has signaled an interest in buying as many as 30 new

---

34 Ivan Safronov, Aleksandra Dzhordzhevich and Sergei Titov, "Il-76 Menyayut poletnoe zadanie," *Kommersant*, June 8, 2017, <https://www.kommersant.ru/doc/3319791>.
35 Lavrov, 2011, 67.
36 Tom Waldwyn, "Russian Military Lift Risks Atrophy," *Military Balance Blog*, Jul. 6, 2017, <https://www.iiss.org/blogs/military-balance/2017/07/russian-military-lift>.
37 Dmitrii Shapkin, "Nazlo torgovym vojnam. Vozdushnyj transport," *Kommersant*, Apr. 12, 2018, <https://www.kommersant.ru/doc/3598275?query=%D0%B2%D0%BE%D0%B7%D0%B3%D0%B0-%D0%B4%D0%BD%D0%B5%D0%BF%D1%80>.
38 Aleksei Tarkhanov, Elizaveta Kuznetsova and Galina Dubina, "Ruslany uvolnyayutsya so sluzhby NATO," *Kommersant*, Apr. 17, 2018, <https://www.kommersant.ru/doc/3605505>.

Il-78M90A AAR aircraft.[39] However, so far there is only a contract for designing and building prototypes, and the project is currently two years behind schedule.[40]

Airborne Early Warning and Control Aircraft

In 2017, Russia possessed no more than 14 Beriev A-50 Airborne early warning and control (AEW&C) aircraft, of which at least 9 were operational. AEW&C aircraft are also considered a centralized resource. Since 2009, four have been upgraded to the A-50U standard.[41] In the longer run, the A-50Us will be augmented and replaced by new A-100 AEW&C aircraft, but so far this project has only been ordered and built as a prototype, meaning that serial production is unlikely to begin before well into the 2020s.[42]

Unmanned Aerial Vehicles

A major finding of the Five-Day War with Georgia in 2008 was that the vintage UAVs of the Russian armed forces did not meet modern standards. There has been a fast acquisition of tactical and operational-level UAVs since then, both domestically built and imported/license built, primarily short- and medium-endurance for the battalion, brigade, and army corps levels. Proliferation of UAVs in the land forces has happened fast. In 2017, a Russian official claimed that there were more than 2,000 in service.[43] Tactical UAVs have apparently been used to great effect in Syria as well as Ukraine for reconnaissance, direction of fire, and battle damage assessment (BDA).

Russia lacks operational UAVs for higher flight levels and longer endurance than the license-built version of the Israeli Searcher, known as

39 UAC Russia on Livejournal, "Tanker obrel telo: na 'Aviastar-SP' idet sborka pervogo rossiiskogo samoleta-toplivozapravshchika," Jan. 2, 2016, <https://uacrussia.livejournal.com/37218.html>.

40 BMPD, "Nachaty letnye ispytaniia pervogo obraztsa samoleta-zapravshchika Il-78M-90A," Jan. 26, 2018, <https://bmpd.livejournal.com/3064340.html>.

41 BMPD, "VKS Rossii poluchili chetvertyi modernizirovannyi samolet A-50U," Mar. 7, 2017, <https://bmpd.livejournal.com/2477303.html>.

42 BMPD, "Pervyi polet opytnogo obraztsa novogo samoleta radiolokatsionnogo dozora i navedeniia A-100," Nov. 18, 2017, <https://bmpd.livejournal.com/2961071.html>.

43 Interfaks-AVN, "Boevaia bespilotnaia aviatsiia v Rossii poluchila kachestvenno novoe razvitie, zaiavliaiut v Sovbeze RF," May 15, 2017, <http://www.militarynews.ru/story.asp?rid=1&nid=451001>.

Forpost, in Russian service. While there are several ongoing development projects, no medium- or high-altitude, long-endurance UAVs are likely to be in Russian service for many years to come.[44]

Precision Guided Munitions

So far, Russia has not adopted the modular approach to building precision-guided weapons by adding guidance kits to existing "dumb" iron bombs, and every precision munition is built as a dedicated weapon, increasing cost and reducing flexibility. Most bombing in Syria has been done with unguided dumb bombs. Despite claims that the advanced SVP-24/"Gefest" bombsight to a degree compensates for this shortcoming,[45] the lack of modern precision weapons has forced the Russian forces in Syria to rely on riskier low-level tactics with attack aircraft and helicopters.[46] Consequently, aircraft in these categories have suffered the largest combat-related losses in Syria, with at least three, possibly four helicopters[47] and one Su-25 attack aircraft destroyed by enemy fire from the ground. All losses of multi-role combat aircraft have been due to accidents, with the exception of the Su-24M fighter-bomber that was shot down by a Turkish F-16 in November 2015.[48]

Maritime Air Power and Power Projection from the Sea

Soviet carrier doctrine designated carrier aviation for fleet air defense, not for force projection.[49] However, when the only Russian carrier *Kuznetsov* deployed to Syria, it was precisely in this role. The necessity of the deployment is debatable, as the Russians had a base at their disposal in Syria, flew sorties from Russia, and at least on two occasions from a base in Iran.

---

44 Pukhov, 2016, 5.

45 Konstantin Sivkov, "Obiekt na vylet," *Novosti VPK* No. 41 (607), Oct. 28, 2015, <http://vpk.name/news/143127_obekt_na_vyilet.html>.

46 Pukhov, 2016.

47 The Russian MoD claims that a Mi-35 helicopter shot down in July 2016, while indeed crewed by Russians, was actually Syrian. Ilia Kramnik, "Krushenie Mi-35: versii," *Lenta.ru*, Jul. 11, 2016, <https://lenta.ru/articles/2016/07/11/mi35version/>.

48 TASS, "Poteri VKS Rossii v Sirii. Dose," Mar. 6, 2018, <https://tass.ru/info/5012726>.

49 See e.g., Robert J. Lee, "A Brief Look at Russian Aircraft Carrier Development," Jan. 9, 1996, <http://www.rjlee.org/rcar.html>.

The deployment of the carrier traded greater cost and complexity for lesser capability, in the form of reduced payloads and playtime, and older and less capable airframes, even though some of the *Kuznetsov's* Su-33 aircraft were upgraded with new systems to give them land attack capability.[50] A substantial number of sorties were flown from the base at Humaymim,[51] the Russians lost two aircraft, and the carrier was withdrawn in January 2017, earlier than originally announced, ostentatiously because it had fulfilled its mission.[52] The deployment of the *Kuznetsov* thus appears to have been, in charitable terms, a limited and perhaps symbolic military success.

The *Kuznetsov* is currently undergoing comprehensive repairs and upgrades, and will return to operational service no earlier than 2021. Aircrew maintain basic carrier operations skills at the land-based training complexes in Eisk and Saki on the Black Sea. While the Russian shipbuilding industry has presented plans for building new carriers, the modernization program for 2018–2027 does not allocate any money for their realization.[53] The utility of maintaining a fleet of aircraft carriers at all was questioned, some years back also by Dmitri Rogozin, then deputy prime minister with responsibility for armaments.[54]

The *Kuznetsov* is retained mostly to maintain competence on carrier operations and for reasons of national prestige. Given the budgetary pressures on Russian defense spending and the complexity of building new carriers, it will likely remain Russia's only carrier at least until the 2030s, and if fiscal sense prevails over prestige, it may just as likely be its last.

Russia's anti-submarine warfare (ASW) fleet relies mainly on Tu-142 and Il-38 Maritime patrol aircraft, between 25 and 50 years old. The Black

50 Aleksei Ramm, "VMF Rossii udarit po terroristam v Sirii s osoboi tochnostyu," *Izvestiia* Sept. 2, 2016, <http://izvestia.ru/news/630104>.

51 S. O'Connor and T. Ripley, "Russian Carrier Jets Flying from Syria, Not Kuznetsov," *IHS Jane's Defence Weekly*, Nov. 30 2016, Vol. 53, No. 48, 5.

52 Russian MoD, "V sootvetstvii s resheniem Verkhovnogo Komanuyushchego Vooruzhennymi Silami Rossijskoj Federatsii V. V. Putina Ministerstvo oborony Rossii pristupaet k sokrashcheniyu gruppirovki Vooruzhennykh Sil v Sirii," Jan. 1, 2017, <https://function.mil.ru/news_page/country/more.htm?id=12107843@egNews>.

53 Ivan Safronov and Aleksandra Dzhordzhevich, "Iadernoi triade opisali perspektivy," *Kommersant*, May 18, 2017, <https://www.kommersant.ru/doc/3299987?utm_source=kommersant&utm_medium=doc&utm_campaign=vrez>.

54 "Potrebnosti v sozdanii avianostsev u Rossii net, zaiavil Rogozin," *RIA Novosti*, Nov. 16, 2013, <https://ria.ru/defence_safety/20131116/977313945.html>.

Sea Fleet is still operating the antique Beriev Be-12 amphibious planes.[55] A few of the older Il-38 airframes are being upgraded with new avionics and sensors, to the Il-38N standard,[56] while the Tu-142s are merely getting full overhauls in order to prolong their service life.[57] There appear to be no plans for building conceptually new maritime patrol aircraft (MPA) on the horizon. This might seem strange for a country that puts so much emphasis on control over sea routes and anti-access capabilities in a wide area around the base areas of its strategic submarines (the so-called bastion areas), as well as on maintaining a presence in the Arctic.[58] It may well be a reflection of a Russian navy preoccupied with submarines and surface vessels over aircraft.[59]

Air Mindedness and Air Power as a Propaganda Tool

The patriotic upbringing of younger generations is mentioned by Valerii Gerasimov as a way to prevent color revolutions.[60] TV shows like *Poligon*, *Samyi-samyi*, and *Voennaya priemka* are devoted to showing off military units, equipment, and operations, and also feature beautifully produced and spectacular presentations of new and modernized aircraft.[61]

The consistent promotion of the Russian military as efficient, modern, and embodying Russian national virtues seems to work. In October 2018, the military was the state institution that Russians trusted the most, according to a poll by the Levada Center, a dramatic improvement since 2012 at

---

55 TASS, "Protivolodochnye samolety-amfibii be-12 proveli uchebnye bombardirovki v Krymu," Oct. 24, 2018, <https://tass.ru/armiya-i-opk/5712769>.

56 BMPD, "Vosmoj modernizirovannyi Il-38N," Dec. 23, 2016, <https://bmpd.livejournal.com/2341327.html>.

57 OAK, "Aviiatsiia VMF RF peredan ocherednoi dalnii protivolodochnyji samolet Tu-142MK kapitalno otremontirovannyi v Taganroge," Nov. 20, 2018, <http://uacrussia.ru/ru/press-center/news/aviatsii-vmf-rf-peredan-ocherednoy-dalniy-protivolodochnyy-samolet-tu-142mk-kapitalno-otremontirovan>.

58 See e.g., Mathieu Boulègue, "NATO Needs a Strategy for Countering Russia in the Arctic and the Black Sea," *Chatham House Expert Comment*, Jul. 2, 2018, <https://www.chatham-house.org/expert/comment/nato-needs-strategy-countering-russia-arctic-and-black-sea>.

59 Maksim Klimov, "Ognennoe nebo rossiiskogo flota," *Nezavisimoe voennoe obozrenie*, Dec. 6, 2018, <http://nvo.ng.ru/realty/2018-12-06/6_1025_sky.html>.

60 Gerasimov, 2016.

61 See, e.g., the three-part series on T-50/Su-57, beginning Nov. 10, 2018: <https://www.youtube.com/watch?v=DN2zCJ5WBn4>.

the start of military reform.[62] This is of course not exclusively associated with air power, as all branches of the military are promoted in the same way, and both land and naval systems are presented in likewise glamorous shows. Nevertheless, air power, with its long-lasting association with cutting-edge technology since the early twentieth century, is also fulfilling a role in the promotion of national pride and trust in Russia's military, industrial, and technological potential, and by extension also in the Russian authorities.

Conclusion

The Russian military, like many others, is currently forced to modernize and renew its inventory as legacy equipment built during the final decades of the Cold War has become obsolete, worn out, or both. The cost of completely new systems is high, and the Russian Air Force is unlikely to expand its nominal numbers by much. However, availability of aircraft and other systems will most likely be better than it was in the 1990s and into the early 2010s,[63] making the Russian Air Force a more effective force, with a combat power more directly corresponding to its nominal number of aircraft and other systems. This is fully in line with the ambitions of the ongoing defense reforms since 2008—creating a leaner, more efficient military.

If the optimistic prognosis of Ruslan Pukhov from 2016 holds, Russia will be the second largest air power nation in the world in 2020,[64] even though it will remain behind the United States in numerical strength. In Europe, it will be the biggest single air power nation. While economic growth remains sluggish, the Russian economy is far more resilient than the Soviet one, and seems to have absorbed the impact of falling oil prices and western sanctions fairly well.

Modernization of the Russian Air Force encompasses both a revitalization and modernization of the strong emphasis on a multilayered, integrated aerospace defense system as well as the development of

---

62 Lada Shamardina, "Pensionnaya reforma perecherknula Krym," *Kommersant*, No. 181, Oct. 4, 2018, <https://www.kommersant.ru/doc/3759662>.
63 No more than 40% of the approximate 2,800 aircraft in the inventory in 2008 were in useable technical condition. Ruslan Pukhov, "Poligon budushchego," *Rossiya v globalnoj politike*, Feb. 16, 2018, <http://www.globalaffairs.ru/global-processes/Poligon-buduschego-17997>.
64 Pukhov, 2016.

modernized force projection and strike capabilities. That is, Russia is building air power capabilities both to protect against, and to carry out the western-style expeditionary warfare of the post–Cold War era. The ambition is clearly to be technologically as close to peer competitor the United States as possible, likely even with an edge in defensive systems.

The priorities and timelines of the modernization of Russian air power reflect doctrinally relatively conservative air power thought. The first priority is to secure control of Russia's enormous airspace, by renewing and rebuilding the integrated air defense system and the fleet of multi-role fighters first. Second is the strategic deterrence capability, nuclear and conventional, while operational support capabilities appear to come last.

The Russian Air and Space Force still lacks large numbers of modern precision weapons, sufficient numbers of AEW&C aircraft, strategic transport aircraft, and air-to-air refueling assets. A lack of advanced long-range reconnaissance assets is likely inhibiting the full exploitation of modernized strike capabilities. The relative scarcity of these supporting resources is also likely to inhabit the amount of advanced training available to Russian pilots to carry out complex, composite air operations, current ambitions for flight hours are not fully up to western standards.

Russia is continuing development of modern (post-Soviet) aircraft designs, with the Su-57 fifth-generation fighter being the most mature program. However, Soviet designs with modernized avionics and weapons will remain dominant in the Russian fleet at least into the middle of this century. The Russian Air Force structure is thus not fully in balance to carry out complex, modern air operations on a large scale, and it will likely take many years before it is. The many different versions of both new-built and older, modernized aircraft are likely to drive up maintenance and running costs. In the end, there is no reason to expect anything else than that Russia will remain a very strong regional air power nation with capacity for limited power projection in theaters abroad, as it has shown in Syria.

References

Babakin, A. "Kadrovyi shtopor aviatsii." *Nezavisimoe voennoe obozrenie*, No. 25, Jul. 25, 2014.

BMPD. "VKS Rossii planiruet imet 700 istrebitelei," Dec. 16, 2016. <http://bmpd.livejournal.com/2326878.html>.

BMPD. "Vosmoj modernizirovannyi Il-38N," Dec. 23, 2016. <https://bmpd.live-journal.com/2341327.html>.

BMPD. "VKS Rossii poluchili chetvertyi modernizirovannyi samolet A-50U," Mar. 7, 2017. <https://bmpd.livejournal.com/2477303.html>.

BMPD. "Ministerstvo oborony poluchilo tretii v 2017 godu polkovoi komplekt zenitnoi raketnoi sistemy S-400," Oct. 27, 2017. <https://bmpd.livejournal.com/2914426.html>.

BMPD. "Pervyi polet opytnogo obraztsa novogo samoleta radiolokatsion-nogo dozora i navedeniia A-100," Nov. 18, 2017. <https://bmpd.livejournal.com/2961071.html>.

BMPD. "Vse, chto vy khoteli znat ob istrebitele Su-35, no stesnialis sprosit," Dec. 31, 2017. <https://bmpd.livejournal.com/3029961.html>.

BMPD. "Nachaty letnye ispytaniia pervogo obraztsa samoleta-zapravshchika Il-78M-90A," Jan. 26, 2018. <https://bmpd.livejournal.com/3064340.html>.

BMPD. "Kontrakt na pervye 12 seriinykh istrebitelei Su-57 ustanovochnoi partii budet zaklyuchen v 2018 godu," Feb. 9, 2018. <https://bmpd.livejournal.com/3087140.html>.

Bocharova, S. and A. Nikolskii. "Dalniaia aviatsiia poluchit 10 Tu-160M2 za 160 mlrd rublei." *Vedomosti*, Jan. 25, 2018. <https://www.vedomosti.ru/technology/articles/2018/01/25/748964-tu-160m2>.

Boulègue, M. "NATO Needs a Strategy for Countering Russia in the Arctic and the Black Sea." *Chatham House Expert Comment*, Jul. 2, 2018. <https://www.chathamhouse.org/expert/comment/nato-needs-strategy-countering-russia-arctic-and-black-sea>.

Dzhordzhevich, A. and I. Safronov. "U trillionov est dva soiuznika – armiia i flot." *Kommersant* No 235 (6229), Dec. 12, 2018. <https://www.kommersant.ru/doc/3500710>.

Evdokimova, A. "Vozdushno-kosmicheskie sily Rossii poluchili bolee 70 novei-shikh RLS." *TV Zvezda*, Jan. 6, 2018/ <https://tvzvezda.ru/news/forces/content/201801060824-5ies.htm>.

Gavrilov, I. "'Sever Arktiki'. Sozdano novoe strategicheskoe komandovanie." *Rossiyskaya Gazeta*, No. 6546, 274, Dec. 1, 2014. <https://rg.ru/2014/12/01/komandovanie-site.html>.

Gerasimov, V. "Tsennost nauki v predvidenii." *Voenno-promyshlennyi kurer*, No. 8, 476, Feb. 27, 2013.

Gerasimov, V. "Po opytu Sirii." *Voenno-promyshlennyi kurer*, No. 9, 624, Mar. 9, 2016.

Ignateva, L. "Модернизация 2.0: Казанский авиазавод замахнулся на 20 миллиардов." *Realnoe vremia*, Aug. 16, 2018. <https://realnoevremya.ru/articles/109582-19-mlrd-na-modernizaciyu-kazanskogo-aviazavod-im-gorbunova.>

Interfaks-AVN. "Boevaia bespilotnaia aviatsiia v Rossii poluchila kachestvenno novoe razvitie, zaiavliaiut v Sovbeze RF." May 15, 2017. <http://www.military-news.ru/story.asp?rid=1&nid=451001>.

Klimov, M. "Ognennoe nebo rossiiskogo flota." *Nezavisimoe voennoe obozrenie,* Dec. 6, 2018. <http://nvo.ng.ru/realty/2018-12-06/6_1025_sky.html>.

Kofman, K. "The Collapsing Russian Defence Budget and Other Fairy Tales." *Russia Military Analysis Blog,* May 23, 2018. <https://russianmilitaryanalysis.wordpress.com/2018/05/23/the-collapsing-russian-defence-budget-and-other-fairy-tales/>.

Kramnik, I. "Krushenie Mi-35: versii." *Lenta.ru,* Jul. 11, 2016. <https://lenta.ru/articles/2016/07/11/mi35version/>.

Lavrov, A. "Reform of the Russian Air Force." M. Barabanov, ed., *Russia's New Army.* Moscow: Centre for Analysis of Strategies and Technologies, 2011.

Lee, R.J. "A Brief Look at Russian Aircraft Carrier Development." Jan. 9, 1996. <http://www.rjlee.org/rcar.html>.

Losey, S. "Pilots Are Flying 17 Hours per Month, But It's Still Not Enough, Air Force Secretary Says." *Airforce Times,* Mar. 1, 2018. <https://www.airforcetimes.com/news/your-air-force/2018/03/01/secaf-air-force-pilots-are-flying-17-hours-per-month-but-its-still-not-enough/>.

Nastin, P. "Shojgu: dve treti letnego sostava VKS poluchili boevoyu praktiku v Sirii." *TV Zvezda,* Aug. 12, 2018. <https://tvzvezda.ru/news/forces/content/201808120023-nfy6.htm>.

Nikolskii, A. "Vozdushno-kosmicheskie sily Rossii obedinili VVS i vojska Vozdushno-kosmicheskoy oborony." *Vedomosti,* Aug. 3, 2015. <https://www.vedomosti.ru/politics/articles/2015/08/04/603322-vozdushno-kosmicheskie-sili-rossii-obedinili-vvs-i-voiska-vozdushno-kosmicheskoi-oboroni>.

Nikolskii, A. "Do kontsa goda Minoborony zakupit ne menee 36 istrebitelei Su-30SM." *Vedomosti,* Aug. 16, 2018. <https://www.vedomosti.ru/politics/articles/2018/08/16/778384-do-kontsa-goda-ne-menee-istrebitelei>.

OAK. "Aviiatsiia VMF RF peredan ocherednoi dalnii protivolodochnyji samolet Tu-142MK kapitalno otremontirovannyi v Taganroge," Nov. 20, 2018. <http://uacrussia.ru/ru/press-center/news/aviatsiya-vmf-rf-peredan-ocherednoy-dalniy-protivolodochnyy-samolet-tu-142mk-kapitalno-otremontirovan>.

O'Connor, S. and T. Ripley. "Russian Carrier Jets Flying from Syria, Not Kuznetsov." *IHS Jane's Defence Weekly*, Nov. 30, 2016, Vol 53, No. 48, 5.

OECD. "Russia." *OECD Economic Outlook*, Vol. 2018, No. 2, Preliminary Version, 178–80.

Peck, M. "Why Russia Won't Go Bankrupt Paying for Its Military Buildup." *The National Interest,* Aug. 8, 2018. <https://nationalinterest.org/blog/buzz/why-russia-wont-go-bankrupt-paying-its-military-buildup-28092>.

Pinchuk, A. "VKS: Na novoi tekhnike k novym vozmozhnostiam." *Krasnaia Zvezda,* No. 159, Sep. 4, 2013, 1.

Pukhov, R. "Poligon budushchego." *Rossiya v globalnoj politike*, Feb. 16, 2018. <http://www.globalaffairs.ru/global-processes/Poligon-buduschego-17997>.

Ramm, A. "VMF Rossii udarit po terroristam v Sirii s osoboi tochnostyu." *Izvestiia*, Sep. 2, 2016. <http://izvestia.ru/news/630104>.

Ramm, R. "Ubijtsam avianostsev popolniat arsenal." *Izvestiia*, Nov. 17, 2017. <https://iz.ru/663570/aleksei-ramm/ubiitcam-avianostcev-popolniat-arsenal>.

RIA Novosti. "Potrebnosti v sozdanii avianostsev u Rossii net, zaiavil Rogozin." Nov. 16, 2013. <https://ria.ru/defence_safety/20131116/977313945.html>.

Russian Ministry of Defence (@mod_russia). Aug. 10, 2017, <https://twitter.com/mod_russia/status/895897917755666434>.

Russian Ministry of Defense. "V sootvetstvii s resheniem Verkhovnogo Komanuy-ushchego Vooruzhennymi Silami Rossijskoj Federatsii V. V. Putina Ministerstvo oborony Rossii pristupaet k sokrashcheniyu gruppirovki Vooruzhennykh Sil v Sirii." Jan. 1, 2017. <https://function.mil.ru/news_page/country/more.htm?id=12107843@egNews>.

Russian Ministry of Defense. "V komandovanii Voenno-vozdushnykh sil VKS podveli itogi za 2018 god." Dec. 4, 2018. <https://function.mil.ru/news_page/country/more.htm?id=12206756@egNews>.

Safronov, I. and A. Dzhordzhevic., "Iadernoi triade opisali perspektivy." *Kommersant*, May 18, 2017. <https://www.kommersant.ru/doc/3299987?utm_source=kommersant&utm_medium=doc&utm_campaign=vrez>.

Safronov, I., A. Dzhordzhevich and S. Titov. "Il-76 Menyayut poletnoe zadanie." *Kommersant*, June 8, 2017. <https://www.kommersant.ru/doc/3319791>.

Shamardina, L. "Pensionnaya reforma perecherknula Krym." *Kommersant*, No. 181, Oct. 4, 2018. <https://www.kommersant.ru/doc/3759662>.

Shapkin, D. "Nazlo torgovym vojnam. Vozdushnyj transport." *Kommersant*, Apr. 12, 2018. <https://www.kommersant.ru/doc/3598275?query=%D0%B2%D0%BE%D0%BB%D0%B3%D0%B0-%D0%B4%D0%BD%D0%B5%D0%BF%D1%80>.

Shoigu, S. "Organizatsiia protiv khaosa." *Voenno-promyshlennyi kurer*, No. 14, 727, Apr. 10, 2018. <https://vpk-news.ru/articles/42095>.

"Shturmany Shpitonkovy." *Krasnaia Zvezda*, Aug. 6, 2015. <http://redstar.ru/index.php/newspaper/item/25216-shturmany-shpitonkovy>.

Simon, L. "Demystifying the A2/AD Buzz." *War on the Rocks*, Jan. 4, 2017. <https://warontherocks.com/2017/01/demystifying-the-a2ad-buzz/>.

Sivkov, K. "Obiekt na vylet." *Novosti VPK*, No. 41 (607), Oct. 28, 2015. <http://vpk.name/news/143127_obekt_na_vyilet.html>.

Sozaev-Guriev, I. "Rogozin i Makarov posporili iz-za novogo bombardirovsh-chika." *Izvestiia*, June 5, 2012. <https://iz.ru/news/526580>.

Tarkhanov, A., E. Kuznetsova and G. Dubina. "Ruslany uvolnyayutsya so slu-zhby NATO." *Kommersant*, Apr. 17, 2018. <https://www.kommersant.ru/doc/3605505>.

TASS. "Poteri VKS Rossii v Sirii. Dose." Mar. 6, 2018. <https://tass.ru/info/5012726>.

TASS. "Protivolodochnye samolety-amfibii be-12 proveli uchebnye bombardirovki v Krymu." Oct. 24, 2018. <https://tass.ru/armiya-i-opk/5712769>.

Tien, N. et al. "Trends in World Military Expenditure." *SIPRI Fact Sheet*, May 2018, 2–3. <https://www.sipri.org/sites/default/files/2018-04/sipri_fs_1805_milex_2017.pdf≥>.

UAC Russia on Livejournal. "Tanker obrel telo: na 'Aviastar-SP' idet sborka pervogo rossiiskogo samoleta-toplivozapravshchika," Jan. 2, 2016. <https://uacrussia.livejournal.com/37218.htm>.

"Voennaya Doktrina Rossijskoj Federatsii." 2014, 14.

Waldwyn, T. "Russian Military Lift Risks Atrophy." *Military Balance Blog*, Jul. 6 2017. <https://www.iiss.org/blogs/military-balance/2017/07/russian-military-lift>.

4 FRANCE
The Light at the End of the Tunnel?

Guillaume de Syon

Introduction

The sudden and public resignation in July 2017 of General Pierre de Villiers, Chair of the French Joint Chiefs, reverberated throughout France. For the first time in France's Fifth Republic (est. 1958), an army leader reporting directly to the President had stepped down because of tensions with his superior. The crisis revealed that French forces were at a crossroad. Whereas in the 1970s through to the end of the Cold War, French military aviation could make a valid claim of preparedness with a modern arsenal in the context of a superpower confrontation, it had to evolve in the 1990s to make do with a new geopolitical landscape.[1] This included not only the appearance of new threats, but also the end of the required draft in 1996 that had furnished an ideal group of young people to recruit professionals from. In addition, the crisis revealed that all the services, including air power, were being required to carry out far more tasks of power projection and military preparation on an ever-shrinking budget than was reasonably possible. This crisis, considered the tip of an iceberg, suggests that even though France's army is one of the biggest in Western Europe, its

---

1   Col. Jean-Pierre Dupont, "L'armée de l'air face au défi de la transformation des capacités," in *Envol vers 2025. Réflexions prospectives sur la puissance aérospatiale*, eds. Grégory Boutherin and Camille Grand (Paris: La Documentation française, 2011), 91. For an overview of the French Air Force during its transition out of the Cold War, see Etienne de Durand, "French Air Power: Effectiveness Through Constraints," in *European Air Power: Challenges and Opportunities*, ed. John Andreas Olsen (Lincoln, NB: University of Nebraska Press, 2014), 3–31.

services, including aviation, found themselves at a crossroad that involved future mission development, the need for clear paths of cooperation, and the need to obtain budgetary guarantees that will allow for ongoing modernization. All these elements became the object of serious attention in the establishment of the 2019–25 Loi de programmation militaire (LPM, military program law),[2] especially as the French Air Force was in the process of releasing a new strategic plan, the first since 2008.

What Does the Current Force Structure, Doctrine, and Threat Environment Look Like for the Nation's Air Force?

French military aviation transitioned out of the Cold War, but retained some its original functions as a result of French nuclear strategy as well as the need to support the national defense industry. Consequently, as Etienne de Durand notes, there is a dearth of doctrinal pronouncements until 2008 when an Air Force concept, paralleling the release that same year of a defensewide White Paper, appeared.[3] The emphasis on quality over quantity and an overall rationalization of services as envisioned under President Nicolas Sarkozy, seemed to put a reemphasized line in place. However, its application became more urgent as the service was required to close several bases and cut thousands of posts. The rationale was that the money saved would be recapitalized into modern equipment, but this happened at a snail's pace, while more cuts under the 2013 White Paper also occurred. In response to this second defense review ordered by President François Hollande, further cuts appeared, all of which affected the publication of a formal strategic air plan by then Air Force chief Denis Mercier, known as *"Unis pour faire face"* (United to face the challenge). The name, a variation on the French Air Force's motto "facing the challenge," masked the further governmental demands for consolidation. Mercier's successor, General André Lanata managed the situation in an able way, but recognized that the new threats and new political demands for swift engagements in distant theaters had turned the Air

2   The LPM involves budget projections over a five-year period with provisions for its conclusion in the sixth year as a new five-year projection begins. Though negotiations between civilian and military sides of the army and the Senate voting the budget, it is reviewed by the finance ministry, and may be subject to cuts in subsequent years because of financial shortfalls or program overruns.

3   Durand, 25.

Force into a full political weapon, but not one that might prove as effective as if it were given the means to support its mission over the longer term.[4]

Budgetary pressures are substantial in any air force, but in France's case, because it follows a five-year plan, they involve more sacrifices to be able to respond to developing threats. French military aviation is also tied closely to the French aeronautical industry. Consequently, it enjoys a certain level of political appeal with civilian leaders, but the budgetary strings are such that some funding approval is carried out so as to support the industrial infrastructure more than the military one.[5] At the end of the Cold War, France's military budget averaged 3 to 4 percent of the gross domestic product (GDP) depending on the overall state of the economy, with one pattern clearly defined: cuts to the budget impacted the actual functioning of the service, and less so equipment purchases.[6] With the end of the draft, this would have more important consequences.

The professionalization of the French armed forces has resulted in a crisis of confidence of the services that are still searching for a new identity, all the while competing with a private sector eager to hire highly trained specialists in all technical fields.[7] The French Air Force, like similar national services/forces, had to contend with these changes as well as the rise of new forms of asymmetrical warfare. Like the United States, France has declared war on terrorism while seeking to maintain its nuclear strike capability and the preservation of various political and strategic interests worldwide. The resulting confusion and frustration at the level of the armed forces has also reached the *armée de l'air* (French Air Force).

Whereas the 2008 White Paper called for some 70 fighters to be ready for immediate, intensive operations, the 2013 iteration lowered the number to 45 (including naval-based fighters).[8] In parallel, however, the French Air Force was called to deploy.

---

4   Interview of General André Lanata, *Air&Cosmos* 2527S (Dec. 16, 2016), 15.

5   Durand, 18; See also the general discussion of the materiel and budgetary shortcomings in Nathalie Guibert, *C'est qui le chef? La vraie crise entre le pouvoir et l'armée* (Paris: Robert Laffon, 2018).

6   Valérie Lelièvre, "Dépenses militaires et contraintes économiques 1971–1995." *Revue française d'économie* 11: 2 (1996), 65–86.

7   Guibert, 270–75.

8   Durand, 28; cf. *French White Paper on Defense and National Security 2013*, 88, <http://www.livreblancdefenseetsecurite.gouv.fr/pdf/the_white_paper_defence_2013.pdf>.

In 2017, newly elected President Emmanuel Macron requested an update on the conclusions reached in the 2013 White Paper. The 2017 Strategic Review stressed among its key points the need for collaborative projects that would limit the stress on resources and reinforce links among allies. The wording of the memorandum does not allow interpretation as to whether this was also a reactive stance to President Trump's negative comments about the North Atlantic Treaty Organization (NATO), though some observers have argued it to be the case.[9] Echoes of the need for greater autonomy appear in French Senate hearings, notably regarding the development of future weapon systems, but the staggering cost projections have also added to the call for cooperation with other nations.[10]

Possible Collaborative Moves

Since the end of the Cold War, French analysts have acknowledged that states threatening French interests now do so with far more capabilities than was the case three decades ago. The only means, however, to respond to all the potential threats involve substantial cooperation with other nations, either as part of NATO or in identifying specific common interests.

Franco-British Cooperation

Until the Brexit vote in 2016, France and the United Kingdom had sought a rapprochement in matters of industrial defense production. Differing strategic outlooks prevented a €200 million study about the possibility of a joint aircraft carrier development from coming to fruition, but other options appear to be underway, thanks in part to the 2010 Lancaster House Treaty. A 2014 joint funding to compare drone approaches (the French "Neuron" and the British "Taranis" unmanned aerial vehicle (UAV) projects) was announced, with possible plans to build a future combat air system (FCAS, also known by its French acronym SCAF). However, a joint Franco-German announcement about the FCAS (see below) in 2018, and the British launch

---

9   *Strategic Review of Defence and National Security 2017*, <https://www.defense.gouv.fr/dgris/presentation/evenements/revue-strategique-de-defense-et-de-securite-nationale-2017>.
10 French Senate, information report 559, May 23, 2017, 35–38.

of the "Tempest" fighter project, perhaps in response to Brexit, suggest that the British commitment is no longer active.[11]

It would be in the interest of both nations to maintain joint development or a drone system, if only for budgetary reasons, that would in turn have good connective capacity with either the F-35 weapons system of the Royal Air Force (RAF), or the Franco-German FCAS. Not everyone in France favors cooperation with Germany, however. Testifying before a senate commission in February 2018, Eric Trappier, chair of the Dassault Corporation, noted that in light of the company's fighter design experience, he would personally prefer continuing cooperation and consultation with the British.[12]

Franco-German Cooperation

Franco-German cooperation, dating back officially to the 1963 friendship treaty (Élysée Treaty), is also entering a critical decade as both nations are defining new fields of cooperation for a possible European army,[13] and common defense vectors, notably medium altitude long endurance (MALE) drones, also known in French as *aéronef piloté à distance* (see below). In terms of training for the Tiger combat helicopter, both armies continue to rely on the joint schooling of pilots and mechanics in Luc (France) and Fassberg (Germany).[14] Both nations also ordered jointly the Airbus A400M Atlas transport, which is slowly entering service after years of delay (see below).

In April 2018, the French and German governments signed an agreement to develop a new, sixth-generation FCAS. Though projected initially to be an equal 50:50 participation between the aerospace industries of both

---

11 Jean-Pierre Maulny, "The French/UK Defense Relationship." *IRIS Analyses*, June 22, 2016, <http://www.iris-france.org/78591-the-frenchuk-defence-relationship/>; Philippe Lemayrie, "L'Entente cordiale au péril du Brexit?" *Le Monde diplomatique, Blog "Défense en ligne*, Oct. 26, 2018, <https://blog.mondediplo.net/l-entente-cordiale-au-peril-du-brexit>.

12 Eric Trappier, testimony before the French Senate Defense Commission, Feb. 28, 2018, <http://videos.assemblee-nationale.fr/video.5624702_5a96a5ff81ec0.commission-de-la-defense--auditions--sur-le-projet-de-loi-de-programmation-militaire-28-fevrier-2018>.

13 Roland Goron, "'L'armée européenne,' nouvelle priorité du couple franco-allemand." *Le Figaro*, Nov. 14, 2018.

14 Margaux Thuriot, interview of generals Pierre de Villiers and Günther Fassberg, Feb. 4, 2013, <https://www.defense.gouv.fr/english/actualites/articles/succes-et-avantages-de-la-cooperation-franco-allemande>.

nations, as of January 2019 the Spanish government has expressed an interest in a spectator role that might lead to active participation once the aircraft parameters are defined. The current vision for 2025 involves identifying both the requirements of the actual fighter aircraft and the communication systems. Unlike other fighter projects currently under consideration, the SCAF would not become operational until 2040, and would be expected to include new advances in artificial intelligence intended to assist the human element. Such interconnectedness, while echoing the direction of U.S. Air Force (USAF) thinking, is not yet fully possible in the French setting without American assistance.

Were such a FCAS project to reach the field successfully, it would imply either a jump in fighter generation or the incorporation of connectivity elements found in the F-35. The Rafale has often been termed a fighter of the "four-and-a-half" generation, comparable in some measures to the F-18 Super Hornet (whereas the F-35 is possibly the best representative of the fifth-generation). While the modernization of the Rafale may help bridge the gap, it is unlikely that Germany's Tornado fighter-bombers, which entered service in the 1970s, will last until 2040, thus raising the spectrum that France's partner could turn to an off-the-shelf, U.S.-built replacement. The question of bridging the gap to 2040, though apparently trivial, even elicited quips in a French Senate hearing in 2017 regarding a French off-the-shelf purchase of F-35s.[15]

The Matter of Space Control

France became the third power in space in 1965, but relied on allied support for much of its spatial military activity. By the 1980s, however, it had fielded the Helios observation system in cooperation with other NATO countries, and the astronauts it contributed to the civilian programs were initially all military officers. In effect, though it had sought cooperation with both the United States and the Soviet Union in the Cold War, the nation made do with the dominance of the United States in the field. By 2008, as was reaffirmed in the 2013 white paper, the French government made clear it saw

---

15 Question of Senator Jean-Pierre Cubertafon to General André Lanata regarding the air force budget for 2018, *Compte Rendu No. 12*, Commission de la défense nationale et des armées, Oct. 18, 2017, 17.

value in the use of space for military observation, but that it would refrain from actual weapons militarization in orbit, in line with the stance of European allies.[16] While this implies that the country will not seek to field space weapons, France has identified two other elements of the space triad that it wishes to maintain and even master: surveillance and connectivity. In a July 2018 speech, however, President Macron announced that he was commissioning a redefinition of French military space policy at the national and European levels.[17] At present, the joint consultation includes multiple military and technological centers, as reflected in the already existing inter-army space command, yet the French space agency, *Centre national d'études spatiales* (CNES), does not appear on the list. This both contradicts the parliamentary decision of 1986 to make the CNES into a dual space agency assisting both civilian and military projects and ignores the CNES as part of the design process for the three *composante spatiale optiques* (optical space components (CSOs)), satellites (see below). Excluding the CNES from future military space projects could weaken France's position, further draining the budgets by duplicating tasks and ignoring already acquired technological know-how. The country does not have the budget, for example, to create a DARPA (Defense Advanced Research Projects Agency)–style agency that would handle classified space projects.[18] It is also unclear whether the French Air Force will manage French military space, although it is a logical assumption since four of the French astronauts are also Air Force officers.

In *December* 2018, five months after President Macron's speech, the first of three CSO military observation satellites was orbited as part of the replacement process for the Helios 1 and 2 surveillance systems currently in place.[19] Whereas CSO-1 would fulfill reconnaissance functions, CSO-2 would be tasked specifically with identification. The first of three CSOs marks a new direction for the French armed forces in that it will

---

16 Xavier Pasco, "De l'utilisation au 'contrôle' de l'espace extra-atmosphérique?", in *Envol vers 2025. Réflexions prospectives sur la puissance aérospatiale*, eds. Grégory Boutherin and Camille Grand (Paris: La Documentation française, 2011), 86. Cf. *Livre Blanc sur la défense nationale 2008*, 135–43.

17 President Emmanuel Macron address at Hôtel de Brienne (Army headquarters), Jul. 13, 2018, <https://www.elysee.fr/emmanuel-macron/2018/07/17/discours-du-president-de-la-republique-emmanuel-macron-a-l-hotel-de-brienne>.

18 Laurent Cabirol, "Spatial militaire: Big bang ou Pschiiit?" *La Tribune*, Nov. 27, 2018, <https://www.asafrance.fr/item/spatial-militaire-big-bang-ou-pschit-en-2019.html>.

19 "CSO/MUSIS," <https://cso.cnes.fr/fr>.

allow direct access to information rather than dealing with a filtering process under U.S. control. However, budgetary constraints already exist in
the 2019–2025 military planning budget, meaning that cooperation with
Belgium and Germany will be essential to putting the CSO-3 into polar
orbit by 2023.

Aside from cooperation with other nations to develop and field new
defense systems, the French Air Force is, of course, charting its own path in
terms of acquisitions. A dual theme is present in the midterm acquisition
and fielding of weapons: interoperability through common platforms, and
polyvalence for use in multiple settings. This appears across the spectrum
of aircraft and their missions.

In terms of jet training, the Swiss-built Pilatus PC-21s began arriving
in fall 2018. A total of 17 aircraft, all operating from a single base (at Cognac),
are expected to replace the aging Socata Epsilon and the Dassault-Dornier
Alphajets that were split between two airfields, though student pilots will
now spend 40 percent of their time in simulators (and they will still switch
to Alphajets for their final training phase).[20] The maintenance cost reduction, estimated to be 10 times less than that for an Alphajet, is to be achieved
in part through a private maintenance contract with a British firm.

France expects to have some 225 fighters (down from a projected
270 in the 2008 White Paper) by 2025, split between Rafale and upgraded
Mirage 2000D. However, the tasks for both types of fighters are evolving,
as is the case in the nuclear strike force. French nuclear posture has evolved
since 1996 into a dyad involving submarine-launched ballistic missiles
(SLBMs) and nuclear bombers.

In 2018, the Mirage 2000N that had replaced, in 1988, the original
Mirage IVA (strategic bombing) and Mirage IIIE (tactical bombing) was
retired in favor of the Rafale B two-seater carrying the ASMP-A cruise missile (that already operated aboard the Mirage 2000N). This shift reflects
not only a long-needed modernization but also a quest for polyvalent missions. Whereas the Mirage 2000N was occasionally deployed in external
operations independent from strategic targets, its limited carrying capacity
prevented it from flying with a laser pod for precision strikes; it also could
not be used in air patrol functions. The Rafale B on the other hand may

---

20 Emmanuel Huberdeau, "Arrivée officielle du PC-21," Oct. 3, 2018, <http://www.air-
cosmos.com/arrivee-officielle-du-pc-21-115603>.

assume such functions as needed, and be used in all other functions that other Rafale air force and naval units (using the Rafale M) are engaged in. As for the modernized Mirage 2000Ds, the upgrades include new air-to-air capabilities, such as the short-range interception, combat and self-defence missile (*missile d'interception, de combat et d'autodéfense*).

To support fighters in their missions, in 2021, France will begin fielding the first of 12 Airbus A 330 multi-role tanker transport (MRTT) Phénix (down from the 14 recommended in the 2013 White Paper). These tankers are scheduled to replace the now half-century-old KC-135Fs as well as three Airbus A310 and two A340 currently operating strategic airlift missions. The Phénix was preceded by the first units of some 70 Airbus A400M airlifters. The A400M Atlas, though facing several years of delay, now seems poised to allay some of its early difficulties.

At the helicopter level, France faces new pressures to replace its existing H 225 Puma and light surveillance Fennec within the 2019–2025 program. The Air Force has only 78 helicopters at its disposal, which emphasizes the need for interoperability among all branches of the service. The missions range from air safety patrols to SAR (sea-air-rescue) and insertion, thus straining availability. For example, only 10 or so of the 23 SAR Puma machines are fully equipped for such missions. Meanwhile, the French Navy retired its Super Frelon and fielded NH-90 helicopters. Problems of corrosion, however, are limiting the efficiency of the new machines.

As for the French army, though it stands to lose the biggest chunk of the budget under the 2013 White Paper, its land army light aviation (*aviation légère de l'armée de terre* (ALAT) helicopter fleet is surviving the budgetary challenge by maintaining some 140 attack helicopters (including 80 Eurocopter Tigers on order) and another 115 prioritized for transport duties.

In addition to fleet renewal, however, the greatest challenge to the French Air Force has come about in the realm of support, maintenance, and training, all with an eye to adapting to new threats. For example, runways at bases where new MRTT tankers will operate do not have the required runway strength to carry planes that are much heavier than the machines they are replacing. Similarly, the A400M transports require new support facilities far greater and more complex than the C-160 Transall planes they are replacing.[21]

---

21 Interview of General Serge Soulet, Air&Cosmos 2527S (Dec. 16, 2016), 56.

Elsewhere, air defense relies on new GM 400 radars and the renovation of the 2215 and 22XX models. A central challenge beyond optimization between all these systems involves the capacity to spot drones. Whereas the French Air Force is able to provide coverage for military-sized drones, strategic sites are, at this time, unable to spot smaller machines. The incidents in 2015 when amateur drones overflew the American embassy in Paris and came into close proximity with the presidential Élysée Palace emphasized the urgency of a solution, which the Air Force is endeavoring to develop.[22]

Another challenge associated with both modernizing and reaching a fleet of 225 aircraft involves training. In a public testimony in 2017, General Lanata noted that he could not supply the number of pilots needed on time because of saturation at the simulator training level.[23] He stressed warnings that General Rony Lobjoit had voiced as head of the French Air Force's human resources office: as a result of the 2008 White Paper downturn, the Air Force lost on average 2,000 personnel per year, all the while facing the need for more complex training as the Rafale entered nationwide service. All told, Air Force personnel fell by almost 30 percent before beginning to increase in 2017. The high level of training required for some personnel, in addition to the pilots, means that ideally, an average aviator would spend up to two decades in the service to pay back the years of training in fields such as cyber systems that face considerable competition from the private sector.[24] Consequently, the service fears that its personnel, though highly trained, will lack sufficient experience and may even face a morale problem when not given sufficient day-to-day resources to fulfill its tasks.

The lack of sufficient staff will affect French Air Force commitments abroad as well as interoperability. At present, the service is engaged in no fewer than four external operations. Though it does cooperate with other services under NATO agreements and other treaties (see above), the French policy of intervention to preserve political interests as needed means that a search for autonomous capacity is essential despite budgetary restrictions. Thus, longer-term projections may call for selective collaboration while avoiding overdependence on U.S. capabilities.

---

22 "Plusieurs drones ont survolé des lieux sensibles de Paris dans la nuit," *Le Monde*, Feb. 24, 2015.

23 Assemblée nationale, compte-rendu No. 12 (Oct. 18, 2017), 33–34, <http://www.assemblee-nationale.fr/15/pdf/cr-cdef/17-18/c1718012.pdf>.

24 Interview of General Rony Lobjoit, Air&Cosmos 2527S (Dec. 16, 2016), 78–79.

Drones

When necessary, France has set aside its domestic industrial priorities to purchase foreign-built materiel. The latest iteration involves the decision to buy off-the-shelf MQ-9 Reaper drones from the United States. The ones fielded in 2014 as part of operation Barkhane in the Sahel region accumulated over 24,000 flight hours and enabled successful cooperation in information sharing between French and U.S. forces in Africa.[25] A total of 12 (four systems of three drones each) of these MALE systems are expected to join French forces by 2019, with the later ones incorporating a weapons capability. The challenge, however, involves training sufficient crew (up to 16 per drone because of extended missions). Eventually, shifting the crew from observation to attack duties will need serious consideration of the risks that have affected USAF crews carrying out strikes (notably, depression).[26]

Though Reaper operations have proven extremely successful, the need for stronger information flows and controls among command posts that differ from their American counterparts has dictated the move toward a European MALE drone design as part of the 2025 operational objectives.

Aircraft Carriers

The French Navy's air wing resources have been constrained ever since it reduced its carrier capacity to one, nuclear-powered boat, the *Charles de Gaulle*. Having served in support of U.S. intervention in Afghanistan as well as French operations in Libya and Syria, the aircraft carrier underwent an 18-month refurbishment in 2017–2018. Due to shifts in U.S. policy interests, it is possible that it will be called upon to provide support in the eastern Mediterranean (Syria) once U.S. withdrawal from the theater of operations is complete.[27] The *Charles de Gaulle* itself is expected to remain in service into the late 2030s when a new-generation aircraft carrier, for which

---

25 Interview of General André Lanata, Air&Cosmos 2527S (Dec. 16, 2016), 19.
26 Interview of General Serge Soulet, Air&Cosmos 2527S (Dec. 16, 2016), 53–57.
27 "Le 'Charles de Gaulle' mobilisable plus longtemps contre Daech si besoin," *Le Marin*, Dec. 31, 2018, <https://www.lemarin.fr/secteurs-activites/defense/33160-le-charles-de-gaulle-mobilisable-plus-longtemps-contre-daech-si>.

study funding was committed in 2018, should enter service.[28] This commitment puts to rest indefinitely the idea of cooperation between the French and British navies over the design of a new aircraft carrier, but raises anew the controversy about having only one aircraft carrier available, as until the late 1990s France could claim better vector projection when it had two conventional aircraft carriers at its disposal. Budgetary restrictions have thus far silenced calls for a second carrier.

Future challenges tied in part to budgetary matters involve the availability of aircraft in service. In December 2017, Defense Minister Florence Parly sounded alarm that, with the exception of aircraft engaged in Operation Chamal, fleet readiness was down by an average of 40 percent. A logical answer would be to subcontract maintenance to the original equipment manufacturer, as the Australian Air Force has done for its NH-90 and Tiger helicopters. However, the military is concerned about how much of the limited budget would then be reserved for private contractors rather than training and fleet renewal, and how proprietary technical information would no longer be shared with the rank-and-file military crews ensuring day-to-day maintenance.[29]

Conclusion

In November 2018, the new Chief of the French Air Force, General Philippe Lavigne, announced the third strategic plan since 2008, simply named *Plan de vol* (flight plan). As in earlier iterations, it did not discard the goals of the preceding plans, but sought to build on them and account for the changing threat assessment situation. Once the LPM of 2019–2025 was approved in July 2018, it was possible to redefine the pace of modernization of the French Air Force under limited budgetary improvements and to project the future needs of the service as well as define potential threats. Five keywords defined the new strategic plan: "Power" to maintain the operational advantage demonstrated in operations in Syria and Libya;

---

28 "Armées: un futur porte-avion pour succéder au Charles de Gaulle," <https://www.gouvernement.fr/armees-un-futur-porte-avions-pour-succeder-au-charles-de-gaulle>.

29 Anne Bauer, "Disponibilité des avions de combat: Florence Parly tape du poing sur la table," *Les Echos*, Nov. 11, 2017; Exchange between Senator Dominique de Legge and Defense Minister Florence Parly, Senate Commission concerning Military programs 2019–2025 (LPM), Apr. 3, 2018, <http://www.senat.fr/rap/a17-473/a17-47312.html>.

"audaciousness," relying on raised morale; "aviator DNA" as a reflection of the need to further the Air Force's identity some two decades after the end of the draft; "agility," a newly found flexibility to face such threats as the rising aggressiveness of Russia, and also the need to field new weapons more smoothly than has been the case to date; and, finally, "connectivity," perhaps echoing the awareness that a modern air force no longer relies solely on its pilots, but also on a series of information technologies and their projection platforms to ensure collaboration among weapon systems. However, in the French context, attracting more and stronger recruits also means linking up with families and youth to ensure that the service can count on a healthy renewal of its personnel.

The resignation of General de Villiers in 2017 tolled the bell for limited budgets that threatened the functionality of all the French services. Since then, however, it appears endeavor on the civilian side combined with an awareness of new threats and the ongoing efforts of the French Air Force to streamline its hardware, training, and polyvalence may offer some light at the end of a long tunnel.

References

Bauer, A. "Disponibilité des avions de combat: Florence Parly tape du poing sur la table." *Les Echos,* Nov. 11, 2017.

Bridey, J.-J. "Compte Rendu: Commission de la defense nationale et des forces armées." *Assemblée Nationale*, compte-rendu No. 12, Oct. 18, 2017, 33–34.

Cabirol, L. "Spatial militaire: Big bang ou Pschiiit?" *La Tribune*, Nov. 27, 2018.

Cabirol, M. "Le futur avion de combat européen va décoller industriellement en janvier 2019." *La Tribune*, Nov. 20, 2018.

CNES. "CSO/MUSIS," 2018. <https://cso.cnes.fr/fr>.

Commission de la défense nationale et des armées. "Question of Senator Jean-Pierre Cubertafon to General André Lanata regarding the air force budget for 2018," *Compte Rendu* No. 12, Oct. 18, 2017, 17.

de Durand, E. "French Air Power: Effectiveness Through Constraints," in *European Air Power: Challenges and Opportunities*, edited by John Andreas Olsen. Lincoln, NB: University of Nebraska Press, 2014.

Défense. "Le 'Charles de Gaulle' mobilisable plus longtemps contre Daech si besoin." *Le Marin*, Dec. 31, 2018.

Dupont, J.-P. "L'armée de l'air face au défi de la transformation des capacités." *Envol vers 2025. Réflexions prospectives sur la puissance aérospatiale*, edited by Grégory Boutherin and Camille Grand. Paris: La Documentation française, 2011.

French Senate, information report 559, May 23, 2017, 35–38.

Goron, R. "'L'armée européenne,' nouvelle priorité du couple franco-allemand." *Le Figaro*, Nov. 14, 2018.

Gouvernement.fr. "Armées: un futur porte-avion pour succéder au Charles de Gaulle." <https://www.gouvernement.fr/armees-un-futur-porte-avions-pour-succeder-au-charles-de-gaulle>.

Guibert, N. *C'est qui le chef? La vraie crise entre le pouvoir et l'armée.* Paris: Robert Laffon, 2018.

Huberdeau, E. "Arrivée officielle du PC-21," Oct. 3, 2018, <http://www.air-cosmos.com/arrivee-officielle-du-pc-21-115603>.

Lelièvre, V. "Dépenses militaires et contraintes économiques 1971–1995." *Revue française d'économie*, 11: 2 (1996).

Lemayrie, P. "L'Entente cordiale au péril du Brexit?" *Le Monde diplomatique, Blog "Défense en ligne,"* Oct. 26, 2018.

Maulny, J.-P. "The French/UK Defense relationship," *IRIS Analyses*, June 22, 2016.

Pasco, X. "De l'utilisation au 'contrôle' de l'espace extra-atmosphérique?" in *Réflexions prospectives sur la puissance aérospatiale, Envol vers 2025*, edited by Grégory Boutherin and Camille Grand. Paris: La Documentation française, 2011.

Sénat. "Exchange between Senator Dominique de Legge and Defense Minister Florence Parly." Senate Commission concerning Military programs 2019–2025 (LPM), Apr. 3, 2018, <http://www.senat.fr/rap/a17-473/a17-47312.html>.

Société. "Plusieurs drones ont survolé des lieux sensibles de Paris dans la nuit." *Le Monde*, Feb. 24, 2015.

Strategic Review of Defence and National Security 2017. <https://www.defense.gouv.fr/dgris/presentation/evenements/revue-strategique-de-defense-et-de-securite-nationale-2017>.

Thuriot, M. "Interview of Generals Pierre de Villiers and Günther Fassberg." Feb. 4, 2013. <https://www.defense.gouv.fr/english/actualites/articles/succes-et-avantages-de-la-cooperation-franco-allemande>.

Trappier, E. Testimony before the French Senate Defense Commission. Feb. 28, 2018. <http://videos.assemblee-nationale.fr/video.5624702_5a96a5ff81ec0.commission-de-la-defense--auditions--sur-le-projet-de-loi-de-programmation-militaire-28-fevrier-2018>.

5 UNITED KINGDOM
The Thin Blue Line

Peter W. Gray

Introduction

British military history, and indeed the wider literary culture of the nation, is deeply imbued with the concept of a thinly spread unit holding fast against a massed attack. The original "Thin Red Line" terminology came from the Crimean War and, in particular, the Battle of Balaclava.[1] It is the title of an 1881 oil painting by Robert Gibb and is found in Rudyard Kipling's famous depiction in "Tommy" of the typical British soldier laboring away under poor conditions and worse leadership and whose sterling qualities were only appreciated when the fighting started in earnest.[2] The term has since entered much broader usage including in the United States where fire departments use it out of respect for firefighters who have lost their lives in the line of duty. It has also widely used in the movie industry.

Even a superficial glance at Britain's current air power lineup will immediately confirm the thin element of the chosen title. Only a slightly closer approach will confirm the blue element from a number of areas including the uniform chosen following the formation of the world's first independent air force over a century ago. From a more philosophical, or poetic, viewpoint it also coincides with Alfred Lord Tennyson's "nations'

---

1 John Selby, *The Thin Red Line of Balaclava* (London: Hamish Hamilton, 1970) and Sir William Howard Russell, *The Great War with Russia: A Personal Retrospect* (London: Routledge, 1895). The campaign has been described as one of the most "ill-managed" in English history and warrants its own chapter in Norman F. Dixon, *On the Psychology of Military Incompetence* (London: Jonathan Cape, 1976).
2 Robert Gibb, *The Thin Red Line*, Oil on Canvas, 1881, and Rudyard Kipling, "Tommy" in *Barrack Room Ballads and other Verses* (London: Methuen, 1892).

airy navies grappling in the central blue."[3] The more pragmatic approach for this chapter is that Britain's air power, both historically and for the future represents an amalgam of the light blue of the Royal Air Force (RAF) and the dark blue of the Royal Navy (RN), especially in the era of the F-35. The phrase that air power is truly joint has been included in so many senior officers' speeches that it has become somewhat hackneyed; the cynic may also say that it has been used to paper over the cracks between the two services. This has been undoubtedly true with the divisions the deepest and most bitter when competition for resources has been most fierce. This chapter will include a brief examination of the joint use of Britain's air power.

Going beyond the brief examination of Britain's air power being "thin," it is tempting to wax lyrical on just how thin it is, with adjectives such as "gossamer" springing to mind. The reality is, of course, more brutal and something like "stripped beyond bare minima" from successive defence cuts is probably more appropriate. As real gaps in capability become increasingly evident it is clear that holes, or even massive tears, have appeared in the very fabric of the military structure.[4] These became increasingly inevitable under the budget cuts that were severely compounded by the heavy usage imposed on various fleets by persistent operational tempo.[5] This again will be examined in more depth in the chapter.

A key underlying theme for this chapter will be to examine whether the "line" has lost its integrity to the point where air power (or air force) critics are again at the stage where they believe that they are able to advocate the reabsorption of the world's oldest air force back into its older predecessor services. The chapter will argue that although it is desperately thin, Britain's air power is actually healthier and more robust than it seems at first sight. In fact, the pendulum is possibly now starting to swing back, and instead of being incrementally reduced (salami slicing in United Kingdom parlance), capability is starting to be slowly added. That said, there

---

3 Alfred Tennyson. "Locksley Hall." <https://www.peotryfoundation.org/poems/45362/locksley-hall>.
4 The most obvious of these was the removal of the maritime reconnaissance capability provided by the Nimrod by its cancellation, and that of its successor in the Strategic Defence and Security Review of 2015. See Cm 9161,"National Security Strategy and Strategic Defence Review 2015," GOV.UK, <https://www.gov.uk/government/publications/national-security-strategy-and-strategic-defence-and-security-review-2015>.
5 The need to replace the worn-out C130 fleet was a classic example of this.

is always a lag in the system between decisions being made, funding allocated, equipment procured and, finally, personnel recruited and trained to make it effective in capability terms.

In producing a chapter such as this about the United Kingdom (U.K.), it is always tempting to use provisos such as "at the time of writing" and so forth. This is always the case in times of change and never more so for the United Kingdom with its exit from the European Union. But rapid change is endemic virtually everywhere and such an approach would ensure that nothing meaningful was ever produced. The chapter will work on the basis that the provision of national security and defence is the first priority of all governments in a nation and are therefore (or should be) above the inter-, and intra, party struggles and bickering. The inertia in the decision-making and procurement systems are usually so sluggish that the pendulum has tended to swing between reasonable limits. The chapter will also attempt to avoid the use of buzzwords and slogans that have so bedeviled the discourse. Phrases such as "combat ISTAR," and the various acronyms around networked capability have been seen in the United Kingdom to come into fashion and then disappear without trace. The underlying fundamental principles (also known as doctrine) evolve slowly and more surely.

This chapter will follow the themes of its neighbors and will examine how the United Kingdom sees its force structure, doctrine, and threat environment. As stated above, it will examine the recent historical budgetary pressures and how these are likely to evolve and with what consequences. It will examine potential changes to doctrinal thinking (as opposed to just the written word, which frequently becomes immersed in inter-service wrangling) along with likely major threats. There will be a brief examination of systems and associated weapons likely to be introduced—or at least fought for. The final section—how the United Kingdom sees its future in coalition and in regional alliance terms—will arguably be the most contentious. The chapter will inevitably draw heavily on official documentation, but its judgments do not reflect official thinking or policy. They are the author's alone.

Threat Environment, Force Structure, and Doctrine

In the ideal world, a nation's defence strategy would be based on the existential threat that it faced. The force structure necessary to counter that threat would be estimated with an added contingency for emergencies and based on the enduring and fundamental principles of the use of force, and doctrine would be written and taught. This structure would almost certainly be changed by that nation's commitments to alliances, partnerships, and treaty obligations. It may also be affected by legacy commitments. In the case of the United Kingdom over the last century and beyond, all of these have been true. But the world is not this simple. The reality is that a defence strategy cannot be considered in isolation from the security of the nation, and this will have both an external and internal element to its composition. This dimension will be further complicated by emerging threats, some with a gradual effect and others with a sudden high impact, such as Al-Qaeda and the attacks of September 11, 2001. The demands of domestic politics also play their part when politicians decide that "something must be done."[6] In this eventuality, and as has been demonstrated many times over the last century, the speed, reach, and availability of air power, along with its powerful media impact, have often made air power the weapon of first political resort, whether it be for instigating a no-fly zone such as in Libya or for casualty evacuation such as for Sarajevo in 1993.

On the other hand, designing and buying military equipment can take years, so defence planners have had to devise a better system, method, or process. Throughout the Cold War, the Warsaw Pact provided the existential threat and the situation was relatively stable. The United Kingdom was part of the North Atlantic Treaty Organization (NATO) alliance and aligned (or more correctly, subjugated) its doctrinal thinking with that of NATO (especially as it was a leading voice in this process). Equipment acquisition was based on an ongoing fight between the services for a better share of a limited budget that was fixed as a percentage of the gross domestic product (GDP). The end of the Cold War saw the defence budget go into freefall as the financiers sought a peace dividend even through

---

6 See Peter Gray, *Air Warfare History, Theory and Practice* (London: Bloomsbury, 2016), 107 and Martin Whitfield and Marcus Tanner, "Sarajevo Rescue Mission Snowballs," *The Independent*, Aug. 14, 1993, <https://www.independent.co.uk/news/sarajevo-rescue-mission-snowballs-britain-offers-further-flights-row-with-un-after-doctors-find-1461141.html>. This coverage shows very clearly the direct intervention of Prime Minister John Major.

Saddam Hussein chose this period to invade Kuwait.[7] The landslide continued until 1997 when the incoming Labour government decided to attempt a balanced and thought-through defence review.[8] The absence of a simple threat necessitated the use of a complex matrix of scenarios at various levels of war and in differing theaters. These were war-gamed and analyzed, and a force structure came out along with a complex set of defence planning assumptions. A key output from this Strategic Defence Review (SDR) was the establishment of the Joint Doctrine Centre, which quickly morphed into a Joint Doctrine and Concepts Centre (JDCC) and is now the Development, Doctrine, and Concepts Centre (DCDC), which styles itself as the Ministry of Defence's (MOD) think tank and "helps inform defence strategy, capability development, operations and provides the foundation for joint education."[9] One of the flaws in the original SDR was that only the British Army had seen fit to generate a coherent body of doctrine on which it could base its force development; this left the Navy and RAF without a sound footing on which to base their work. One of the aims of the JDCC was to produce agreed and joint doctrine.

The current DCDC draws its staff from the civil service, the three military services, and from partner nations. It reports to the Central Staff and therefore both forms and informs policy, but is in no way independent. It takes views from academics from various institutions, but is not bound to reflect their views, not least because policy has to be affordable.[10] For the purposes of this chapter, the key document to begin the cycle is the "Global Strategic Trends: The Future Starts Today" (GST) paper. In classic Whitehall language, the opening wording states that the document does not repre-

---

7   The author was a student on the Joint Services Defence College course at Greenwich in 1991 and the final exercise was to make and justify high percentage defence cuts: in actual fact, the students could not keep pace with the swinging cuts being made in Whitehall. The cuts were made under *Options for Change* in 1990 and *Front Line First* in 1994.

8   "Modern Forces for the Modern World," Strategic Defence Review, July 1998, <https://webarchive.nationalarchives.gov.uk/20121018172816/http://www.mod.uk/NR/rdonlyres/65F3D7AC-4340-4119-93A2-20825848E50E/0/sdr1998_complete.pdf>.

9   See "Development, Concepts and Doctrine Centre," GOV.UK, <https://www.gov.uk/government/groups/development-concepts-and-doctrine-centre>.

10 The 1997 SDR had a series of Supporting Essays internally produced but with external consultation. *Global Strategic Trends* commissioned "over 70 pieces of academic research" from 42 institutions, p. 7. See "Global Strategic Trends: The Future Starts Today," GOV.UK, DCDC, 6th edn, last modified Oct. 15, 2018, <https://www.gov.uk/government/publications/global-strategic-trends>.

sent official U.K. government or MOD policy—just that of the DCDC. It is, nevertheless, signed off by the Chief of Defence Staff and the Permanent Secretary of the MOD. The document aims to provide a strategic context for those in the MOD and wider government involved "in developing long-term plans, policies and capabilities."[11]

In terms of defining a future threat environment, the subtitle of the GST is fairly apposite in that it would be fair to take the actual threat for today and extrapolate into the future. The GST, for example, acknowledges that China will continue to grow in economic terms and in military capability to reach at least parity with the United States and eclipse Western Europe and Russia.[12] What the document does not say, and realistically cannot predict, is what the United Kingdom would choose to do in the event of a rise in tension in China's sphere of influence. The chances of military intervention by the United Kingdom seems highly improbable in most likely scenarios; the obligation normally accepted by the United Kingdom to join coalitions for military intervention would be considerably lower than in more traditional areas of influence. In conventional military terms, the classic eventuality requiring military action would be with NATO in the event of a transgression from a resurgent Russia. In particular, the GST emphasizes the importance of the Arctic to Russia, both in potential economic terms and for military growth; the document highlights the reacquisition of military bases and capabilities that were lost after the end of the Cold War.[13] The GST also acknowledges the traditional Russian concerns around internal and external security and cites the continuing growth in military expenditure above that for other areas of government spending (5.6 percent of GDP in 2016). Major modernization programs include the Russian strategic deterrent capability, precision weapons capability, and the means of delivery. Specialist areas of development include cyber, robotics, and potentially autonomous systems.[14] The recent use of the extremely toxic nerve agent Novichok by the Russian agents in the United Kingdom

---

11 Ibid., 7.
12 Ibid., 12.
13 Ibid., 228.
14 Ibid., 229.

confirms their retention of this capability and, more worryingly, their total disdain for the legal and ethical implications of deploying it.[15]

The GST confirms the view that Southwest Asia will remain volatile and that conventional military capabilities will continue to receive major investment.[16] The dominant military powers in the region will be Turkey, Iran, Israel, and Saudi Arabia. Iran and Saudi will likely continue to mount coalitions and to use proxy actors. The scope for major conflict in the region, including the use of nuclear weapons (or their by-products) cannot be ruled out.[17] Asymmetric conflict, including through terrorist organizations such as Islamic State and potentially its successors, is highly likely. This is where the traditional military threat environment overlaps into the security scene with the proliferation of religious zealotry translating into a willingness to commit deadly attacks on domestic targets. From the era of the Provisional Irish Republican Army attacks in Northern Ireland, the British mainland and on-service personnel in Germany and the Netherlands, the standard U.K. mantra has been along the lines of military aid to the civil power with domestic police primacy. This will remain the case in the event of further Islamic (or other) extremist attacks in the United Kingdom, but the ever closer linkages between defence and security policy will continue to influence decisions to intervene abroad and, therefore, force structures.

Notwithstanding a never-ending succession of defence spending cuts, reviews, efficiency measures, and restructuring, the United Kingdom has consistently attempted to maintain a balanced air portfolio retaining the capability to do a bit of everything. Or at least this was true until the "Strategic Defence and Security Review" of 2010 (SDSR 2010).[18] This Review was of the more radical rounds of cuts in that, at a stroke, it removed the maritime reconnaissance capability provided by the Nimrod force. Not only

---

15 See "Novichok Nerve Agent Use in Salisbury: UK Government Response," GOV.UK, Mar. to Apr. 2018, last modified Apr. 18, 2018, <https://www.gov.uk/government/news/novichok-nerve-agent-use-in-salisbury-uk-government-response>.
16 "Global Strategic Trends: The Future Starts Today," GOV.UK, 165.
17 Ibid.
18 CM 7948 *Securing Britain in an age of Uncertainty: The Strategic Defence and Security Review*, 19 Oct. 2010. See "Securing Britain in an Age of Uncertainty: The Strategic Defence and Security Review," GOV.UK, <https://www.gov.uk/government/publications/the-strategic-defence-and-security-review-securing-britain-in-an-age-of-uncertainty> and <https://assets.publishing.service.gov.uk/government/uploads/system/uploads/attachment_data/file/62490/Factsheet8-RoyalAirForce.pdf>.

was the decision taken to scrap the MRA4 acquisition, but also the entire existing fleet was axed. Part of the fig-leaf logic was that the anti-submarine role could be adequately performed by the ships and submarines of the RN. This may be partly true, but it is probable that the removal of the air capability did little to enhance the security of the nuclear deterrent boats as they deployed to their patrol areas. SDSR 2010 also rationalized the United Kingdom fast-jet fleet by removing the Harrier fleet, just leaving the Eurofighter Typhoon and the Tornado, which in due course would be replaced by the F35 Lightning II.[19] The removal of the Harrier had major implications for the RN's retention of a fast-jet capability as it had been a key partner in Joint Force Harrier. The demise of the Nimrod may well have been influenced by cost overruns in the project. The capability was nominally reinstated in SDSR 2015 with the purchase of nine Boeing P8 maritime patrol aircraft that would have an overland capability.[20]

Following the cuts and rationalizations, the RAF was left with its aging legacy fleet of Tornado GR4 aircraft until its withdrawal from service on March 31, 2019. This was a highly capable ground attack aircraft that also had an important reconnaissance capability. It could carry Paveway III and IV GPS/laser-guided bombs, Brimstone air-to-ground missiles, Storm Shadow cruise missiles, an internal 27mm Mauser cannon, and a Litening II targeting pod. It could also take RAPTOR, Sky Shadow, and BOZ countermeasure pods. The Tornado had been on operations throughout the post–Cold War period from no-fly-zone operations over Iraq to more recent work against Islamic State (IS)/Daesh. The ability to use precision weaponry as part of a coalition air tasking order has made the GR4 a valuable asset and the aim is for this capacity to continue with the F-35.

The Typhoon FGR Mk4 is described as a "highly capable and extremely agile fourth-generation multi-role combat aircraft."[21] For much of its early career in the RAF it was only able to operate in the air defence role, but a succession of upgrades means that it now has a genuine precision attack capability. It can carry the usual air defence suite of weapons, including infrared and radar guided missiles (ASRAAM and AMRAAM respectively)

---

19 Ibid.

20 This aircraft is reported to be due in 2020—see RAF, "About the P-8A," <https://www.raf. mod.uk/aircraft/p-8a/>.

21 RAF, "About the Typhoon FGR4," <https://www.raf.mod.uk/aircraft/typhoon-fgr4/>.

and Paveway II and IV GPS/laser-guided bombs; Brimstone and Storm Shadow will follow. The RAF currently has six operational squadrons at RAF Coningsby in Lincolnshire and RAF Lossiemouth in Scotland. The future force structure will comprise the Typhoon and the F-35.

The Intelligence Surveillance Target Acquisition and Reconnaissance (ISTAR) fleet has also seen extensive operational service over the last 30 years. RAF Waddington in Lincolnshire is the ISTAR fleet hub and has two squadrons of E-3D Sentry aircraft plus the operational conversion unit.[22] Battlefield surveillance is provided by two squadrons of Sentinel RMk 1 and a one squadron of Shadow R1 aircraft.[23] Dedicated electronic surveillance is provided by the RC-135 River Joint squadron.[24] The MQ-9A Reaper is a remotely piloted, medium-altitude, long-endurance aircraft and is designed for a range of ISTAR missions as well as precision attack. The RAF has two squadrons, one at Waddington and the other at Creech Air Force Base in the United States.[25] Again, they are in constant operational use in a variety of theaters.

The Air Transport and air-to-air refueling hub is at RAF Brize Norton in Oxfordshire. It has two squadrons of Airbus Voyager aircraft operating in both the transport and tanking roles.[26] The Voyager is supplemented by one squadron of Atlas A400M transport aircraft, along with a test and evaluation unit and a conversion unit.[27] Brize Norton also hosts the remnants of the C130 fleet with a similar structure to the A400M.[28] At RAF Northolt, closer to London, there is one squadron of BAE 146 aircraft for VIP duties.[29]

22 RAF, "About the E-3D," <https://www.raf.mod.uk/aircraft/e-3d/>.

23 RAF, "About the Sentinel R1," <https://www.raf.mod.uk/aircraft/sentinel-r1/> and RAF, "About the Shadow R1," <https://www.raf.mod.uk/aircraft/shadow-r1/>.

24 RAF, "About the RC-135W Rivet Joint," <https://www.raf.mod.uk/aircraft/airseeker-rc-135w-rivet-joint/>.

25 See RAF, "39 Squadron," <https://www.raf.mod.uk/our-organisation/squadrons/39-squadron/>. For an account of life as an operator on these squadrons see Peter Lee, *Reaper Force: Inside Britain's Drone Wars* (London: John Blake, 2018), and for an analysis of drone warfare see the University of Birmingham Policy Commission Report VI, *The Security Impact of Drones: Challenges and Opportunities for the UK*, Oct. 2014, <https://www.birmingham.ac.uk/Documents/research/policycommission/remote-warfare/final-report-october-2014.pdf>. The author was one of the commissioners on this report. Although the military loathes the word, everyone else calls them drones.

26 RAF, "About the Voyager," <https://www.raf.mod.uk/aircraft/voyager/>.

27 RAF, "About the Atlas (A400M)," <https://www.raf.mod.uk/aircraft/atlas-a400m/>.

28 RAF, "About the C-130J Hercules," <https://www.raf.mod.uk/aircraft/c-130j-hercules/>.

29 RAF, "About the BAE 146," <https://www.raf.mod.uk/aircraft/bae146-ccmk2-and-cmk3/>.

The RAF has also had to slim down its command structure in parallel with the other two services and the head office in the MOD. The old of system of multiple commands has been amalgamated into one—Air Command at RAF High Wycombe in Buckinghamshire—the old wartime headquarters of Bomber Harris. The various groups, again all at High Wycombe, have been reduced to six. No 1 Group covers the fast jet and ISTAR fleets; 2 Group, the air transport and force protection forces; 22 Group is responsible for training all RAF personnel; 38 Group is responsible for engineering, logistics, communications, and medical support; 83 Expeditionary Air Group commands the RAF air capability in the Middle East; and finally, 11 Group has recently reformed to cover multi-domain operations and new and evolving threats—it brings air, cyber, and space command and control under one commander.[30]

Beyond the single service aspects of U.K. air power, the rotary wing world is in a truly joint organization called, not surprisingly, Joint Helicopter Command. It brings under one command the battlefield assets of the RAF, the Army Air Corps, and the RN. It was formed in 1999. Command rotates between the three services. Aircraft types include the Chinook, Puma, Gazelle, Lynx, Apache, and Merlin.

As discussed above, air power doctrine writing has only seriously come of age in the last 20 years. For much of the Cold War, the RAF relied exclusively on NATO doctrine, primarily at the tactical level. The few service-specific manuals left on the shelves were outdated and merely used as turgid fodder for promotion examination candidates. There were some early attempts to formulate real doctrine, usually by the incumbent of the RAF Director of Defence Studies post.[31] As discussed above, the real catalyst for taking doctrine seriously was following the 1997/78 Defence Review coincident with the formation of the JDCC. The first result was the publication of a new, third edition of AP 3000, *British Air Power Doctrine*.[32] This

---

30 RAF, "Groups," <https://www.raf.mod.uk/our-organisation/groups/>. For more on 11 Group see G. Jennings, "RAF Reforms 11 Group to Counter New and Evolving Threats," *Jane's 360*, Nov. 8, 2018, <https://www.janes.com/article/84412/raf-reforms-11-group-to-counter-new-and-evolving-threats>.

31 See, for example, Air Publication (AP) 3000, 2nd edn, 1993, and Group Captain Peter Gray, "Air Power and Joint Doctrine: An RAF Perspective," *RAF Air Power Review*, Vol. 3, No. 4, Winter 2000.

32 AP 3000 *British Air Power Doctrine*, Air Staff, 1999.

and its successor volumes were single service productions and published through the air staff, albeit with joint consultation. The first version of the *Air Power Doctrine* to come out under the joint banner was in 2013 with the Joint Doctrine Publication (JDP) 0-30 *UK Air and Space Doctrine*.[33] The second edition was published in December 2017.[34] The joint publications aim for consistency with higher-level doctrine and with the sister documents for the other two environments.[35] Arguably, this is to be applauded if the aim is for all of the arguments to be corralled into a homogenous mass. The risk, however, is that they could become too bland to be of use either to guide the armed forces in the execution of their mission or to be taught to any reasonable level of student. The underlying aspect of doctrine is that fundamental principles on which doctrine is based are just that; they should not be manipulated to suit others' interpretation of what may look consistent.

The current edition of JDP 0-30 reemphasizes the four key roles of air power as control of the air; intelligence, surveillance, and reconnaissance; attack; and air mobility.[36] Air power itself is now defined as "the ability to use air capabilities in and from the air to influence the behaviour of actors and the course of events."[37] These roles and the definition have been consistent over the various iterations. The main change was the drop from five roles to four; the one that was excised was the use of air power for strategic effect. Anecdotally, this was due to apparent difficulty in explaining what this meant to land audiences. In reality, it had more to do with a long-standing (predominantly khaki) phobia around air power "doing it alone." The meat of the publication is in Chapter 3, and covers the foundations of air power. In addition to just setting out the detail behind the roles and functions of air power, the wording, and in particular the footnoting, attempts to reconcile the various environments' differing terminology. The provision of close air support for example, equates in the land environment to "close support fires" of which there are two subsets—close air support and

---

33 JDP 0-30 *UK Air and Space Doctrine*, DCDC 2013.
34 JDP 0-30 *UK Air and Space Doctrine*, DCDC 2017.
35 The higher level is in JDP 0-01 *UK Defence Doctrine*, DCDC 2014. See also JDP 0-10 *UK Maritime Doctrine*, DCDC 2017 and JDP 0-20 *UK Land Doctrine*, DCDC 2017.
36 JDP0-30, 4.
37 Ibid., 8.

close combat attack.[38] There are several other examples from which one may conclude that there has been a harmonization of doctrine writing, but not an agreement over terminology or practice. At least the practitioners have somewhere authoritative to find the divergences.

Cash Is King: Future Budgetary Pressures

Although this slang expression is invariably used in the commercial sector, its equivalent for this chapter runs along the lines that money, not policy, is king. This is a view that is widely shared across the government procurement system and in the various military headquarters. The big problem now, and for the foreseeable future, is that it is in short supply. The standard pattern in the United Kingdom is for the various government departments to negotiate with the Treasury (the Finance Ministry), hoping to agree a settlement for at least the coming financial year. These negotiations are then brought together in what is termed a "spending review." One of the key problems in this process is that departmental plans are notoriously optimistic, especially in their assumptions.[39] Furthermore, they focus on the short term to the detriment of long-term responsible planning. In layman's terms, this means that departments can get away with being able to balance the books next year without promising anything for the future. In equipment procurement terms, for MOD this means that all the department needs to do is be able to pay for what will be delivered in the next year or two, and delay other programmes beyond the review radar range. This also means that the equipment program is unaffordable in the absence of a miraculous uplift in funds, which never happens. The shortfall, in turn, leads to bickering between the services as to which pieces of equipment will be funded and which are delayed. Of course, cash can be found from elsewhere, including manpower; the army tends to lose out on this front for obvious reasons of density. The equation on spending is compounded by the need to acknowledge the interests of the domestic defence industry and the vast chain of small suppliers. Regrettably, the Treasury also only

---

38 Ibid., 34.
39 See, for example, the report by the Institute for Government, *The 2019 Spending Review: How to Run It Well*. <https://www.instituteforgovernment.org.uk/sites/default/files/publications/IfG_2019_%20spending_review_web.pdf>, 4–5.

focuses on the short-term figures rather than on sorting out bad accounting practice.[40]

In 2006, NATO set its member countries a target of spending at least 2 percent of GDP on defence. This was seen as an attempt by the alliance to get the European nations to lessen their reliance on the United States as by far the major player. There is no real significance in the 2 percent; it neither represents a necessary and sufficient amount of money, nor does it reflect a needed capability.[41] The United Kingdom has achieved this target, along with only two other nations (Greece and the United States), every year since 2010.[42] For the last five years, expenditure in real terms has been stable at £36 billion, but is expected to rise to £37.6 billion in 2019/20. It should be noted that expenditure on operations and interventions is above and beyond the defence budget and is met directly by the Treasury.[43] This includes those urgent operational requirements (UORs) deemed necessary actually to fight the campaign. In the past these have included weapons upgrades and unmanned systems. The problem with this arrangement is that eventually the systems are either scrapped or have to be brought into the core program with all of the sustainability costs. When dealing with something like the Reaper program, the black hole in the defence spending budget can only get worse. There have also been accusations that the MOD has sought Treasury UOR funding for upgrades that should have been part of the normal program.

Over the period 2016–2026, the MOD estimates an expenditure of £18 billion on combat air, including the Typhoon, Tornado, and F-35; £16.6 billion on air support, including Voyager, A400M, and C130; £10.6 billion on helicopters, including Merlin, Apache, and Chinook; and £4.6 billion on ISTAR.[44] The MOD estimates £13.5 billion on weapons, but has not broken the figure down between naval and air systems. A clear difference between Cold War and recent operations is the actual use of weapons with

40 The so-called black hole has been estimated to be about £21 billion. See Deborah Haynes, "Going to War on a £21bn Hole in the Defence Budget," *The Times*, May 25, 2018.

41 See Neil Dempsey, *UK Defence Expenditure*, House of Commons Briefing Paper CBP 8175, Nov. 8, 2018, 3.

42 Ibid.

43 Ibid., 11. In 2017/18, counter-Daesh operations cost about £0.5 billion.

44 "A Breakdown of Planned Defence Expenditure 2018," Defence Contracts Online, <https://www.contracts.mod.uk/blog/breakdown-planned-defence-expenditure-2018/>.

associated costs. As stated above, these costs should be met directly from Treasury but are not always covered in full.[45]

Although the acquisition of F-35s is underway and the two Queen Elizabeth air carriers are close to completion, the actual number of airframes to be bought remains uncertain. The upcoming series of spending reviews will be critical as will any potential government appetite for tackling the so-called black hole in the equipment program. In the meantime, in the United Kingdom, public and media interest in public spending focuses on the National Health Service, education, and social services; there are no votes in defence.

Future Doctrine and Threats

If, as discussed above, money reigns supreme above policy, then doctrine will come third by a long way. Although the United Kingdom now has a well-established doctrine production system with the DCDC, it is firmly under official policy ownership and there is no prospect of changes in doctrine, no matter how "fundamental" the principles, changing policy, or even sparking a meaningful dialogue.[46] In terms of high-level doctrine, any significant changes or even trends are therefore unlikely. Although the doctrine manuals are produced under a joint banner, they very much reflect the views of each service headquarters; resulting wording has to brokered around compromise. The lower level doctrine surrounding how air operations are planned and executed is almost invariably based on a Combined Air Operations Centre set of processes. With the exception of part of the Libyan campaign, these have all been U.S. led and the U.K. air commanders will follow those procedures for the foreseeable future.

Arguably more importantly than doctrine for British air power, the RAF continues to invest in intellectual capital. This is normally described under the conceptual component of fighting power alongside the moral

---

45 Nick Hopkins, "Libya Conflict May Cost UK £1.75bn," *The Guardian*, Sep. 25, 2011, <https://www.theguardian.com/politics/2011/sep/25/libya-conflict-uk-defence-bill>.
46 This is based on doctrine being the fundamental principles of war collated for the guidance of senior commanders and their folk.

and physical.[47] The guardian for this quest for intellectual progress is the RAF Director of Defence Studies (D Def S).[48] Among other routine tasks for the air staff, D Def S acts as editor for the RAF *Air and Space Power Review*, which is a peer-reviewed air power journal aimed at publishing the latest thinking on the subject from serving personnel and key academics.[49] The RAF also has a formal fellowship scheme (Chief of the Air Staff's (CAS) Fellowships) aimed at encouraging air power thinking and education at post-graduate level beyond the staff college system. Courses include full- and part-time doctoral study (the Portal Fellowship named after the wartime CAS), master-level degrees including in international relations at Cambridge (the Tedder Fellowship), and distance learning offerings. The RAF has opened up these programs to all ranks—not just the officer corps. These courses are highly commendable and go some way to countering the natural tendency of air forces to focus on the tactical level of air power as practitioners. The extent to which anti-intellectualism is apparent in the RAF is a delicate subject and one that is outside the scope of this chapter.

The future threat environment has largely been discussed in the context of the GST above. The reality is that unless there is a sudden maverick attack on a British dependency that has to be countered—as in the 1982 Argentine invasion of the Falkland Islands—or a major attack on NATO necessitating an Article 5 response, the United Kingdom will have the discretion on whether it needs to be involved. Even in the NATO Article 5 case, the scale of the response will be discretionary. Despite claims from the British Army otherwise, the most likely U.K. response will be an air deployment in preference to "boots on the ground." It is outside the scope of this chapter to discuss the drawbacks in land deployments, but it would be fair to say that the speed of deployment, and reach and variety of precision attack weaponry makes air the weapon of first political choice. Adding precision attack to a coalition order of battle is always welcome, especially when it has its own organic tanker and transport support and ISTAR assets.

---

47 See JDP 0-01 *UK Defence Doctrine*, <https://assets.publishing.service.gov.uk/government/uploads/system/uploads/attachment_data/file/389755/20141208-JDP_0_01_Ed_5_UK_Defence_Doctrine.pdf>.
48 The author held this post from 1999 to 2002.
49 See <https://www.raf.mod.uk/what-we-do/centre-for-air-and-space-power-studies/publications/>.

The Wish List—Future Weapons and Systems

The U.K. weapons and aircraft programs, subject to affordability, center on continuing improvements and upgrades to the Typhoon. Primarily, these will be carried out under the auspices of Project Centurion and will involve incorporation of existing Tornado weapons systems, such as the Storm Shadow deep-strike cruise missile; the Meteor, beyond visual range air-to-air missile; and the Brimstone precision attack weapon.[50] Beyond these upgrades, it is likely that the Typhoon will receive the next-generation standoff missile, the Spear Cap 3. It is also planned to obtain a new radar, new helmet, and expendable active decoys.[51]

The bright shining hope beyond that, as with many air forces, is the introduction of the F-35. It will replace the Tornado fleet and become the mainstay of the RAF manned-strike capability. It will also return a fixed capability to the RN Fleet Air Arm. Despite a prime ministerial announcement to the contrary, the United Kingdom will be purchasing the F-35B with its vertical- and short-lift capability. The planned buy is 138 aircraft.[52] Following the demise of the Nimrod maritime patrol capability, the RAF has ordered nine Boeing P8 aircraft, which will be known as the Poseidon MRA Mk1.[53]

Long-term RAF hopes rest on the recently announced Tempest program. This will nominally be a manned aircraft, although options are open for a remotely piloted version. It is to replace the Typhoon from 2035. The U.K. Defence Secretary unveiled a model of the Tempest at the 2018 Farnborough International Airshow as part of the Future Combat Air Strategy.[54] It is to be designed by a consortium of firms led by BAE Systems with Leonardo, Rolls Royce, and MBDA. Part of the rationale behind the program is

---

50 Air Force Technology, "RAF to Upgrade Eurofighter Typhoon with Tornado Capabilities," Sep. 5, 2018, <https://www.airforce-technology.com/news/raf-upgrade-eurofighter-typhoon-tornado/>.

51 Ibid.

52 See "What You Need to Know About the F-35B," Forces Network, May 21, 2019, <https://www.forces.net/news/what-you-need-know-about-f-35b-lightning-ii>.

53 Allison George, "RAF Announce that the P-8A Maritime Patrol Aircraft Will Be 'Poseidon MRA Mk1' in UK Service," UK Defence Journal, Aug. 22, 2018.

54 GOV.UK. "Combat Air Strategy: An Ambitious Vision for the Future." Jul. 16, 2018. <https://www.gov.uk/government/publications/combat-air-strategy-an-ambitious-vision-for-the-future> and Callum Jones, "RAF Unveils £2bn Plan for New Tempest Fighter Jet (and It Has Lasers)," The Times, Jul. 16, 2018.

to retain a sovereign U.K. design capability in advanced military aircraft. This will involve building the capacity to generate U.K.-based intellectual property with the associated people skills and facilities. This will allow the United Kingdom to "secure operational advantage and freedom of action or export opportunities."[55] There is clearly a long way to go in this ambitious program, especially given the likely costs. It is very much seen as a Typhoon replacement rather than an alternative to F-35. The rhetoric on show at Farnborough was very clear on this, emphasizing the future mix. It is evident that the restrictions imposed by the United States on sharing aspects of the F-35 technology package continues to cause frustration in the aerospace industry. This is particularly the case with firms that shared their intellectual property with the United States in the first instance to aid the development of F-35.

Coalition Operations

The United Kingdom has consistently sought to support key allies, and especially the United States, by providing small, but worthwhile capabilities to armed interventions. These deployments have been invariably supported by the United Kingdom's own tanker, transport, and ISTAR elements. It is highly likely that this will continue for the foreseeable future. The developing mix of Typhoon and F-35 followed by F-35 and Tempest will ensure that the United Kingdom has a genuine precision attack capability that will add value to any attack option (ATO). The ATO is particularly important in coalition operations for political, symbolic, and practical reasons. Too many countries have restricted their offerings to air defence assets. One only has to think of operations against Serbia where the Adriatic was the best-defended piece of airspace in the world, against no viable threat. Although the United Kingdom could offer niche capabilities, it seems unlikely that it will opt to restrict a coalition offering in this way.

---

55 MOD, "Combat Air Strategy: An Ambitious Vision for the Future."

Conclusion

The United Kingdom has traditionally willingly deployed air assets to support coalition operations and will continue to do so. The critical mass may not be huge—a thin blue line indeed—but it has always been highly capable and has integrated well with allies. The air and ground crews are highly professional and capable. The United Kingdom has also demonstrated an excellent record of fitting seamlessly into command and control organizations at all levels. In terms of overall force structure, two decades of defence spending cuts have taken the force to unprecedented lows, even removing a whole capability. It is now fair to say, however, that the pendulum is swinging back the other way, which augurs well for U.K. air power. The procurement of the Poseidon P8 and the announcement of the Future Combat Air Strategy are excellent omens for the future. Beyond the physical manifestations of British air power, the willingness of the U.K. government to work with the military forces and with defence industry and its suppliers or to regenerate the growth of intellectual capital is also an excellent sign for the future. Within the RAF this has remained a constant theme for the last 40 years with every encouragement, and indeed formalization and expansion, of the CAS Academic Fellowship scheme to allow service personnel of all ranks to gain masters and research degrees. This fostering of the intellectual component of fighting power has also to be seen as a positive sign. In short, the blue line may be thin, but will soon be intact and ready to flourish across U.K. defence and, in particular, as the RAF enters its second century.

References

Air Force Technology. "RAF to Upgrade Eurofighter Typhoon with Tornado Capabilities." Sep. 5, 2018.

Air Publication (AP) 3000 *Air Power Doctrine*, 2nd edition. London: Royal Air Force, 1993.

Directorate of Air Staff. Air Publication (AP) 3000 *British Air Power Doctrine*. London: HMSO, 1999.

Defence Contracts Online. "A Breakdown of Planned Defence Expenditure 2018." <https://www.contracts.mod.uk/blog/breakdown-planned-defence-expenditure-2018/>.

Dempsey, N. *UK Defence Expenditure*, House of Commons Briefing Paper CBP 8175. Nov. 8, 2018, 3.

Dixon, N.F. *On the Psychology of Military Incompetence*. London: Jonathan Cape, 1976.

Forces Network. "What You Need to Know About the F-35B." May 21, 2019.

George, A. "RAF Announce that the P-8A Maritime Patrol Aircraft Will Be 'Poseidon MRA Mk1' in UK Service." *UK Defence Journal*, Aug. 22, 2018.

Gibb, R. *The Thin Red Line*. Oil on Canvas, 1881.

Gray, P. "Air Power and Joint Doctrine: An RAF Perspective. "*RAF Air Power Review*, Vol. 3 No. 4, Winter 2000.

Gray, P. Air Warfare History, Theory and Practice. London: Bloomsbury, 2016.

GOV.UK. "Novichok Nerve Agent Use in Salisbury: UK Government Response, March to April 2018." Last modified Apr. 18, 2018. <https://www.gov.uk/government/news/novichok-nerve-agent-use-in-salisbury-uk-government-response>.

GOV.UK. "Combat Air Strategy: An Ambitious Vision for the Future." Jul. 16, 2018. <https://www.gov.uk/government/publications/combat-air-strategy-an-ambitious-vision-for-the-future>.

GOV.UK. "Global Strategic Trends: The Future Starts Today." DCDC, 6th edn. Last modified Oct. 15, 2018. <https://www.gov.uk/government/publications/global-strategic-trends>.

GOV.UK. "National Security Strategy and Strategic Defence Review 2015." <https://www.gov.uk/government/publications/national-security-strategy-and-strategic-defence-and-security-review-2015>.

GOV.UK. "Development, Concepts and Doctrine Centre." <https://www.gov.uk/government/groups/development-concepts-and-doctrine-centre>.

GOV.UK. "The Strategic Defence and Security Review: Securing Britain in an Age of Uncertainty." Oct. 19, 2010. <https://www.gov.uk/government/publications/the-strategic-defence-and-security-review-securing-britain-in-an-age-of-uncertainty>.

Haynes, D. "Going to War on a £21bn Hole in the Defence Budget." *The Times*, May 25, 2018.

Hopkins, N. "Libya Conflict May Cost UK £1.75bn." *The Guardian*, Sep. 25, 2011.

Institute for Government. "The 2019 Spending Review: How to Run It Well." <https://www.instituteforgovernment.org.uk/summary-2019-spending-review-how-run-it-well>.

Jennings, G. "RAF Reforms 11 Group to Counter New and Evolving Threats." *Jane's 360*, Nov. 8, 2018.

Jones, C. "RAF Unveils £2bn Plan for New Tempest Fighter jet (and It Has Lasers)." *The Times*, Jul. 16, 2018.

John, S. *The Thin Red Line of Balaclava*. London: Hamish Hamilton, 1970.

Joint Doctrine Publication 0-30 *UK Air and Space Doctrine*, DCDC 2013.

Joint Doctrine Publication 0-01 *UK Defence Doctrine*, DCDC 2014.

Joint Doctrine Publication 0-30 *UK Air and Space Doctrine*, DCDC 2017.

Joint Doctrine Publication 0-10 *UK Maritime Doctrine*, DCDC 2017.

Joint Doctrine Publication 0-20 *UK Land Doctrine*, DCDC 2017.

Kipling, R. "Tommy" in *Barrack Room Ballads and other verses*. London: Methuen, 1892.

Lee, P. Reaper Force: Inside Britain's Drone Wars. London: John Blake, 2018.

RAF. "About the Typhoon FGR4." <https://www.raf.mod.uk/aircraft/typhoon-fgr4/>.

RAF. "About the E-3D." <https://www.raf.mod.uk/aircraft/e-3d/>.

RAF. "About the Sentinel R1." <https://www.raf.mod.uk/aircraft/sentinel-r1/>.

RAF. "About the Shadow R1." <https://www.raf.mod.uk/aircraft/shadow-r1/>.

RAF. "About the RC-135W Rivet Joint." <https://www.raf.mod.uk/aircraft/airseeker-rc-135w-rivet-joint/>.

RAF. "39 Squadron." <https://www.raf.mod.uk/our-organisation/squadrons/39-squadron/>.

RAF. "About the Voyager." <https://www.raf.mod.uk/aircraft/voyager/>.

RAF. "About the Atlas (A400M)." <https://www.raf.mod.uk/aircraft/atlas-a400m/>.

RAF. "About the C-130J Hercules." <https://www.raf.mod.uk/aircraft/c-130j-hercules/>.

RAF. "About the BAE146." https://www.raf.mod.uk/aircraft/bae146-ccmk2-and-cmk3/>.

RAF. "Groups." <https://www.raf.mod.uk/our-organisation/groups/>.

RAF. "About the P-8A." <https://www.raf.mod.uk/aircraft/p-8a/>.

Russell, W.H. *The Great War with Russia: A Personal Retrospect*. London: Routledge, 1895.

Strategic Defence Review. "Modern Forces for the Modern World." Jul. 1998.

Tennyson, A. "Locksley Hall." <https://www.poetryfoundation.org/poems/45362/locksley-hall>.

University of Birmingham Policy Commission Report VI. *The Security Impact of Drones: Challenges and Opportunities for the UK*. Birmingham Policy Commission, Oct. 2014.

Whitfield, M. and Tanner, M. "Sarajevo Rescue Mission Snowballs." *The Independent*, Aug. 14, 1993.

6 ITALY
At One Hundred, and Beyond

Gregory Alegi

Introduction

The Italian Air Force (ITAF) is among the world's oldest independent air forces, a mere five years younger than the United Kingdom's Royal Air Force. It is a combat-proven, balanced, well-equipped, well-trained, medium-sized air force, capable of delivering autonomous packages as well as contributing meaningful capabilities to alliance and coalition operations within the so-called expanded Mediterranean, which Italy sees as its natural area of strategic interest.

The ITAF mission is to "defend Italy, its territory, its sovereignty and vital interests, ensuring the preparation, operational effectiveness and the employment of air forces in the framework of the national and international security system."[1] To carry it out, it has an annual budget of roughly €2.4 billion and 41,000 personnel.[2] It is structured around three functional commands reporting to the Capo di Maggiore Aeronautica (Chief of Air Staff (CAS)). The Comando della Squadra Aerea (Force Command (CSA)) comprises the Comando Operazioni Aeree (Air Operations Center (COA)), the Comando Forze di Combattimento (Combat Forces Command (CFC), with three fighter wings, three attack-reconnaissance wings, a missile

---

1 CAS Lt. Gen. Pasquale Preziosa, text prepared for the hearing at the House and Senate defense committees, June 23, 2013. For the subsequent discussion see <http://documenti.camera.it/leg17/resoconti/commissioni/stenografici/html/04c04/audiz2/audizione/2013/06/26/indice_stenografico.0005.html>.
2 *Coccarde Tricolori 2018*, 214–16. Published mid-year since 2003, the bilingual *Coccarde* is a convenient source of open data, including capsule descriptions of the equipment used by the Italian armed forces, from aircraft to light weapons and vehicles.

support wing, and the national aerobatics team), the Comando Forze di Supporto e Speciali (Support and Special Forces Command (CFSS), with a helicopter wing, two air transport wings, a mixed tanker-airborne early warning (AEW) wing, and two infantry wings). The Comando Scuole (Training Command) comprises both ground instruction schools and three flight-training wings. The Comando Logistico (Logistics Command, COM-LOG) provides support functions, but also controls the high-quality Centro Sperimentale Volo (Flight Test Center (CSV)) and Sardinia test ranges.

The service prides itself as being the most technological and innovative among the Italian armed forces, which makes it particularly sensitive to emerging domains and complex new environments.[3] This can be traced to its mid-1990s awareness that in order to remain relevant the ITAF would need to incorporate space, unmanned aircraft systems (UAS), and "global information," effectively establishing its goals for the next quarter-century.[4]

The ITAF's latest slogans, "We exist because we are useful" (2018), "Centered around the squadron" (2017), and "Air Force 4.0" (2016), express its key tenets of contributing to the national community, having an operational mindset, and embracing innovation in every aspect. But they also express the crucial challenge of correctly identifying future threats and preparing to counter them with the proper technologies and within sustainable budgets.

The Rebirth of Italian Air Power

In 1911 an Italian Army captain flew the world's first operational sortie. During World War I the country fielded a large Army air service and a much smaller naval one. On March 28, 1923, these were separated from their original service and merged into the new Regia Aeronautica (Royal Air Force (RA)), the third branch of the Italian military. Contrary to popular belief, Fascism did not favor RA over the senior services—rather, during

---

3 Gregory Alegi, "Il futuro poggia su tradizione e innovazione," interview with CAS Lt. Gen. Alberto Rosso, *Aeronautica*, December 2018, 4 (Rosso interview A). Similar concepts in Alessandro Cornacchini, "Intervista al capo di Stato Maggiore dell'A.M. gen. s.a. Alberto Rosso," *Rivista Aeronautica*, December 2018. (Rosso interview B).
4 CAS Lt. Gen. Mario Arpino, "L'Aeronautica Militare del Terzo Millennio," lecture delivered at the Centro Alti Studi Difesa (CASD), June 19, 1998, 13. Arpino served as CHOD from 1999 to 2001, only the third ITAF officer to hold the post since it was created in 1925.

the 1926–1933 peacetime years, RA funding was one-fourth that of the Army and one-half that of the Navy.

Also contrary to popular belief, administrative independence did not equate with the pursuit of strategic air power advocated by Giulio Douhet, the ardent, universally known but unheeded prophet of the strategic use of aircraft. If anything, the RA contribution to victory in Ethiopia (1935–1936) and Spain (1936-1939) adhered to the close air support precepts of Amedeo Mecozzi, the vocal opponent of Douhetism.

After World War II the armed forces struggled with the lack of a solid defense culture in Italy.[5] Defeat on the battlefield and Cold War ideological divisions combined to deprive the armed forces of both political and popular support. The various Christian Democrat-led coalition governments tended to view defense as a necessary evil, while the avowed pacifism of the Communist-dominated opposition was tainted by its explicit subservience to Moscow. As W.B. Yeats wrote, "The best lack all conviction, while the worst/are full of passionate intensity."[6]

In this context, the Army continued to dominate the Italian defense establishment in terms of both resources and thinking. In fact, while an embryo joint staff (now known as the Stato Maggiore della Difesa (Defense Staff (SMD)) was created as early as 1925, the Capo di SMD (Chief of Defense Staff (CHOD)) post was held in succession by 11 Army generals. The position was opened to all services in 1972, under an Army, Navy, Army, Air Force rotation. Between 1972 and 2018, the Air Force held it four times, or for 20.2 percent of the period, about one-half of the time of the Army and two-thirds of the time of the Navy respectively.[7]

As the RA morphed into the ITAF, it narrowly avoided disbandment by emphasizing cooperation with surface forces, laced with air defense. The context was hardly conducive to original thinking and the service contented itself with adhering to the U.S.-inspired North Atlantic Treaty Organization (NATO) Air Doctrine, a body of force employment norms the

---

5 Virgilio Ilari, *Storia militare della prima repubblica 1943–1993* (Ancona: Nuove Ricerche, 1994). A short description in Gregory Alegi, "Forze Armate," in *Dizionario del liberalismo italiano*, Vol. I, 473–78 (Soveria Mannelli: Rubbettino, 2011).

6 W.B. Yeats, "The Second Coming," *The Dial*, 1920.

7 CAS Lt. Gen. Enzo Vecciarelli became the fifth Air Force CHOD on November 5, 2018. Including his projected three-year term adds to the ITAF share approximately five points, equally split between Army and Navy.

key advantage of which was to drive the integration of allied air forces.[8] Years later, ITAF leadership would deplore having abandoned the field to academics or inside staff meetings and lectures.[9]

When the United States gave some European countries a role in its deterrent, under so-called dual key arrangements, the ITAF added first a short-lived ballistic missile unit and then a fixed-wing strike component. With U.S. defense aid over, numbers dwindled, and obsolescence crept in, until a 1977 law allowed the fleet to recapitalized and the tri-national Tornado combat aircraft to go into production. Reacting to growing regional instability, Italy eventually reconsidered its strategic priorities and defense model, publishing a White Paper (1985), defining its area of interest, and eventually styling itself as a medium-sized regional power.[10]

The ITAF traces its rebirth to its participation in the first Gulf War, which marked the service's return to real operations. Despite its overall success, Operazione Locusta (Operation Locust, as the mission was known) brought to light the need to radically overhaul ITAF assumptions, organization, and equipment. Among the immediate lessons identified was the fallacy of believing that war was unlikely and that the United States would make up for all shortcomings;[11] the fallacy that a modern service could operate without tankers, precision guided munitions, secure communications, and deployability; awareness that rapid reaction and expeditionary roles required abandoning the territorially organized command structure; and finally, the need to reexamine doctrine—indeed, to return to thinking in terms of doctrine rather than of conducting operations.

---

8 Arpino, "L'Aeronautica Militare del Terzo Millennio," 7. Even the recently created Air Warfare Center within the ITAF Air Operations Center at Poggio Renatico is concerned with the development and validation of tactics, procedures, and concepts of air operations and not with fresh theoretical approaches to air power.
9 Stato Maggiore Aeronautica, *SMA 9 Potere Aereo-spaziale – Fondamenti*, June 2011 edition, 7 (SMA 9).
10 Luigi Caligaris and Carlo M. Santoro, *Obiettivo difesa. Strategia, direzione politica, comando operativo* (Bologna: Il Mulino, 1986), particularly 9–109; Carlo M. Santoro, *La politica estera di una media potenza. L'Italia dall'Unità ad oggi* (Bologna: Il Mulino, 1991).
11 For an example see of this mentality see Bruno Servadei, *Ali di travertino.* (Ravenna: SBC, 2011), particularly 176–84.

Upon becoming CAS, Lt. Gen. Mario Arpino articulated the new ITAF vision.[12] The doctrine he outlined in 1997 called the ITAF to defend against "direct threats to [the nation's] territory, structures or population" and operate "out of area, alone or with other countries, to protect or re-establish international rule of law and stability, or to protect national interests within the so-called *areas of strategic interest.*"[13] The recent conflict in Bosnia had shown air power to be a force multiplier with political-strategic value, but organizations would need to be streamlined, starting at the NATO level.[14] For Arpino, who had managed Locusta from the coalition headquarters in Riyadh, limited funding would not be allowed to lower the quality of either equipment or training.[15] His vision of the future encompassed information warfare, information technology, and simulation (to lower costs), but rejected the idea of specializing individual European air forces around specific capabilities on the basis that air defense equated with sovereignty.[16] From this stemmed the need to maintain a complete range of capabilities, albeit in small numbers, and a radical restructuring from air regions to a new functional model.[17]

The strategic concept published in 2005 by CHOD Admiral Giampaolo Di Paola identified Italy's area of strategic interest as the expanded Mediterranean, defined to comprise NATO (which would obviously include the Atlantic), the European Union (EU), the Balkans, Eastern Europe, North Africa, East Africa, and reaching as far as the Persian Gulf.[18] Simultaneously the ITAF, under Lt. Gen. Leonardo Tricarico, CAS between 2004 and 2006, carried out a comprehensive force structure review aimed not only

---

12 Stato Maggiore dell'Aeronautica [SMA], *La dottrina dell'Aeronautica Militare.* (Rome, 1997); Arpino, "L'Aeronautica Militare del Terzo Millennio." According to Arpino, this vision owes much to Lt. Gen. Basilio Cottone, CAS in 1983–86, whose privately published 2015 memoir *Volando e sorvolando* does not however mention them.

13 SMA, *La dottrina,* 11.

14 SMA, *La dottrina,* 28 ff.

15 Arpino, "L'Aeronautica Militare del Terzo Millennio," 15.

16 SMA, *La dottrina,* 52–56; Arpino, "L'Aeronautica Militare del Terzo Millennio," 16–17.

17 SMA, *La dottrina,* 57–59; Arpino, "L'Aeronautica Militare del Terzo Millennio," 17; SMA, *La nuova struttura dell'Aeronautica Militare* (Rome, 1999).

18 CHOD Adm Giampaolo Di Paola, *Il concetto strategico del capo di Stato Maggiore della Difesa* (Rome, 2005). <http://www.difesa.it/SMD_/CaSMD/ConcettoStrategico/PublishingImages/87587_Documento%20completo%20(File%20Pdf%202,48%20Mb)pdf.pdf>. The 2015 White Paper confirmed the Euro-Atlantic and Euro-Mediterranean areas as "fulcrum," "national interests," and "absolute priority" (paragraphs 42, 44–50, 64).

at identifying a leaner structure for a more sustainable future, but also at mapping out the various steps to the stand-down, in order to avoid the expense and trauma of emergency decisions. The bold review led ITAF to transfer little-used airports and air traffic control to civilian agencies, and also to concentrate air defense on just two bases in Tuscany and Apulia. The overall transformation plan was described by CAS Lt. Gen. Vincenzo Camporini in 2007. [19] On the whole, the project succeeded and positioned the service to mitigate the cuts instigated by a 2012 military establishment review law.[20]

After holding an international air power conference, in June 2011, under CAS Lt. Gen. Giuseppe Bernardis, the ITAF published Stato Maggiore dell'Aeronautica (Air Force General Staff (SMA)) 9, its first formal comprehensive doctrinal document.[21] Titled *Fundamentals of Aero-space power*, SMA 9 found "the great airpower theorists (Douhet, Mitchell, Mecozzi, etc.)" obsolete because of their insistence on the use of violence against "classic centers of gravity."[22] The document posited the "perception of airpower usefulness" to be "in direct proportion to the ability to react to crises by making the required capabilities quickly and truly available" and stressed the centrality to modern air forces of a "true expeditionary capability."[23] SMA 9 listed air power capability pillars as command and control / battle management (Air C2/BM), mobility and projection, situational awareness (Intelligence Surveillance Target Acquisition and Reconnaissance (ISTAR) and tactical-operational military intelligence), ability to generate strategic effects (information operations), force application (attack, defense, kinetic or not, lethal or not), and support within coalitions in reaching air superiority in some degree.[24] The five-year plan published in 2013 under CAS Lt. Gen. Pasquale Preziosa repackaged these capabilities in four macrosets (command and control, combat, combat support, and combat service support), excluding information operations but mentioning access to space as

19 SMA, *La trasformazione.* (Rome, 2007).
20 Law no. 244, Dec. 31, 2012.
21 SMA 9 published shortly after joint publication PID/S-1, *La dottrina militare italiana*, never publicly distributed.
22 SMA 9, 47.
23 SMA 9, 29.
24 SMA 9, 45–46.

something to be closely monitored.[25] Importantly, the document explicitly linked capabilities to the duration of the effort.

The depth and originality of the doctrinal renaissance is difficult to assess, as the published texts draw upon international doctrine even for terminology. SMA 9 quotes openly from the British AP 3000 and NATO AJP-33 publications, and a seminal 2014 "Food for thought" paper mentions approvingly British JDP 0-30, US AFDD 3-2, and Joint Air Power Competence Centre (JAPCC) "Start paper project on NATO AP future vector."[26] On the other hand, talented ITAF staff contribute to doctrinal development at European Air Chiefs (EURAC), JAPCC, SMD, and other forums, so that an Italian contribution to their formulation is far from impossible.

Whatever the source, the ITAF shares the view that countries face a broad range of threats, frequently asymmetric and aimed at various centers of gravity, including information.[27] While it is fully aware that crises cannot be managed solely by the military or solely by air power, it also argues that the same applies to land or sea power.[28] This is in part the consequence of accepting that threats will certainly include non-state actors and hybrid forms, as well as split operations that might expose its domestic territory—for instance, through attacks on home-based control stations for UAS operating over distant theaters, or satellite ground stations.

Between Domain Competence, Jointness, NATO, and the EU

For several years now, the Italian defense establishment has pushed for interservice integration and cooperation. The process started in 1997, when legislation introduced by Defense Minister Beniamino Andreatta redesigned the higher echelons for both effectiveness and efficiency.[29] Twenty

---

25 Aeronautica Militare, *Verso il 2018. Linee guida e d'indirizzo strategico* (Rome, 2013), 40–42.

26 Brig. Gen. Luigi Del Bene, "Potere Aereo-Spaziale nei conflitti di 4ª e 5ª generazione. Ruolo e implicazioni capacitive." Unpublished paper.

27 CHOD Gen. Vecciarelli, "It is no longer the case of defending borders, as it might have once been. Today we must defend our citizens' flows, their freedom of movement, of starting business and industrial activities in many corners of the world, the security of energy flows, and I believe the armed forces cam make a great contribution to all this," quoted in Flavia Giacobbe, "A difesa di Vecciarelli. Le parole della Trenta contro una polemica inutile." Nov. 12, 2018, <https://formiche.net/2018/11/difesa-vecciarelli-trenta/>.

28 Del Bene, "Potere Aereo-Spaziale nei conflitti di 4ª e 5ª generazione."

29 Law no. 25, Feb. 18, 1997, later incorporated in the Codice dell'Ordinamento Militare (Military Code, Legislative Decree No. 66, Mar. 15, 2010 (COM).

years later, the 2015 White Paper called for a combined total of 150,000 military and 20,000 civilian staff.[30] By 2017 the multi-year planning document envisioned a National Integrated Force at once joint, interoperable, and able to guide coalition or alliance missions in the Euro-Mediterranean area.[31]

The 1997 law subordinated service Chiefs to the CHOD, whose decisions would no longer be recommendations but binding orders. The 1998 creation of the new Comando Operativo di vertice Interforze (Joint Operations Command (COI)), reporting directly to the CHOD, centralized control of major operations and exercises and opened the door to other joint commands.[32]

The new accounting principles introduced in 2008 brought further clarity by grouping military expenditure under four broad "missions" and eleven "programs" that cut across the different services. The main mission was defined as "Defense and security of [national] territory" and included seven programs, namely preparation and employment of each of the four services (ITAF, Army, Navy, and Carabinieri, which had been granted independence from the Army in 2000) plus overall planning of the armed forces and their procurement, military security, and functions not directly connected to military defense tasks. In the 2010 Code of Military Organization (COM) Parliament attributed to the military establishment the task of defending "the nation's territory and its maritime and air lanes of communication" and defined ITAF as "the air operations component of the military defenses of the State."[33]

The four missions were confirmed without change in the 2015 White Paper, which aimed to achieve "more with less" by envisaging greater jointness.[34] This included making COI the Deputy CHOD for operations, with responsibility extending to single-service operations, and bringing force support under the Direttore Nazionale degli Armamenti (National Director for Armaments (DNA)) and adding logistics, turning the office into a Direttore Nazionale degli Armamenti e della Logistica (National Director for

---

30 *Libro bianco,* para 72. The figure, which is increasingly viewed as inadequate, pegged the ITAF at 33,000.

31 DPP, 21.

32 Arpino, "L'Aeronautica Militare del Terzo Millennio," 13.

33 Legislative decree no. 66, March 15, 2010, Articles 88 and 139.

34 *Libro Bianco per la sicurezza internazionale e la difesa* (Rome, 2015), <www.difesa.it/Primo_Piano/Documents/2015/04_Aprile/LB_2015.pdf>.

Armaments and Logistics (DNAL)) which would carry out its new responsibilities through a new Comando Logistico della Difesa (Defense Logistics Command (CLD)). The Segretariato Nazionale della Difesa (National Secretariat for Defense (SG)) and its technical-administrative procurement activities would be separated from DNAL and, significantly, be mandatorily held by civilians. The changes were supposed to be introduced within six months. In truth, the goals proved somewhat elusive, because the process was delayed by factors ranging from culture to the pressure of extended asymmetric operations to short-term, single-service priorities. Upon becoming Deputy Chief of the Defense Staff in 2015, Lt. Gen. Enzo Vecciarelli described the situation blithely: "In more recent times, still under everyone's eyes, in order to achieve a somewhat imaginary supremacy of one component over the other, no energies have been spared in developing full-fledged, more or less open internal clashes, perhaps successful in some areas but clear losers in the overall Defense Organization complex."[35]

In the ITAF view, jointness and dual use are served most effectively by placing assets in the hands of the service closest to the respective domain and then making the capability available to all services and, indeed, institutional users. In other words, ITAF believes that integration revolves around leveraging *competenza ambientale* ("domain competency").[36] The concept is similar to the "complexity of the cultural dimension" found in SMA 9,[37] and can be summarized as rooted in the familiarity with the aspects that enable an individual service to extract the best from a given domain, minimizing the cost and delay of learning. This does not imply domains should be exclusive, but recognizes that in most cases lead services are naturally identified. Service capabilities, on the other hand, are to be made available across the national defense and security community in a joint and dual perspective.

In theory, this concept is far from revolutionary. The technical-administrative directorates of the current DNA/SG have long been designed around it, making the Army-led Direzione degli Armamenti Terrestri (Land Armaments Directorate, TERRARM) the focal point for all light weapons

---

35 Lt. Gen. Enzo Vecciarelli, Open Letter of the Deputy Chief of Staff of Defense, Mar. 2, 2015. Years before Gen. Camporini had joked that the measure of his lack of success as CHOD had been the failure to merge even such minor things as the various foreign language schools.
36 Rosso interview B, 11.
37 SMA 9, 31.

and military vehicles, the Navy-led Direzione Armamenti Navali (Naval Armaments Directorate, NAVARM) the reference for warships, and the Air Force-led Direzione Armamenti Aeronautici e Aeronavigabilità (Aeronautical Armaments and Airworthiness Directorate, ARMAEREO) for the procurement of air combat systems.

Extending domain competence to the operational realm, however, brings integration to a much higher level and requires careful examination of individual activities. For instance, the ITAF justified first entering the special forces club by transforming its skydiving team into the 17° Stormo Incursori (17th Wing), and in the land forces arena, by converting the airfield protection force into the 16° Stormo Fucilieri dell'Aria (16th Air Infantry Wing), in terms of specific operational requirements, including opening, securing, and operating airfields in the early phases of expeditionary missions, target acquisition, and personnel recovery behind enemy lines. The 17th Wing was also placed under the overall command of the Comando Operativo delle Forze Speciali (Special Forces Operational Command (COFS)), thereby contributing to Italy's joint special forces effort.

Factors played differently in the unmanned air systems arena. When the rapid spread of UAS became apparent, ARMAEREO drew up a pioneering certification model that served as template at the European level. With uniform regulations in place, including the requirement for licensed military pilots to operate aircraft over 150 kg maximum take-off weight (MTOW), individual services were largely free to adopt any compliant micro, mini, and light UAS. Higher-end systems, including Predator A and B, were concentrated in a Defense UAS Center of Excellence operated by the ITAF on behalf of joint users. In fact, the original 28° Gruppo (28th Squadron), recently complemented by the 61° Gruppo (61st Squadron), has also contributed its capabilities to various police and civil protection operations, indicating that "dual use" is more than a catchword.[38]

Experience in the Gulf and over the Balkans convinced the ITAF that future operations would be much more international than envisaged under Cold War scenarios. By 1999, CAS Lt. Gen. Andrea Fornasiero declared it "imperative to think of the ITAF in terms of the real ability to operate (with

---

38 CAS Lt. Gen. Enzo Vecciarelli, hearing at the House and Senate defense committees, May 18, 2016, 9. Vecciarelli also offered to make available the "extraordinary multi-sensor capabilities of the F-35" for dual-use information gathering.

its own assets and logistics) together with the other armed forces of Italy, of European countries and NATO."[39] By 2010 the "multinational and European dimension" Fornasiero envisaged for the future was enshrined in the COM, which called for the Italian military to be "easily integrated in multinational complexes" and to be able to "participate in peace support actions, also on a multinational basis."[40] This reflected operational reality, as well as the political need for international mandates to overcome the Italian constitutional repudiation of "war as a way to offend the freedom of other people and as means to resolve international controversies" by leveraging the provision in favor of international organizations aiming at "ensuring peace and justice among Nations."[41]

With the single exception of air defense, Italy has been an active proponent of European defense, joining with greater sincerity than other countries initiatives in both procurement (European Defense Agency (EDA), Organisation for Joint Armament Co-operation (OCCAR)) and "sharing and pooling" (European Air Transport Command in 2014 and European Personnel Recovery Center in 2015). By 2016 the process was considered to have achieved an "extraordinary integration capability" ready for "a European dimension of defense," should the political will for it exist.[42]

"We Exist Because We Are Useful"

In an attempt to change the image of the military as a burden for the country or as an expensive inward-looking toy, the ITAF, from 2016, made "usefulness" a key element of its identity and message. The ITAF had of course always served Italy in a number of non-military roles, ranging from air traffic control to emergency services, from weather forecasting to disaster relief, from medical flights to firefighting. But with the exception of earthquakes and other national emergencies, the Air Force contribution to the nation has gone largely unperceived. Even when the end of the Cold War encouraged a more open relationship with society and the press, ideological opposition to Italian participation in so-called peace missions led to a

---

39 CAS Lt. Gen. Andrea Fornasiero, foreword to SMA, *La nuova struttura*.
40 Legislative decree no. 66, March 15, 2010, Art. 88.
41 Constitution of the Italian Republic, Art. 11.
42 Vecciarelli hearing, 10.

different kind of secrecy being imposed. [43] In this light, the armed forces' persistent "inability to explain not just our needs but also the depth of our loyalty and spirit of service to our country" is less "inexplicable" than Lt. Gen. Vecciarelli lamented in 2015.[44] Because of its reliance on technology and small numbers, the ITAF has a small footprint and suffers disproportionately from this overarching misperception of how it protects citizens and of its value as a political and diplomatic asset. This, in turn, affects the ability of the Ministry of Defense to obtain budgets adequate to Italy's level of ambition on the world scene and thus to influence outcomes in industrial, diplomatic, and political issues.

The ITAF has for some time worried that "Air Power runs the risk, in some way, to fall victim to its own success," a position echoed in the so-called NATO paradox.[45] In other words, the ITAF fears that other services or agencies unreasonably expect that future operations will enjoy air supremacy as in recent conflicts in the Balkans, Afghanistan, Iraq, and Libya, and that someone will make this supremacy available at no cost. Whether conscious or not, this mindset downplays the contribution and usefulness of air power to the battlespace, leading to the loss of capabilities crucial for air forces but, more importantly, for joint operations.

Although they are expensive, the ITAF argues that air force systems are crucial enablers and force multipliers for situational awareness and to allow simpler assets (helicopters, tactical UAS, or even lighter than air vehicles) to coexist in increasingly crowded airspace. They are also vital in allowing land operations to be carried out with minimal casualties on both sides, thus contributing to acceptance by public opinion. This line of reasoning, remindful of the saying "If you think education is expensive, try ignorance," becomes even more important in an information dominance framework.

---

43 Some chroniclers of Italian out-of-area operations have expressed bitter contempt for this communication policy in the very title of their books, e.g., Gianandrea Gaiani, *Iraq Afghanistan – Guerre di pace italiane* (Venezia: StudioLT2, 2008), and Meo Ponte, *Eroi di una guerra segreta* (Milano: Longanesi, 2018).

44 Vecciarelli, Open letter.

45 SMA 9, 31; Del Bene, "Potere Aereo-Spaziale nei conflitti di 4a e 5a generazione," "at the political level decision-makers continue to stress the growing importance of the aero-space instrument as fundamental enabler to solve crises; but the same decision-makers seem unable or unwilling to work collectively to ensure that the same instrument remains relevant and effective."

Air defense, the first and foremost ITAF mission, is arguably also among the most difficult to communicate in terms of usefulness. The apparent lack of visible threats to Italian territory, together with the general challenge of explaining prevention-oriented services, has already resulted in sharp reductions in the missile component. Somewhat encouragingly, the manned fighter element, which also provides air defense to Albania and Slovenia (plus rotational air policing of certain NATO member countries), has regained its Trapani base, reprieved following the 2011 war against Libya, and is widely expected to base a so-called mini squadron at Treviso-Istrana, to protect the north-eastern quarter. Slow-moving interception, with both helicopters and armed trainers, is much publicized during major public events, and the service is studying how to expand its capabilities against small drones. Given Italy's exposure to threats originating in the unstable Mediterranean region, air defense will remain the central ITAF mission. In October 2018, the newly appointed CAS, Lt. Gen. Alberto Rosso, acknowledged the ITAF need for a successor to its current Eurofighter Typhoon but declined to weigh in about competing so-called sixth-generation fighter concepts from France-Germany colla and the U.K. By March 2019 the think tank, Istituto Affari Internazionali (IAI), had produced a paper recommending that Italy join the U.K. Tempest programme, albeit leaving capability requirements aside and the analysis based on largely industrial grounds.[46] The ITAF looks to expand its air defense role in space, initially through the EU Space Situational Awareness program, including hosting the five-nation European consortium at the Pratica di Mare air base.[47]

The ITAF has no doubt that space is useful to twenty-first century air forces—indeed, its centrality to the contemporary way of life, economy, well-being, and military operations (for communications, observation, and precision) has already contributed to make space a center of

---

46 Stefano Pioppi, "La tabella di marcia del generale Alberto Rosso," <http://www.airpressonline.it/36088/rosso-aeronautica/>.

47 Alessandro Marrone and Michele Nones (eds.), *Europe and the Future Combat Air System* (IAI, 2019), <https://www.iai.it/en/pubblicazioni/europe-and-future-combat-air-system>. For U.K. assumptions and vision behind Tempest/FCAS see MoD, "Combat Air Strategy: An Ambitious Vision for Future," <https://www.gov.uk/government/publications/combat-air-strategy-an-ambitious-vision-for-the-future>; <http://www.esa.int/Our_Activities/Operations/Space_Situational_Awareness/About_SSA>; Vecciarelli hearing, 20. The SSA system should eventually be integrated in the Poggio Renatico Air Operations Center.

importance and therefore vulnerable.[48] The increasing need for a multi-user prospective in the governance of Italian space and aerospace policy has led to concentrating space and aerospace policy under the Prime Minister's office, with the stated purpose of changing the direction of the Agenzia Spaziale Italiana (Italian Space Agency (ASI)).[49] The reform is of direct interest to the Ministry of Defense, whose Innovation Center (CID) in 2009 produced the classified *Space Environment Concept* (EC-001).[50] Although so far the ministry has refrained from making the ITAF the lead service for the emerging space domain, it has given it responsibility for military human space flight (which also produced universally recognized spokespersons for the service), aerospace medicine, space weather, hypersonic and sub-orbital flight, and space access by air launch;[51] in turn, the ITAF has spun out its space planning from its A-3 department to the new Ufficio Generale Spazio (Space General Office (UGS)), with offices for Aerospace Innovation and Space Policy and Operations.[52]

With new suborbital and hypersonic technologies closing the gap between air and space, the ITAF has already adopted an ambitious vision based on a continuous, unhyphenated aerospace environment suitable for piloted flight activities up to about 100 kilometers (62.5 miles).[53] ITAF leadership views this continuum as a natural extension of the service's traditional atmospheric environment. The ITAF is translating this broad outlook into reality through initiatives ranging from training aircrew for suborbital flight to links with other initiatives maturing on the commercial and civil side (including a possible Virgin Galactic suborbital spaceport in

---

48 Del Bene, "Potere Aereo-Spaziale nei conflitti di 4a e 5a generazione."

49 Law no. 7, Jan. 11, 2018. ASI previously reported exclusively to the Ministry of Education, University and Research (MIUR). See Gregory Alegi, "Il futuro dell'Asi e uno sguardo al passato," *Air Press*, No. 95 (Dec. 2018), 34–35.

50 A draft *Environmental Concept* Air, slated to become EC-004, existed in 2013 but is not listed among published CID documents,
<http://www.difesa.it/SMD_/Staff/Reparti/III/CID/Dottrina/Pagine/default.aspx>.

51 Vecciarelli hearing, 20.

52 <http://www.aeronautica.difesa.it/organizzazione/loStatoMaggiore/organigramma/Pagine/UGS.aspx>. At the time of writing, a similar office was expected to be created within SMD.

53 Aeronautics, operating up to about 20 kilometers (12.5 miles), have traditionally been separated from space, which by convention begins at about 100 kilometers, by an empty 80-kilometer band used solely to move between the two environments. This lack of continuity led to the two environments being linked, as well as separated, by a hyphen: "aero-space."

Italy), from research agreements to launch small satellites, and from fighters to the development of stratospheric platforms.[54]

Another argument for ITAF usefulness is its impact on the economy, as a driver of technology, exports, and skilled jobs. In the past, this allowed the various branches of the military to obtain additional funding through broad industrial-support laws (1975–1977) or through support for individual programs through other government departments, lately the Ministry for Industry and Economic Development (MISE). In recent years, MISE funding has been the precondition for many ITAF programs, whether of interest to the service or industry. In turn, this has enabled Italy to remain one of a few nations with an extensive design and manufacturing presence in the aerospace, defense, and security industry.

Finally, the ITAF is useful because, as the only Italian government organization with flight as its core business, it maintains a complete range of aviation-related capabilities that others prefer to acquire ready-made from other providers, whether domestic or international, commercial or governmental. These capabilities, which range from flight-testing to accident investigation, are in high demand from other departments, which without the ITAF would have to invest considerable resources in developing and maintaining non-core skills.[55]

While these examples confirm that the ITAF is indeed useful, they do not in themselves guarantee that the message is reaching the general public—or even the defense community.

---

54 Lt. Col. Walter Villadei, "L'aeronautica militare Qualifica 1 primi ufficiali per il volo suborbitale," Nov. 29, 2018, <http://www.aeronautica.difesa.it/comunicazione/notizie/Pagine/NASTAR.aspx>; Flavia Giacobbe, "Accesso allo spazio e nano-satelliti. Così l'italia fa sistema," *Air Press* online, Dec. 13, 2018, <http://www.airpressonline.it/36904/spazio-sistema-aeronautica-lanciatore/>.

55 For instance, the Ente Nazionale Aviazione Civile (National Civil Aviation Agency, ENAC) and the Agenzia Nazionale Sicurezza Volo (National Flight Safety Agency, ANSV) borrow ITAF pilots, flight inspectors, accident investigators, and air traffic controllers. On a structural level, the Italian Navigation Code establishes Armaereo as the authority for military aircraft type certification and continued airworthiness (Art. 745). UGS sponsored an unclassified national workshop on stratospheric platforms, held in Rome on April 5, 2019.

Air Force 4.0

In 2016, CAS Lt. Gen. Enzo Vecciarelli launched the "Aeronautica 4.0" slogan to indicate that in the future the Air Force would be driven by automation, smart technologies, virtual reality, and modular, flexible, and fully interoperable cyber systems.[56] It could be argued that Vecciarelli was merely restating in twenty-first century parlance the continuing centrality of technology to air forces, but the term quickly morphed into a buzzword for a modern and cohesive ITAF committed to all around innovation, including new organizational models to cope with the legislative mandate to cut personnel to 33,000 by 2024.[57] To this end, the ITAF submitted itself to an external audit by the University of Florence, in part through the SHRMxL lab run with Bordeaux, Brussels, Madrid, and Montreal universities. Early outputs are expected to include the mapping of top-level processes, an "integrated knowledge portal," and the "Strategic portfolio model."

Organizational change is also behind the 2017 "Centered around the squadron" slogan, a metaphor that leverages the fond memories of comradeship and no-nonsense focus on operations.[58] In an approach blending the pragmatism of shrinking resources with the idealism of a new approach, Lt. Gen. Vecciarelli asked everyone to act as a single, great squadron, unencumbered by the layers of command and indecision that pilots often lament in large central organizations. The decision implied moving away from the heavy USAF model to more nimble structures, capable of achieving greater resource/output ratios. Over the decades, inspiration has come from other national services, including the Royal Netherlands Air Force and the Royal Air Force. In recent years the Heyl Ha'Avir (Israel Defense Force / Air Force) has also provided inspiration, its streamlined structure, informal climate, extreme efficiency, and reliable strength appearing as attractive

---

56 Stefano Cosci, "Aeronautica 4.0," interview with Lt. Gen. Vecciarelli, in *Rivista Aeronautica*, No. 6/2016 (Dec. 2016), 14 (Vecciarelli interview).

57 Stefano Pioppi, "Aeronautica 4.0, innovazione e investimenti per la difesa," <http://www.airpressonline.it/35316/aeronautica-4-0-innovazione-e-investimenti-per-la-difesa/>.

58 The ITAF and its predecessors have traditionally named "Stormo" the regiment-level units roughly equivalent to other air forces' "Wing." While the slogan "Lo Stormo al centro" translates literally into "Centered around the Wing," the more universal term "squadron" avoids the impression that everything revolves around aircraft and flying. For various interpretations of the slogan see Associazione Pionieri dell'Aeronautica (APA), *Lo Stormo al centro. Storia, riflessioni, prospettive* (Vicchio: LoGisma, 2017).

alternatives to stifling bureaucracy.[59] Even while leveraging the value of the metaphor, an ITAF project group is rethinking operational structures to uncouple squadrons and other force elements from their parent Wings and permanent bases.[60] While studies have been carried out, including comparisons with the USAF, Luftwaffe, Armée de l'Air, and the RAF, at the time of writing no decisions have been announced.

Still, it would appear that the ITAF shines in implementation and optimization but lags somewhat in formulating fresh thought, as Lt. Gen. Vecciarelli stated with unusual vigor in 2016; in his opinion, the pursuit of technical capabilities had placed the ITAF among "the most advanced air forces," but had also made it "less adequate in confronting the often deep changes which had occurred in the meantime."[61] What was required, he argued, was to reset assumptions and challenge received ideas, engaging personnel in dialogue and soliciting suggestions from all quarters.

The heavy reliance on digitization feeds into the cyber domain, another step in the ITAF multi-domain transformation. Although recent operational experience has seen Western forces enjoy the advantages of largely uncontested cyberspace, the ITAF believes that increasing reliance on exchanges of information will also multiply vulnerabilities, including those arising from "split operations." SMD has for some time recognized the importance of net-centric warfare and the 2015 White Paper envisaged forming a Comando Interforze Operazioni Cibernetiche (Joint Cyber Operations Command (CIOC)).[62] ITAF analysts and engineers are in high demand by other administrations and agencies and when CIOC was formed on September 30, 2017, the ITAF Brigadier General Francesco Vestito was selected to lead it.[63] This demand can be ascribed to the information tech-

---

59 The relationship can be traced to the 1998 Frecce Tricolori display for the 50th anniversary of the birth of Israel, progressing gradually to joint exercises in Italy and Israel, to the purchase of equipment such as bomb guidance kits and targeting pods, and finally to the 2012 purchase of two G550 CAEW systems and the Optsat 3000 observation satellite, in exchange for Israeli acquisition of Alenia Aermacchi M-346 advanced trainers.

60 Lt. Gen. Alberto Rosso, "Lo Stormo di oggi e di domani," in APA, 84.

61 Vecciarelli interview, 14.

62 Stato Maggiore della Difesa, *La trasformazione net-centrica.* (Rome, 2009); "The Joint Cyber Command Is Born," *Informazioni Difesa*, No. 3, 2017 (June 2017), <http://www.difesa.it/InformazioniDellaDifesa/periodico/Periodico_2017/Documents/Numero3/cyber_defence.pdf>.

63 Analisidifesa, "Difesa cyber nazionale: parla il generale Vestito," June 2, 2018, <https://www.analisidifesa.it/2018/06/difesa-cyber-nazionale-parla-il-generale-vestito/>.

nology (IT) and cyber experience gathered by the ITAF by virtue of the preponderant weight of IT and electronics in combat aircraft like the Eurofighter Typhoon,[64] the need to protect data exchanges between air defense and air traffic management systems, electronic combat reconnaissance aircraft, the massive computing power required for weather forecasting, the development of Virtual, Live, and Constructive training environments, and dedicated exercises like Virtual Flag and Cyber Eagle.

The Industrial Dimension

Because air force capabilities revolve around technology more than in any other service, theorists recognize that "Airpower includes not only military assets, but aerospace industry and commercial aviation"—or, as Napoleon might have said, air forces fly on technology.[65] At the same time, service goals and planning are often framed as "shopping lists" of equipment that is lacking. In the past, this often translated into military staffs and procurement branches becoming surrogate industrial planners, when not rubber-stamping industrial decisions or social measures. As Lt. Gen. Arpino observed in 1998, "the laudable intent of helping national industry develop" sometimes resulted in the tendency to "a semi-autarchic vision, which in the short term provided work for industry and its employees but in the medium term substantially penalized the service's operational effectiveness."[66] Twenty years later, the diagnosis still applies to business lines that, like combat aircraft, remain dependent on government orders, particularly to support research and development (R&D).

The defense industry is estimated to have contributed in 2017 about one percent of the Italian gross domestic product (GDP), with €14 billion in sales (including €9.5 billion in exports), €1.5 billion in R&D, and €1 billion

---

64 Rosso interview A, 5.
65 Philip D. Meilinger, "10 Propositions on Airpower," *Airpower Journal*, Spring 1996, 14–15, <http://www.au.af.mil/au/awc/awcgate/au/meil.pdf>.
66 Arpino, "L'Aeronautica Militare del Terzo Millennio," 6.

in taxes, with over 50,000 direct employees.[67] Studies estimate the aerospace multiplier in Italy at 3.28 in economic return and 3.06 in employment. The dominant player is Leonardo (formerly Finmeccanica), the €11.5 billion revenue company (2017) that is among the world's 10 largest players and active in many fields crucial to contemporary aerospace power, including aircraft (but with a separate division for aerostructures), helicopters, electronics, space, and cyber security; it also holds a 25 percent share in European missile group MBDA. Outside Leonardo, Avio is a world leader in light launchers and Tecnam in general aviation, while Piaggio Aero Industries is active in turboprop business airplanes. The largest capability gap lies in engines: neither Aero Avio, the GE-owned leader in gearboxes, nor the Piaggio engine division have complete design and manufacture capabilities.

For the Italian defense community, the mismatch between the breadth of industrial capabilities and procurement funds poses specific challenges. While its largest player is much smaller than competitors like Boeing (€81.6 billion in 2017), Lockheed-Martin (€44.8 billion in 2017), or Airbus (€40 billion in 2018), it is too large to thrive on the Italian market alone. [68] To help industry focus its resources, the 2015 White Paper called for SG/DNA to define national goals and means, including employment impact, in an official industry and technology strategy (SIT).[69] The paper was never published and its role, at least in terms of technologies, would appear to have largely shifted to the interdepartmental aerospace committee created in 2018.

---

67 Data from sources including The European House – Ambrosetti, *La filiera italiana dell'aerospazio, della difesa e della sicurezza* (Milan, 2018), and Col. Mario Toscano, "La politica industriale della Difesa," May 10, 2017, <http://www.fnsi.it/upload/70/70efdf2ec9b086079795c442636b-55fb/2cd894c7d7f2f054bedd80ffff6848f7.pdf>. Older data for comparative purposes in Prometeia, *Il contributo di Finmeccanica all'Italia. Tecnologia, crescita, investimenti.* (Bologna, 2014), <http://www.oxfordeconomics.com/my-oxford/projects/276233>, and Prometeia, *Il sistema industriale della difesa per il sistema Paese*, slides presented in Rome on Jul. 13, 2015.

68 In reply to this, over the past 15 years Leonardo has expanded in the United Kingdom (by acquiring Westland helicopters and the BAE Systems electronics division), the United States (by acquiring the electronics group DRS), and Poland (by acquiring PZL Swidnik, another helicopter manufacturer), to the point where it now considers these countries domestic markets.

69 If the SIT was completed, it has not been published or released. The official military think tank CEMISS has, however, published a paper on this subject written by a Finmeccanica researcher (<http://www.difesa.it/SMD_/CASD/IM/CeMiSS/DocumentiVis/Rcerche_da_pubblicare/AI-SA-27_La_politica_industriale_nel_settore_della_Difesa.pdf>).

Prioritizing capabilities would prevent the duplication of "assets, functions, institutional actors and users" that the ITAF fears might occur in the UAS field, and which already points to the multidimensional interplay of military, political, and industrial issues.[70] The challenges for the ITAF medium altitude long endurance (MALE) systems' aspiration are linked in part to the difficulties encountered by Piaggio Aerospace Industries. The century-old company unveiled its P.1HH Hammerhead MALE project in 2013, with sensor suite and electronics provided by Selex (now Leonardo Divisione Elettronica). The company eventually secured an eight-aircraft launch order from the United Arab Emirates, conditional upon the ITAF placing its own launch order to guarantee product quality. In 2015, Airbus, Dassault, and Alenia Aermacchi (now Leonardo Divisione Velivoli) unveiled their own EuroMALE proposal, putting Italy in the unusual position of participating in two largely overlapping projects. Deconflicting these projects was predicated on different timeframes; in fact, it was argued that industry would leverage its P.1HH experience to gain more meaningful roles in EuroMALE. Unfortunately, the P.1HH prototype was lost in unclear circumstances and the company's attempt to leapfrog by announcing an improved P.2HH failed to secure the vital defense contract, pushing Piaggio into receivership. Italian MALE capability will thus likely come via participation in the NATO Alliance Ground Surveillance (AGS) initiative, which should see five Northrop Grumman RQ-4 Global Hawks begin operations from Sigonella in 2019. On the other hand, armed drones, which the ITAF has long seen as the next logical step for its UAS component, make a good case for the importance of a strong domestic industrial base. In 2009 the U.S. Congress did not approve arming the U.S.-made General Atomics MQ-9A Predator B for the ITAF, leading the service to look to European solutions.[71]

Industry factors also played a significant role in the domestic opposition to the ITAF selection of the F-35. Understandably, the electronics sector resented its small work share in its information age system, in terms of both immediate development and long-term support. Although the Cameri final

---

70 Vecciarelli hearing, 19.

71 <http://www.ilsole24ore.com/art/notizie/2013-05-18/mistero-armi-droni-italia-140640. shtml>. As a paradox the aircraft incorporated weapon-carrying provisions at delivery. The notion of armed UAS later met with domestic political resistance and remains stalled to this day, largely outside the control of the military.

assembly and check-out (FACO) facility mostly placated the airframe business, some championed additional batches of the part-domestic Typhoon, which offered greater margins on new builds and upgrades as well as supporting the software engineering capabilities. Industry sources were in all likelihood behind press speculation about the possible ITAF selection of the Leonardo M-346FA, the light attack variant of the T-346A lead-in fighter trainer, touted as a low-cost complement to the F-35A, but for which no ITAF doctrinal concept or operational requirement is known to exist.[72]

Given these broad premises, in the past 20 years the overall ITAF landscape has been generally positive, with the service successfully acquiring and fielding virtually everything it had declared necessary in 1998, including UAS, AEW, and the F-35.[73]

Italy identified the F-35 to succeed the Tornado very early, a crucial decision considering that these 1970s fighter-bomber and electronic combat reconnaissance jets have participated in every ITAF out-of-area operation. Italy signed the Concept Development Phase memorandum of intents in 1998.[74] Over the following 20 years, it skillfully navigated both outright opposition and lukewarm half-praise to select, order, acquire, build, and introduce the F-35A into service, thereby securing high-end attack and strike capabilities for at least another generation. In November 2018, the ITAF declared initial operational capacity (IOC); in December, it appeared to be winning over parts of the originally anti-F-35 new government. Speaking at a public event, the new Undersecretary for Defense Angelo Tofalo went as far as describing the F-35 as "an aircraft which has an excellent technology,

---

72 With Leonardo close to completing deliveries against known M-346 orders, past experience suggests that the possibility of an industrial-continuity order cannot be completely ruled out.
73 Arpino, "L'Aeronautica Militare del Terzo Millennio," 31–32. The two main exceptions are the MEADS missile system, the sad story of which is described in the text, and the NH-90 helicopter, which planners replaced with the AW149 that has yet to enter production.
74 Michele Nones, Giovanni Gasparini, Alessandro Marrone, *Il programma F-35 Joint Strike Fighter e l'Europa*, Quaderni IAI No. 31 (Oct. 2008); Leonardo Tricarico and Gregory Alegi, *La linea aerotattica dell'aeronautica militare. Dall'esperienza operativa alle esigenze attuali per le operazioni future* (Rome: ICSA, 2014), <http://www.fondazioneicsa.info/2014/10/02/la-linea-aerotattica-dellaeronautica-militare/>; RID, *Speciale x-Trà F-35*, Oct. 2015; *L'F-35 negli scenari asimmetrici* (Rome: ICSA-CESMA, 2016); Corte dei Conti, *Relazione speciale sulla partecipazione italiana al Programma Joint Strike Fighter F-35 Lightning II*, 2017.

possibly the best in the world at this moment."[75] The remaining hurdles are delivery rates and F-35B basing. In order to reduce annual spending, the government is widely expected to roughly halve delivery rates.[76] The reduction would drive up both unit and program costs, but the medium- to long-term increase would be trumped by the short-term political benefit. F-35B basing is a largely interservice squabble arising from the Italian decision to equip both Air Force and Navy with the STOVL variant. The ITAF explains its decision by pointing to the lack of long runways in likely expeditionary theatres, but the Navy fears that losing the STOVL monopoly might lead to a joint F-35B force and even the loss of its jet element. Despite the savings to be gained from not duplicating facilities within a few dozen miles from each other, the Navy resists co-locating its Grupaer with the ITAF 13° Gruppo (13th Squadron) at Amendola.[77] Unofficial naval sources have spread the idea that similar savings could be achieved by standardizing the ITAF F-35 lineup on the A model alone, which would of course grant the Navy a greater role in future out-of-area operations.[78] Although at the time of writing the ITAF is upgrading Amendola to support the F-35B, the issue remains officially undecided.

The missile component of the air defense system, in both its long- and short-range elements, is probably the largest remaining capability gap created by budgetary constraints and external events. The gap—or, more accurately, the indecisive approach to filling it—has important implications for the Italian role in MBDA.[79] MBDA suffers from the mismatch between

---

75 Gianluca De Feo, "5Stelle, Tofalo e la riabilitazione degli F35: Hanno un'ottima tecnologia, forse la migliore al mondo," <http://www.repubblica.it/politica/2018/12/19/news/tofalo_1_f35_croce_e_delizia_la_marcia_indietro_dei_grillini_sul_caccia_delle_polemiche-214640220/>.
76 Emilio Pucci, "F35, c'è il via libera all'acquisto di 90 aerei: missione del sottosegretario grillino in Usa," *Il Messaggero*, Jan. 22, 2019.
77 Flavia Giacobbe, "Rosso, l'F-35 e il caccia del futuro," *Air Press,* No. 95 (Dec. 2018), 5.
78 For a recent example see Giuseppe Ligure (presumably a pseudonym for retired Vice Admiral Giuseppe Lertora), "Il paradosso della Difesa nelle acquisizioni dell'F-35B STOVL," Nov. 28, 2018, <http://www.liberoreporter.it/2018/11/dossier/il-paradosso-della-difesa-nelle-acquisizioni-dellf-35b-stovl.html>.
79 The consolidation of Europe's missile companies began in 1996, when parts of Matra Defence and BAe Dynamics merged their missile activities to form Matra BAe Dynamics (MBD). Matra BAe Dynamics represented half of Matra Hautes Technologies' missile business; the other half was Matra Missiles, which became Aérospatiale-Matra Missiles (AMM). In December 2001, MBD (including AMM) and the missile and missile systems activities of AMS merged, creating MBDA. "Airbus, BAE and Leonardo Complete MBDA Merger." *Defense Daily International*, Dec. 21, 2001.

the Leonardo share and the limited number of domestic industrial programs critical to maintaining its Italian design and manufacturing base. Long equipped with the Nike-Hercules system, including its nuclear-tipped variant, the ITAF had envisaged replacing the missile component with the U.S.-built Patriot but the acquisition was canceled by political fiat in the wake of the fall of the Berlin Wall, just as the first Gulf War highlighted the threat from ballistic missiles. To plug the capability gap, Italy joined Germany in selecting the Medium Extended Air Defense System (MEADS), a Lockheed-Martin design for the U.S. Army. MEADS passed several tests, including an extensive demonstration at the ITAF test center at Pratica di Mare in July 2014 and a six-week test at Pratica and Freinhausen, in Germany, in September 2015, but the unilateral U.S. program cancellation again left the ITAF unable to rebuild its missile defense.[80] A possible solution would reuse certain items funded and developed by European MEADS partners, in conjunction with the MBDA Sol-Air Moyen Portée/Terrestre (SAMP/T) missile.

For short-range defense, the ITAF has long relied on the national Spada system, developed and variously upgraded by Selenia, whose missile business was later merged into MBDA. Its Aspide missile, common to the Navy's Albatros and Army's Skyguard systems, was planned to be replaced with the MBDA Common Anti-air Modular Missile Extended Range (CAMM-ER), the development of which the United Kingdom joined in 2016.[81] The program was made somewhat urgent by age-related issues with the Aspide, which have been addressed through waivers now set to expire. However, the 2019 budget prepared by the new Five Star (M5S)-Lega government cancelled CAMM-ER funding, adding to the ITAF land-based air defense capability gap and creating another headache for MBDA Italia as sole developer and successful exporter of the Aspide.

With the withdrawal of the Piaggio PD.808GE and Aeritalia G.222VS aircraft coinciding with the growing importance of electronic warfare, the ITAF needed to plug the resulting gaps and, if possible, acquire new capabilities. Budget realities have prevented acquiring a Joint Airborne

---

80 *Verso il 2018*, 39; Vecciarelli hearing, 8; DPP, 31.

81 Gianandrea Gaiani, "A rischio il programme per la difesa aerea CAMM ER?," Analisidifesa, Sep. 28, 2018, <https://www.analisidifesa.it/2018/09/a-rischio-il-programma-per-la-difesa-aerea-camm-er/>.

Multimission Multisensor Systems (JAMMS), but the ITAF hoped to meet the need by leasing specific platforms, including the Lockheed Martin AML Dragon Star (2012–2016) and a Beechcraft King Air SPYDR (from 2016).[82]

Backbone of the ITAF rotary-wing component for about 35 years, the AB.212 helicopters, roughly equivalent to the Bell UH-1N, are ripe for replacement. Out-of-area operations accelerated fleet wear and tear; more importantly, they highlighted the need to shift the role of the ITAF helicopter units from peacetime search and rescue (SAR) to more active concepts. The Air Staff initially envisaged replacing both HH-3F and AB.212 with a single NH-90 fleet.[83] This evolved into a two-tier concept, with the 15-ton AW101 (HH-101A in Armaereo parlance) replacing the HH-3F for combat SAR and personnel recovery and the 8-ton AW149 succeeding the AB.212 in utility roles. After AgustaWestland (now Leonardo Helicopters) lost bids in Turkey (2011) and Poland (2015), it put the AW149 on hold, leading the ITAF to adopt a militarized 6-ton AW139 civil helicopter as a temporary solution, albeit unsuitable for operational environments. In the future, the ITAF is looking to replace the TH-500E fleet of its 72° Stormo (72nd Wing), possibly redesigning the helicopter school around a public-private joint venture that would sell courses to other parties and thereby reduce overhead. A recent agreement with the Army to designate the 72nd Wing as a joint training establishment goes in the same direction.

Another temporary solution is the P-72A Surveyor maritime patrol aircraft (MPA), delivered in small numbers to the 41° Stormo (41st Wing) to replace the Breguet Atlantic anti-submarine warfare platform. The Navy, which under the compromise hammered out in the 1950s retains operational control of the MPA force, has voiced its concern for the loss of anti-submarine warfare (ASW) capabilities and would like the P-72A to be replaced with a more potent type. No funds have been allocated so far, however.

Italy was an early participant in the Airbus A400M, withdrawing for reasons ranging from high cost, unrealistic Airbus planning numbers

---

82 <https://theaviationist.com/2013/06/07/aml-at-work/>;<http://www.ghidica.it/2017/ -09/11/laeronautica-militare-e-lintelligence-elettronica/>. According to the *Documento Programmatico Pluriennale 2017–19*, the ITAF intends to continue the SPYDR lease through 2019 while launching a permanent acquisition program for two SPYDR aircraft (<http://www. difesa.it/Content/Documents/DPP/DPP_2017_2019_Approvato_light.pdf>, 35).

83 Arpino, "L'Aeronautica Militare del Terzo Millennio," 31.

(44 aircraft, providing about five times the capacity of the single C-130J Hercules squadron), limited industrial returns (with participation in the Airbus Military subsidiary, rather than its parent company), inadequate performance, and unreliable timing—a much criticized assessment subsequently vindicated by program history. It later considered joining the NATO C-17 force with a one-aircraft quota, but the plan was cancelled in 2011–12 under the government of Mario Monti. Although the current KC-767A tanker-transports can carry troops and supplies over considerable distances, they are unable to load heavy Army equipment; more importantly, sustained transport use would hamper their irreplaceable role as tankers. The perceived level of priority was probably reduced by the gradual downgrading of Italian presence in the Gulf and the European Air Transport Command (EATC).

Looking ahead, it is likely that the ITAF will increasingly express its needs in terms of capabilities to meet probable threats and maintain qualitative parity with benchmark air forces. Technological sophistication and budget realities suggest that it will become increasingly difficult for it to become the sole, or even main, supporter of every segment and field in which Italian industry operates at present. In this respect, the ITAF will probably drive industry integration at the continental level, whether at a functional level (i.e., pushing for Italian industry to align itself with a single European project in each field, such as the sixth-generation fighter) or in wide-ranging redesign (e.g., exchanging non-leading business across borders to create strong homogeneous entities capable of competing on world markets). Its ability to help strengthen Italian industry, however, might be limited by inadequate government and industry understanding of the importance and implications of the new European Defense Fund (EDF).[84] Whatever the ITAF vision, without appropriate levels of national co-funding for the EDF and strong political leadership, mandatory Italian EDF contributions might very well result in supporting foreign competitors.[85]

---

84 <http://europa.eu/rapid/press-release_IP-18-4121_en.htm>.
85 Stefano Pioppi "Aerospazio e difesa. Il grande salto richiesto alla politica e all'industria," Dec. 7, 2018, <http://www.airpressonline.it/36814/aerospazio-difesa-icsa-volpi/>.

Conclusion

Over the past 20 years the ITAF has followed a surprisingly consistent and stable path of transformation into a modern force, qualitatively equivalent and in some areas superior to other European air forces, as indicated by the selection of Amendola to host the Tactical Leadership Program Flying Course 2018-4. Media attention centered around the European debut of the fifth-generation Lockheed-Martin F-35A and its interactions with fourth-generation aircraft—and, in fact, midway through the exercise the ITAF declared its 13° Gruppo (13th Squadron, belonging to the locally based 32° Stormo/32nd Wing) had achieved IOC with the F-35A, the first European-based F-35 unit to reach this milestone on the new fighter. But from another perspective, the measure of change probably lay in the overall ITAF lineup, which depicted a service that had achieved many of its long-standing goals, including tankers and early-warning aircraft, and recapitalized crucial air defense, attack and training components, albeit often with suboptimal numbers. Together with less visible assets and capabilities, like the Joint Force Air Command (ITA-JFAC), this explains the proud assessment offered by Lt. Gen. Rosso in his first extended interview upon taking office:

> If I look at the capabilities and assets we now have and compare them with what we had when I was a squadron pilot, I see a huge difference. Flying the F-104 made me very happy in romantic terms, of flying the aircraft, but in international exercises our role and contribution were virtually zero. We were considered those who wanted to participate in modern scenarios despite having a vintage aircraft ... Now everything has changed and it is the others who look to us with admiration and respect, because we have modern equipment but especially because we have personnel trained to be competent, credible, authoritative and capable of fully leveraging the technologies available to them.[86]

---

86 Rosso interview A, 12–13. Rosso entered the ITAF Academy in 1978 and served with 4° Stormo (4th Wing) in various capacities between 1984 and 1996 and 2002 and 2004.

The ITAF appears to have a clear concept of its future, in relation to both conventional and emerging threats, a realistic appreciation of its ability to influence events and to what extent, a genuine commitment to its role as joint enabler, and a solid understanding of the required capability portfolio. Security threats are, of course, subject to change, and there is no guarantee that unforeseen paradigm shifts or asymmetrical conflicts will not arise to challenge the ITAF roadmap for continued and sustainable relevance. At this time, however, the most evident threat to ITAF capabilities and operations must be inadequate financial resources.

By tradition, the Italian military have been chronically underfunded, with net defense expenditure hovering around or below one percent of GDP—as generals often lamented, "less than what the Americans imposed upon Japan to keep them from rearming." Cuts continued even after Italy committed to both NATO and EU defense funding targets.[87] The quality of the budget now differs sharply from the hypothetical ideal breakdown between personnel costs (50 percent), operations (25 percent) and investment (25 percent). Between 2002 and 2016, the ITAF saw its budget decrease by approximately one-quarter, from €4.114 billion to €2.807 billion, before climbing up slightly again to €3.096 billion.[88] With the cost of personnel growing by 48 percent (from €1.627 billion to €2.412 billion), the drop was absorbed largely by operations and investment. Salaries now account for about 80 percent of the budget, roughly twice as much as 15 years ago, a situation all the more remarkable given the 16 percent reduction in staffing levels over the same time span.

The situation was somewhat ameliorated by generous ad hoc funding for out of area operations (*missioni*) and MISE support for industrial programs (including in the 2017–22 period T-346A, T-345A, Typhoon, HH-101, and Tornado), particularly as new acquisitions began to include related infrastructure and a number of years of logistic support. Each role has expanded its capabilities into adjacent areas to maximize efficiency and cost effectiveness while adding capabilities and leveraging aircrew training.[89] Ground attack aircraft carry air-to-air armament on routine sorties,

---

87 DPP, viii.
88 Author's analysis based on *Coccarde Tricolori* figures.
89 Rosso interview A, 4.

which enables them to be directed toward potential intruders if necessary;[90] similarly, trainers and helicopters train to intercept slow movers that might prove difficult targets for fast jets.

To reduce overhead in flight training, in which it has developed considerable expertise, the ITAF is looking to launch an international fighter training school in a joint venture with industry, with the aim of attracting other air forces through a combination of competitive cost and quality training.[91] Savings are already provided by increasingly comprehensive simulation capabilities, cutting down on the number of aircraft required as well as on expensive flying hours, without compromising training outcomes. In the same way, distance learning and teleconferences cut down on travel costs and increase personnel availability. Smaller cost-cutting measures are used in many areas, from the lease of excess capacity to pay for the maintenance of the extensive pipelines built during the Cold War to deliver petrol, oil, and lubricant (POL) securely to air bases to EU energy-saving programs to fund building upgrades. But stable policy and prudent management can compensate for diminishing resources only up to a point, and increasing difficulty in funding support contracts is already a limiting factor. In 2016, Lt. Gen. Vecciarelli identified the main challenge for the airlift component as the "chronic underfunding of logistic support (spares and maintenance)" drastically limiting fleet availability.[92]

To a certain extent, budget uncertainty is as bad as outright cuts.[93] While all reductions are painful and curtail investment, mid-program changes have a negative impact on three different levels. By stretching out the development phase, sudden reductions delay full production, leaving programs mired in non-recurring costs before meaningful outputs are delivered. In addition, assets are often complementary or build upon each other, so that relatively minor cuts can translate into significant capability losses. Finally, staffs must devote large amounts of time to

---

90 Vecciarelli interview, 16.
91 Craig Hoyle, "New Italian Training School to Take Off with M-346," Jul. 17, 2018, <http://www.flightglobal.com/news/articles/farnborough-new-italian-training-school-to-take-off-450399/>.
92 Vecciarelli hearing, 10.
93 Fondazione ICSA, "La funzione difesa in tempi di crisi economica: riflessioni e prospettive," Camera dei Deputati, Osservatorio di Politica Internazionale, Approfondimenti, No. 67 (Nov. 2012).

continuous replanning, which distracts them from their overall function. For these reasons, the Italian military has constantly requested multi-year budget-planning and stability. The principle was enshrined in the COM and the resulting 2017 three-year plan,[94] but the latest 2019 budget cuts indicated that governments and parliament do not feel bound by them. Should the trend continue, ITAF concepts and plans for its second century of existence would become largely irrelevant. Worse yet, its ability to contribute to the National Integrated Force would disappear.

References

"Airbus, BAE and Leonardo Complete MBDA Merger." *Defense Daily International.* Dec. 21, 2001.

Aeronautica Militare. "Verso il 2018. Linee guida e d'indirizzo strategico." Rome, 2013.

Alegi, G. "Il futuro dell'Asi e uno sguardo al passato." *Air Press*, No. 95, Dec. 2018.

Alegi, G. "Forze armate," in *Dizionario del liberalismo italiano,* Vol. I, Soveria Mannelli, Rubbettino, 2011, 473–78.

Analisidifesa. "Difesa cyber nazionale: parla il generale Vestito." June 2, 2018. <https://www.analisidifesa.it/2018/06/difesa-cyber-nazionale-parla-il-generale-vestito/>.

Associazione Pionieri dell'Aeronautica (APA). "Lo Stormo al centro. Storia, riflessioni, prospettive." Vicchio, LoGisma, 2017.

Caligaris L and C. M. Santoro. *Obiettivo difesa. Strategia, direzione politica, comando operative.* Bologna: Il Mulino, 1986.

Cenciotti, D. "Lockheed Martin's 'Dragon Star' Airborne Multi-Intelligence Laboratory at Work." *Aviationist*, June 7, 2013.

Cornacchini, A. "Intervista al capo di Stato Maggiore dell'A.M. gen. s.a. Alberto Rosso." *Rivista Aeronautica*, Dec. 2018.

Cosci, S. "Aeronautica 4.0," interview with Lt. Gen. Vecciarelli, *Rivista Aeronautica*, No. 6, Dec. 2016.

De Feo, G. "5Stelle, Tofalo e la riabilitazione degli F35: Hanno un'ottima tecnologia, forse la migliore al mondo." *Repubblica*, Dec. 19, 2018.

Di Paola, G. "Il concetto strategico del capo di Stato Maggiore della Difesa." Rome, 2005. <http://www.difesa.it/SMD_/CaSMD/ConcettoStrategico/PublishingImages/87587_Documento%20completo%20(File%20Pdf%20 2,48%20Mb)pdf.pdf>.

European Commission. "EU Budget: Stepping Up the EU's Role as a Security and Defence Provider." June 13, 2018.

Fondazione ICSA. "La funzione difesa in tempi di crisi economica: riflessioni e prospettive." Camera dei Deputati, Osservatorio di Politica Internazionale, Approfondimenti, No. 67, Nov. 2012.

Gaiani, G. "Mistero sulle armi per i droni: l'Italia alza la voce con Washington e minaccia scelte europee." *Il sole 24 ore*, May 18, 2013.

Gaiani, G. "A rischio il programma per la difesa aerea CAMM ER?" Analisidifesa, Sep. 28, 2018. <https://www.analisidifesa.it/2018/09/a-rischio-il-programma-per-la-difesa-aerea-camm-er/>.

GhiDiCa. "L'Aeronautica Militare e l'intelligence." Sep. 11, 2017. <http://www.ghidica.it/2017/09/11/laeronautica-militare-e-lintelligence-elettronica>.

Giacobbe, F. "A difesa di Vecciarelli. Le parole della Trenta contro una polemica inutile," Nov. 12, 2018.

Giacobbe, F. "Accesso allo spazio e nano-satelliti. Così l'italia fa system." *AirPress online*. Dec. 13, 2018. <https://www.airpressonline.it/36904/spazio-sistema-aeronautica-lanciatore/>.

Giacobbe, F. "Rosso, l'F-35 e il caccia del futuro." *AirPress online*, No. 95. Dec. 2018.

Hoyle, C. "New Italian Training School to Take Off with M-346." *Flight-global.com*, Jul. 17, 2018. <https://www.flightglobal.com/news/articles/farnborough-new-italian-training-school-to-take-off-450399/>.

Ilari, V. *Storia militare della prima repubblica 1943–1993*. Ancona: Nuove Ricerche, 1994.

Informazioni Difesa. "The Joint Cyber Command Is Born," No. 3, June 2017.

Lertora, G. "Il paradosso della Difesa nelle acquisizioni dell'F-35B STOVL." Nov. 28, 2018. <http://www.liberoreporter.it/2018/11/dossier/il-paradosso-della-difesa-nelle-acquisizioni-dellf-35b-stovl.html>.

Meilinger, P.D. "10 Propositions on Airpower." *Airpower Journal,* Spring 1996.

Nones, M., G. Gasparini and A. Marrone. "Il programma F-35 Joint Strike Fighter e l'Europa." *Quaderni IAI* No. 31, October 2008.

Pioppi, S. "Aeronautica 4.0, innovazione e investimenti per la difesa." *Air-Press online*, Sep. 19, 2018. <https://www.airpressonline.it/35316/aeronautica-4-0-innovazione-e-investimenti-per-la-difesa/>.

Pioppi, S. "La tabella di marcia del generale Alberto Rosso." *AirPress online*, Nov. 1, 2018. <https://www.airpressonline.it/36088/rosso-aeronautica/>.

Pioppi, S. "Aerospazio e difesa. Il grande salto richiesto alla politica e all'industria." *AirPress online*, Dec. 7, 2018. <https://www.airpressonline.it/36814/aerospazio-difesa-icsa-volpi/>.

Prometeia, "Il contributo di Finmeccanica all'Italia: Tecnologia, crescita, investimenti." Bologna: Oxford Economics, 2014.

Pucci, E. "F35, c'è il via libera all'acquisto di 90 aerei: missione del sottosegretario grillino in Usa." *Il Messaggero*, Jan. 22, 2019.

Santoro, C.M. *La politica estera di una media potenza. L'Italia dall'Unità ad oggi.* Bologna: Il Mulino, 1991.

Servadei, B. *Ali di travertino.* Ravenna: SBC, 2011.

Stato Maggiore dell'Aeronautica. "La dottrina dell'Aeronautica Militare." Rome, 1997.

Stato Maggiore Aeronautica. "SMA 9 Potere Aereo-spaziale – Fondamenti." June 2011 edition.

Stato Maggiore della Difesa. "La trasformazione net-centrica." Rome, 2009.

Tricarico, L. and G. Alegi. "La linea aerotattica dell'aeronautica militare. Dall'esperienza operativa alle esigenze attuali per le operazioni future." Rome: ICSA, 2014.

Villadei, W. "L'aeronautica militare qualifica i primi ufficiali per il volo suborbital." Nov. 29, 2018. <http://www.aeronautica.difesa.it/comunicazione/notizie/Pagine/NASTAR.aspx>.

Yeats, W.B. "The Second Coming." *The Dial.* 1920.

7 GERMANY
The Future of the Luftwaffe

Richard Shimooka[1]

Introduction

As with many western states, Germany's position in the international system is in flux, undergoing significant (and ongoing) reorientation over the past several decades. Some of the major developments that it has had to deal with include the collapse of the Soviet Union, the rise of instability on Europe's periphery, and Russia's reassertion of its sphere of influence over the former Soviet Union states. It is also in a period of unprecedented technological change, possibly creating revolutionary changes to military operations.

These external challenges alone would be difficult enough to navigate. However, added to them are fundamental shifts in the internal politics of Germany, Europe, and the Atlantic alliance. Where Germany was once able to maintain a complementary relationship between Europe and the United States to promote its interest, the emerging dynamics suggest this may no longer be possible. The Bundeswehr, and the Luftwaffe in particular, sit within this challenging nexus, which will have significant consequences for the future of German air power.

This chapter will explore the Luftwaffe's approach to dealing with the future security of Germany. The first section discusses political, budgetary, and strategic considerations: these are foundational determinants

1 The author would like to thank Christoph Bergs, PhD candidate, King's College London, for his assistance in translating German documents, as well as several anonymous reviewers. Nevertheless, any errors contained within are the author's alone.

of how the Bundeswehr, and by extension, the Luftwaffe, will be expected to operate in the future. The second and third sections will discuss present strategy and doctrine, as well as force structure and readiness, respectively. These sections are critical for understanding how air power is perceived and operates today within the Luftwaffe and in the near future. The fourth section covers the future developments. Building on the previous sections, it will provide a detailed examination of where German air power may go in the medium to long term.

Political, Industrial, Budgetary, and Strategic Considerations

There have been two continuities running through German defence and security policy since the end of World War II. The first is a strong pacifist strand in German society largely a result of the country's role in the World War II. Today's Germany is now frequently referred to as a civilian power—one that has a strong aversion to the use of military power for anything except national defense. This was evident in Germany's general reluctance to undertake any out-of-area operations, such as in the Balkans in the 1990s or Afghanistan in the 2000s.

A much less contentious guiding principle of German defense and security policy since the end of World War II has been multilateralism. From the inception of the Bundeswehr in 1955, and even earlier, German foreign and defense policy has remained heavily wedded to multilateralism, with the North Atlantic Treaty Organization (NATO) as its cornerstone. For much of the Cold War, the German military was nearly inseparable from the alliance structures, to the extent that they could not operate independently without them.

One area of particular political controversy is Germany's participation in a nuclear sharing arrangement with the United States. This policy has its origins in the Cold War, which allowed the German leadership to hold some level of control in the use of nuclear weapons, while signaling the continued U.S. commitment to defending Europe.[2] The German public is generally unsupportive of the state's participation in the scheme, placing

---

2 Douglas Stuart, "Introduction of European Policies and Opinions Relating to Tactical Nuclear Weapons," in T. Nichols, D. Stuart, and J.D. McCausland (eds.), *Tactical Nuclear Weapons in NATO* (Carlisle: Strategic Studies Institute (2012), 229–33.

various governments in the uncomfortable position of defending a pub-licly unpopular nuclear deterrent.[3] This lack of support is despite the recent rise in tensions with the Russian Federation, which has arguably renewed deterrence's military and political value. The government of Angela Merkel has retained significant ambiguity on the topic, refusing to clearly state a position in spite of the need to make a decision on the replacement of its current fleet of aircraft, the Tornado.

Prior to 1993, the European Union (EU) and its preceding organi-zations had only a latent security focus. This began to change with the creation of the European Union. The first substantive effort to involve the European Union in defense matters was in 1998, with the St. Malo Dec-laration between the United Kingdom and France.[4] Although progress has been slow on building up EU-based capabilities, developments since 2014 have seen some significant progress. A key creation has been the Permanent Structured Cooperation (PESCO), a comprehensive series of multinational programs designed to address key gaps in European military capabilities.[5]

Nevertheless, Germany has remained a steadfast supporter of the transatlantic relationship with the United States. Its policies have roughly followed the "Three Ds" as outlined by Madeleine Albright in 1998.[6] These were no decoupling of NATO from Europe, no duplication of existing structures or capabilities that would waste limited European resources, and no discrimination against non-EU NATO members. In particular, Germany met significant success triangulating policies that support both NATO and the European Union simultaneously and synergistically.

However, the election of U.S. president Donald Trump in 2016 and his subsequent statements questioning the utility and cost of NATO have frayed the transatlantic relationship to some degree. These concerns

---

3 Götz Neuneck, "European and German Perspectives," T. Nichols, D. Stuart, and J.D. McCausland (eds.), *Tactical Nuclear Weapons in NATO,*(Carlisle: Strategic Studies Institute, 2012) 257–78.
4 Maartje Rutten, "From St. Malo to Nice: European Defence: Core Documents," Joint Decla-ration issued at the British-French Summit, St. Malo, Dec. 3–4, 1998, European Union Institute for Security Studies (2001) 8–9.
5 "Permanent Structured Cooperation (PESCO) Updated List of PESCO Projects – Overview." Council of the European Union, Nov. 19, 2018, <https://www.consilium.europa. eu/media/37028/table-pesco-projects.pdf>.
6 Madeleine Albright, "The Right Balance Will Secure NATO's Future," *Financial Times*, Dec. 7, 1998; Rutten, "From St. Malo to Nice: European Defence: Core Documents," 10–12.

have been exacerbated by the United Kingdom's 2016 referendum to exit the European Union, which has raised questions about its reliability as an ally. While Germany has not abrogated Albright's Three Ds formulation in light of these events, Berlin is making clear moves to prioritize developing its European-centric defense capabilities and military industrial development. As we will see later in this chapter, it may eventually facilitate this outcome in the end.

Germany's external security challenges have considerably affected overall defense policy. Prior to 2008, Germany's primary security concerns largely centered on the instability caused by weak states. This was underlined by the crisis in the former Yugoslavia in the 1990s followed by stabilization efforts in Afghanistan during the 2000s. The Bundeswehr's reforms during this period reoriented the force structure from one primarily focused on a massive superpower conflict to a more flexible one that could deal with this new reality. This was particularly the case after the 9/11 terrorist attacks in the United States.[7] Nevertheless, military spending in this period lagged significantly and the force size ultimately contracted. As Patrick Keller suggested in 2012, Germany was never an enthusiastic supporter of the shift toward out-of-area operations, even within a NATO context.[8] Consequently, the force structure retained significant elements of a conventional warfare capability, although perhaps with greater flexibility to deploy abroad.

Between 1990 and 2008, successive German governments attempted to maintain good relations with Russia, with mixed results.[9] This started to shift in the aftermath of the conflict in Georgia in 2008, with increasingly provocative moves along Russia's periphery, including the annexation of Crimea, direct and indirect support for insurgents in the Donbas region of Ukraine, and threatening actions against the former Soviet Baltic republics of Estonia, Latvia, and Lithuania. The 2016 *White Paper on German*

---

7 Christian Anrig, *The Quest for Relevant Air Power: Continental European Responses to the Air Power Challenges of the Post–Cold War Era* (Maxwell: Air University Press, 2011), 51, 154, <https://www.airuniversity.af.edu/Portals/10/AUPress/Books/B_0125_ANRIG_QUEST_RELEVANT_POWER.pdf>.
8 Patrick Keller, "Germany in NATO: A Status Quo Ally," *Survival*, Vol. 54, No. 2 (2012).
9 Christopher Chivvis and Thomas Rid, "The Roots of Germany's Russia Policy," *Survival*, Vol. 51, No. 2 (2009).

Security Policy and the Future of the Bundeswehr highlights this as a major challenge:

> This is reflected, for example, by an increase in Russia's military activities along its borders with the European Union (EU) and the North Atlantic Alliance (NATO). In the course of extensively modernising its armed forces, Russia appears to be prepared to test the limits of existing international agreements. By increasingly using hybrid instruments to purposefully blur the borders between war and peace, Russia is creating uncertainty about the nature of its intentions. This calls for responses from the affected states, but also from the EU and NATO.[10]

While the German government continues its attempt to create a balanced Russian policy based on constructive engagement, since 2014 there has been a growing move to reinforce the country's security. This includes participating in NATO's Enhanced Forward Presence in the Baltics in order to deter any coercive actions by Russia. In addition to the immediate challenge provided by the Russian Federation, there remains the challenge of failed and failing states, with continued deployments in Afghanistan and Iraq as well as operations in Mali alongside France.

In light of the emerging unstable regional environment, the German government has made tentative steps to improve the state of its military forces. In 2014, it agreed to meet the NATO Wales Summit goal of spending 2 percent of the gross domestic product (GDP) on defense within a decade, rising from 1.2 percent in 2013.[11] The Merkel government made significant increases to the budget, with it rising 6.9 percent in 2016–2017. There is, however, some concern as to the long-term efficacy of that spending, with much destined to be deferred spending on existing procurement projects and exercises.[12] Rather it would largely assist in raising the readiness and availability of the Bundeswehr, a topic that will be discussed in later sections.

---

10 "White Paper 2016: On German Security Policy and the Future of the Bundeswehr," Federal German Republic (2016), 31.

11 "Wales Summit Declaration." *NATO* (Sep. 5, 2014) <https://www.nato.int/cps/en/natohq/official_texts_112964.htm>.

12 International Institute for Strategic Studies, *Strategic Survey 2015* (Routledge, 2015), 148.

A final area of consideration is industrial development and capacity, which is considered a major pillar of German security. The German defense industry has long been a major factor in the high qualitative capability in the Bundeswehr. In the past several decades, there has been a growing move toward multinational European programs, such as the A400M transport, Eurofighter and Tiger helicopter. The importance of these efforts is clearly stated in the 2016 Defense White Paper, which states that a "independent, strong and competitive defence industry in Europe, including the national availability of key technologies, is essential."[13]

Previously, the development of an indigenous European defense industry was viewed as complementary to strengthening the transatlantic relationship. However, this policy may be reaching its limits. The primary driver for the emerging air force procurement strategy has been the Franco-German relationship and industrial considerations. It has created a rival approach to the efforts of other European states, creating a major point of friction. Italian representatives have already voiced concerns that the Franco-German partnership has largely monopolized decision-making to the exclusion of their interests.[14] A good example is the announced partnership between Germany's MTU and France's Safran to produce an engine for a next-generation combat aircraft, which means that other national manufacturers may be locked out of lucrative work.[15]

This is likely one factor behind the United Kingdom's decision to pursue an independent manned fighter capability to replace their Eurofighters, known as the Tempest. With BAE designated the prime contractor and Rolls Royce in charge of engine development, the desire to preserve a strong workshare for national industrial champions is quite evident.[16] Although Airbus representatives have stated their desire for the United Kingdom to join their efforts to develop such a capability with Dassault,

---

13 "White Paper 2016: On German Security Policy and the Future of the Bundeswehr," 31.

14 Tom Kington, "Taking Sides: Italian Defense Industry Rep Attacks Franco-German Fighter Deal," *Defense News*, Feb. 15, 2019, <https://www.defensenews.com/global/europe/2019/02/15/taking-sides-italian-defense-industry-rep-attacks-franco-german-fighter-deal/>.

15 Michael Gubisch. "France and Germany sign first FCAS development contracts." *Flight Global*, Feb. 7 2019, <https://www.flightglobal.com/news/articles/france-and-germany-sign-first-fcas-development-contr-455591/>.

16 "Team Tempest, From Rolls Royce Came Thrust," *European Defence Review*, Jul. 17, 2018, <https://www.edrmagazine.eu/rolls-royce-the-tempest>.

the divergent directions between the Franco-German and U.K. approaches are becoming increasingly evident. Airbus and BAE have already clashed over assisting countries outside Europe with their defense abilities, the latest being Turkey with its TFX program and Japan.[17]

Similarly, American representatives have raised the potential for PESCO projects to exclude non-European NATO members from participating in a common defense framework.[18] However, this has abated somewhat over the last year.[19] Nevertheless, this is a potential area of friction, which may dilute Germany's ability to contribute to a common defense of Europe.

Present Strategy and Doctrine

The Bundeswehr, including the Luftwaffe, is expected to meet a broad spectrum of threats and challenges facing Germany. Its primary role is the defense of the state and its alliance treaty partners, most notably NATO and the European Union. The 2018 Konzeption der Bundeswehr described the capability profile as geared toward NATO and EU planning objectives and demands on the armed forces. Its main priorities are the close integration and progressive integration of European armed forces, strengthening of the European pillar in NATO, and the more coherent interaction between NATO and the EU are priority.[20]

It goes on to explain that the German state's defence, the alliance's defence (NATO and the European Union), and crisis prevention are the

---

17 Douglas Barrie, "New Franco-German Combat-Aircraft Programme: A Reminder of UK's Uncertain Position," *IISS Military Balance Blog*, Jul. 14, 2017, <https://www.iiss.org/blogs/military-balance/2017/07/new-franco-german-combat-aircraft>; Chris Pocock, "France and Germany to Develop New Fighter without UK?" *AIN Online*, Aug. 2, 2017, <https://www.ainonline.com/aviation-news/defense/2017-08-02/france-and-germany-develop-new-fighter-without-uk>.

18 "EU's New Defense Union Must Not Undermine NATO, Warns Stoltenberg," *DW*, Feb. 13, 2018, <https://www.dw.com/en/eus-new-defense-union-must-not-undermine-nato-warns-stoltenberg/a-42570471>.

19 Martin Banks, "From Cyber to Military Mobility: EU Members Endorse New Defense Objectives," *Defense News*, Nov. 20, 2018, <https://www.defensenews.com/global/europe/2018/11/20/from-cyber-to-military-mobility-eu-members-endorse-new-defense-objectives/>.

20 "Konzeption der Bundeswehr," German Ministry of Defence, Jul. 20, 2018, 5, <https://www.bmvg.de/resource/blob/26544/9ceddf6df2f48ca87aa0e3ce2826348d/20180731-konzeption-der-bundeswehr-data.pdf>. Translated by the author.

most ambitious and complex tasks facing the military.[21] The Luftwaffe is also expected to support peace and stabilization operations, particularly in failed and failing states. However, prevailing interpretations of German national law require any out-of-area operations be conducted within a multilateral framework based on collective security, such as the United Nations (UN) and NATO.[22]

As noted above, NATO remains the cornerstone of Germany's defense policy; interoperability is implicitly and explicitly stated to be a fundamental consideration for the Luftwaffe, at all levels of operations.[23] This has had a marked effect on the Luftwaffe's force structure and operational ability, as it relies heavily on alliance organization and capabilities for its day-to-day operations. Germany is a major participant in the allied air command structure, hosting a Combined Air Operations Centre (CAOC) in Uedem, maintaining a Deployable Air Command and Control Centre, and providing a significant portion of the personnel for the overall air command at Ramstein.[24]

In recent years, however, there have been efforts to introduce greater flexibility in both the national and multinational frameworks for the Luftwaffe. The most fundamental reforms were intended to improve national command and control (C2) structures. In 2011, the Bundeswehr introduced a military-wide restructuring. This was a slightly scaled-back version of various reforms implemented over the previous decade, intended to furnish the German government with the ability to plan and implement air operations autonomously.[25] The structural reforms also focused on improving the Luftwaffe's internal interoperability, with its constituent aerial divisions consolidated in geographically based multi-capability units.

The creation of new organizational reforms (known as Structures 5 and 6) assisted in creating a national C2 system, and also improved the Bundeswehr's ability to contribute to allied structures. For example,

---

21 Ibid.

22 Russell Miller, "Germany's Basic Law and the Use of Force," *Indiana Journal of Global Legal Studies*, Vol. 17, No. 2 (2010), <https://www.repository.law.indiana.edu/cgi/viewcontent.cgi?article=1419&context=ijgls>.

23 "White Paper 2016: On German Security Policy and the Future of the Bundeswehr," 23.

24 "Deployable Air Command and Control Centre (DACCC)," NATO, <https://ac.nato.int/about-aircom/deployable-air-command-and-control-centre> (Accessed Sep 12., 2019).

25 Anrig, *Quest for Relevant Airpower*, 163.

the lieutenant general in charge of the Luftwaffe command was dual hat-
ted as the commander of the CAOC at Uedem. These efforts were further
assisted by broader efforts to improve European organizational capabilities
to assist both NATO and the European Union. The manifestation of these
efforts is the NATO Framework Nations concept, which was spearheaded
by Germany and adopted as alliance policy in 2014. This approach boosts
interoperability between NATO and EU countries in order to operate more
effective multinational formations for specific tasks. This was one of Germany's primary objectives with regard to NATO articulated in the 2016
Defense White Paper:

> Increase European capability development and the interlinking
> of European armed forces by means of the Framework Nations
> Concept in order to strengthen NATO's European pillar. In this
> context, Germany is prepared to pave the way and assume a
> wide range of responsibilities as the framework nation. At
> the same time, Germany will provide its partners with key
> capabilities in a sustainable manner.[26]

Germany's framework primarily focuses on security on the eastern
periphery of Europe. The Luftwaffe is at the forefront of this effort, with it
contributing up 75 percent of the Multinational Air Group's capabilities in
a crisis.[27]

The Luftwaffe's primary tasks are:

> As the first force to deploy, the air component can exert political
> influence at various levels of escalation and de-escalation at an
> early stage, even with limited commitment and sometimes even
> without the need to deploy to the mission area. It can thus make
> a decisive contribution to Germany's ability to pursue foreign
> policy objectives and deterrence in an Alliance context.[28]

---

26 "White Paper 2016: On German Security Policy and the Future of the Bundeswehr," 23.
27 Rainer L. Glatz and Martin Zapfe, "NATO's Framework Nations Concept," CSS Analyses
in Security Policy, No. 2018, Center for Security Studies, ETH Zurich, Dec. 2017, 2. <http://
www.css.ethz.ch/content/dam/ethz/special-interest/gess/cis/center-for-securities-
studies/pdfs/CSSAnalyse218-EN.pdf>.
28 "Air Power Development Strategy 2016," Federal Ministry of Defence 2015, 10.

Furthermore, the air component is also responsible for providing reconnaissance, to better inform operational, strategic, and political decision-making, and for ensuring force protection and air superiority for the freedom of movement of other combat arms.

Considering its objectives and Structure 6, the Luftwaffe can operate as an independent capability to achieve these aims. However, it has also developed a fairly robust cooperation arrangement with the Army to operate in conjunction with it. When the Luftwaffe is tasked with air interdiction in support of the Army, operations are organized within what is called the "fire support coordination line" that ensures proper coordination occurs near friendly ground forces. For close air support, joint fire support teams and groups work with other indirect assets to provide demanded effects. The system is fairly well regarded among Army officers to deliver effects when required.

Force Structure and Readiness

The Luftwaffe can be considered among the more technologically advanced western air forces, with a large and diverse fleet of combat aircraft. In total, it has 211 tactical fighter aircraft. The force structure is based on 143 Eurofighters, a multinational, advanced fourth-generation aircraft. Originally designed for air-to-air operations, the second and third tranches added an air-to-ground capability. In November 2018, the German government announced the purchase of 33 additional aircraft to replace the least capable first-tranche aircraft, of which delivery was complete in 2008.[29] The Eurofighter was once considered among the more costly fourth-generation fighters currently in service, with a cost of €90 million per airframe in 2009.[30] That cost was projected to decline to €75 million in future years.[31] This can be attributed in large part to the highly inflexible workshare arrangement

---

29 Thomas Wiegold, "Aufgestockter Verteidigungshaushalt soll große Rüstungsprojekte ermöglichen," *Augen Geradeaus!* Nov. 5, 2018, <https://augengeradeaus.net/2018/11/aufgestockter-verteidigungshaushalt-soll-grosse-ruestungsprojekte-ermoeglichen/>.
30 "Haushaltsausschuss des Deutschen Bundestages billigt Bundeswehrprojekte." Federal Ministry of Defence, June 9, 2009, <https://www.bundeswehr.de/resource/resource/MzEzNTM4MmUzMzMyMmUzMTM1MzMyZTM2MzEzMDMwzAzMDMwMzAzMDY3NmE2ODDc1NjEzNTZkNzAyMDIwMjAyMDIw/MIP85>.
31 Wiegold, "Aufgestockter Verteidigungshaushalt soll große Rüstungsprojekte ermöglichen."

among the four member states. Each country produces one of the four major sub-assemblies, as well as maintaining a final assembly line for its national aircraft. This system ensures each country receives a large amount of the workshare, but at a far higher cost than if a more efficient approach was adopted. As a cost-saving measure, early-tranche Luftwaffe Eurofighters do not carry the PIRATE Infrared search and track system.

In addition to the Eurofighter, the Luftwaffe also operates a substantial fleet of Panavia Tornado that entered the fleet in the late 1970s and 1980s. The fleet underwent a major mid-life upgrade in the 2000s, while being supplanted by the Eurofighter in some air-to-ground roles.[32] Although the aircraft are facing obsolescence and retirement in the coming decade, the Tornado provides a unique set of capabilities within the force and NATO. The German fleet is separated into two types. The first 68 are the interdictor/strike (IDS) version, which are primarily used for conventional air-to-ground operations. These aircraft have recently undergone the upgrade that introduced new cockpit displays and greater networking capabilities, which should maintain their operational relevance until 2025.[33]

The Tornado also has a secondary role as part of NATO's dual-key nuclear response, being one of the only European aircraft cleared to carry the U.S. B61 nuclear bomb. Germany's participation in the nuclear arrangement has engendered significant controversy. For many years it seemed that the dual-key role would be phased out with the retirement of the Tornado, with the Bundestag even voting for its removal in 2010.[34] However, increased Russian assertiveness on its periphery has shifted perceptions within Germany and forced the government to reassess its position. The 2016 Defense White Paper maintains the strategic ambiguity, both acknowledging the need for a credible nuclear deterrent and Germany's role in the dual-key arrangement, while reaffirming Germany's commitment to nuclear disarmament.[35]

---

32 Leland M. Nicolai, "Aircraft Update Programmes: The Economical Alternative?" *Defense Technical Information Center Compilation Report,* June 28, 2010, <https://apps.dtic.mil/dtic/tr/fulltext/u2/p010301.pdf>.

33 Oliver Pieper, "ASSTA 3.1: Der Modernste Tornado der Luftwaffe," German Airforce Association (2018), <http://www.idlw.de/assta-3-1-der-modernste-tornado-der-luftwaffe>.

34 Jeffery Larsen, "Future Options for NATO Nuclear Policy," Atlantic Council, Aug. 30, 2011, <https://www.files.ethz.ch/isn/132730/083011_ACUS_FutureNATONuclear.pdf>.

35 "White Paper 2016: On German Security Policy and the Future of the Bundeswehr," 65.

In addition to the IDS aircraft, the Luftwaffe also operates 29 electronic combat and reconnaissance (ECR) variants of the Tornado, which were delivered in the early 1990s. These are primarily tasked with suppression of air defenses and intelligence surveillance and reconnaissance (ISR) missions. While somewhat limited in their capability compared to a dedicated platform such as the Boeing EA-18G Growler, they still provide a critical role in ensuring the continued viability of other platforms in increasingly contested threat environments. They are an essential force multiplier within the Luftwaffe's force structure for now and into the near future.

The Bundeswehr as a whole possesses several other air capabilities relevant to the application of air power. The Army possesses 67 Tiger attack helicopters, which are under their command and are focused on supporting ground maneuver units. However, they may be tasked to the Luftwaffe to undertake operations in support of broader objectives. In addition, the Luftwaffe also controls ground-based air defense capabilities under the 2011 restructuring, which assists in meeting the overall objective of air superiority. This includes three batteries of Patriot missiles of 30 launchers in total. This system is tentatively planned to be replaced by the Medium Extended Air Defense System (MEADS), a development of the Patriot system with an open architecture battle management system intended to improve interoperability with other platforms.[36]

In addition to its national capabilities, the Bundeswehr also contributes to several key multinational programs. The most visible is the NATO Airborne Warning and Control System, which is a key C2 and sensor platform for the air defense of European NATO territory. Germany is a key participant, providing a substantial portion of the personnel and funding. It also jointly operates its transport fleet within a EU framework, pooling aircraft in order to gain greater flexibility and capacity for operations.

At present, the greatest challenge facing the Luftwaffe is the low availability rates for its capabilities. The Eurofighter has been singled out for this problem, with several news outlets suggesting that fewer than 10 of over 125 aircraft are considered operational at any one time.[37] Although

---

36 "Medium Extended Air Defense System," Lockheed Martin (2016) <https://www.lockheedmartin.com/content/dam/lockheed-martin/mfc/pc/meads/mfc-meads.pdf>.

37 Matthias Gebauer, "Luftwaffe hat nur vier kampfbereite 'Eurofighter'," *Der Spiegel*, May 2, 2018, <http://www.spiegel.de/politik/deutschland/bundeswehr-luftwaffe-hat-nur-vier-kampfbereite-eurofighter-a-1205641.html>.

the situation is not as dire as some have stated, it nevertheless points to some serious logistical and fleet management shortcomings. The Luftwaffe may have around 40 aircraft available to fly at any single time: a total fleet availability rate of roughly 25 percent. This does not compare well to other western countries: the U.S. military's mission availability rate spans 50 to 80 percent, and Canada's 30-year-old CF-18 fleet has a total fleet availability of 50 percent.[38] The generally low availability of the Eurofighter is a well-established problem among all of the aircraft's users. A key chokepoint has been the collaborative spares management program established for the aircraft, which has been described as deficient by a 2016 RAND Corporation report.[39] Late spares delivery and long maintenance times were frequently cited as issues leading to low availability. These issues have been compounded recently by a fault with the aircraft's Defensive Aid Sub-System that renders it inoperable without a refit, leading to only 10 of the 40 flyable aircraft being considered operationally capable.[40]

The issues with the Eurofighter are not unique to that fleet: they are observable with other systems such as the A400M strategic lift aircraft now entering service with the Luftwaffe. In 2017, *Der Spiegel* reported that only one out of eight of the aircraft currently in service was available for operations.[41] Other aircraft in service have experienced similar technical issues, such as the Tiger attack helicopter.

When she was Germany's minister of defense, Ursula von der Leyen identified this issue as one that requires urgent redress, and increased funding on areas to boost readiness, including expanding inventories of spare parts, catching up on deferred maintenance, and modernizing

---

38 Aaron Metha, "Mattis Orders Fighter Jet Readiness to Jump to 80 percent — in One Year," *Defense News,* Oct. 8, 2018, <https://www.defensenews.com/air/2018/10/09/mattis-orders-fighter-jet-readiness-to-jump-to-80-percent-in-one-year/>; Dennis Roberts, "Buying Old F-18 Jets Is a Bad Idea. It Just Makes Our Existing Problem Worse," *National Post,* Nov. 20, 2017, <https://nationalpost.com/opinion/dennis-roberts-buying-old-f-18-jets-is-a-bad-idea-it-just-makes-our-existing-problem-worse>.

39 Mark A. Lorell and James Pita, "A Review of Selected International Aircraft Spares Pooling Programs," RAND Corporation (2016), 20, <https://www.rand.org/content/dam/rand/pubs/research_reports/RR900/RR999/RAND_RR999.pdf>.

40 Gebauer, "Luftwaffe hat nur vier kampfbereite 'Eurofighter'."

41 "Luftwaffe kann nur einen von acht A400M einsetzen," *Der Spiegel,* Feb. 11, 2017, <http://www.spiegel.de/politik/deutschland/airbus-a400m-luftwaffe-kann-nur-einen-der-acht-transporter-einsetzen-a-1134022.html>.

capabilities.[42] Some successes have emerged as a result.[43] Unfortunately, these problems are deep-seated and are unlikely to be easily resolved with short-term budget increases, such as the case of the Eurofighters' spares support arrangements.

Future Efforts

With the uncertain international environment, rapid technological progress and commitment to enhancing security, the Bundeswehr has established a comprehensive roadmap to develop, procure, and employ its future force structure. However, it is also tempered by German interests in industrial development and politics.

The key guiding principle is a systems-of-systems concept. This vision of future operations is similar to that suggested by major western military powers. It foresees a battlefield dominated by a large number of highly interoperable capabilities that provide information gathered by reconnaissance capabilities.[44] To this end, the roadmap states:

> For the system-of-systems concept to be implemented properly, systems must be able to communicate in near real time and the required data bandwidth (for encrypted data too) must be provided and protected.[45]

This is a critical statement, highlighting the importance of information networks to the future operational capability of the Bundeswehr. The statement also highlights how the new opportunities provided by networks are also viewed as a potential vulnerability, with the "the robustness and IT security of weapon and employment systems are of fundamental importance."[46]

---

42 Holger Hansen, "Germany Should Boost Defense Spending: Defense Minister Tells Parliament." *Reuters,* June 8, 2018, <https://www.reuters.com/article/us-germany--idUSKCN1J41R6>.
43 Nicholas Fiorenza, "Germany Raises Equipment Readiness," *Janes Defense Weekly*, Mar. 12 (2019), <https://www.janes.com/article/87162/germany-raises-equipment-readiness>.
44 "Air Power Development Strategy 2016," 14.
45 Ibid., 13.
46 Ibid., 13.

One potentially problematic aspect of the German system-of-systems concept is its effort to incorporate flexibility into new weapons designs

> ... in which platforms are mainly equipped with modular components (particularly armaments, sensors, and mission equipment), ensures considerable flexibility and increases the potential of multinational cooperation, e.g. by providing an additional "opt-in" option for smaller partners. A system of systems thus makes it possible to meet changing requirements despite prolonged armaments cycles.[47]

Alone, this may not be a serious problem. It differs somewhat from other interpretations of system of systems, such as in the United States, which emphasizes a greater diversity of capabilities providing distributed functionality inside a networked environment.[48] However, the German approach seems to focus on fewer platform types that can be modified to meet a wide variety of mission sets, as the following passage notes:

> Wherever requirements permit, it is useful to combine several capabilities on a single platform (multirole design). Capabilities provided by aircraft should also overlap.[49]

Developing a modular platform able to meet a wide range of contingencies, like the above suggests, is likely to be a very costly approach to defense capability and susceptible to delays. This may have significant ramifications for the Luftwaffe of the future, as discussed below in regard to the Future Combat Air System (FCAS) program.

One area where the Bundeswehr remains wedded to maintaining a separate capability is in reconnaissance, which exists within the rubric of the Airborne Reconnaissance Network. Its importance is discussed in detail in the 2016 Air Power Development Strategy.

The central place that reconnaissance inhabits within the Luftwaffe's concept of operations is reflected by the developed nature of its

---

47 Ibid., 13.
48 "Operating in Contested Environments," Defense Advanced Research Projects Agency, Mar. 30, 2015, <https://www.darpa.mil/news-events/2015-03-30>.
49 "Air Power Development Strategy 2016," 13.

recapitalization plans. It lays out a fairly demanding series of requirements for future capabilities:

- They must operate at long range, deep in the theater, and at short and medium range.
- They must have a high availability rate and a long endurance and must be able to respond rapidly.
- They must operate in all weather at any time of the day or night.
- They must have many different systems, must be adaptable, and have new sensors.[50]

A key shift outlined is the greater use of unmanned aerial vehicles (UAVs) due to their persistence and disposability versus manned systems. The first major step was the acquisition of four MQ-4 Tritons, which are likely to be retrofitted with an Airbus signals intelligence package.[51] This would address what the Bundeswehr described as an urgent capabilities gap, as these situations provide valuable insights at the strategic, operational, and even tactical level of operations.[52] Other reconnaissance capabilities are expected to join the force in the coming years and augment the Luftwaffe's capabilities. On the tactical level, the German Army will replace its existing Kleinflugger Zielortung UAV with the Luna-NG, which possesses greater range and improved sensor and networking capabilities.[53] At the broader battlefield level, around 2025 the Luftwaffe will deploy the European Medium Altitude Long Endurance (EuroMALE) UAV, currently being co-developed by Airbus, Dassault, and Leonardo for the German, French, and Spanish governments. In the interim, the Luftwaffe may purchase a "bridging solution" in the form of an existing platform such as the MQ-9 Reaper.[54] Finally, there are a several naval systems for maritime reconnaissance.[55] In sum, these programs will seemingly provide persistent coverage

---

50 "Air Power Development Strategy 2016," 21.
51 Dave Majumdar, "Germany Is Buying 4 Northrop Grumman MQ-4C Triton UAVs," *National Interest*, Apr. 8, 2018, <https://nationalinterest.org/blog/the-buzz/germanys-buying-4-northrop-grumman-mq-4c-triton-uavs-25262 >.
52 "Air Power Development Strategy 2016," 22.
53 "LUNA NG Tactical Unmanned Aircraft System," EMT, No. 08 (2012) <https://www.emt-penzberg.de/uploads/media/LUNA-NG-en_02.pdf>.
54 "Luftwaffe Preparing for Future Challenges," *Joint Air Power Competence Centre Journal*, No. 20, Jul. 16, 2015, <https://www.japcc.org/interview-lieutenant-general-mullner/>.
55 "Air Power Development Strategy 2016," 23.

at all levels of operations, as outlined above. Given the effort to develop secure battlefield networks, the Bundeswehr has proposed a credible concept of operations for reconnaissance in the future.

FCAS, the centerpiece of German future efforts, is a joint effort with France and Spain to develop the next generation of air power capabilities. The program is intended to field a variety of systems, with potential roles including quick reaction alert, defensive and offensive counter air missions, air interdiction, close air support, joint time-sensitive targeting, anti-surface warfare, and ISR.[56] FCAS will incorporate existing platforms such as the Eurofighter for several decades, often in expanded roles.[57] Existing platforms will be supplanted by UAVs, which Airbus envisions will utilize technologies such as automation, mission systems, and swarming.[58] Here a major platform is likely to be a derivative of the Dassault nEUROn Unmanned Combat Aerial Vehicles (UCAV) program. This system will be required to operate in defended airspace as older aircraft become less survivable due to advances in adversaries' anti-access and area denial platforms, such as surface-to-air missiles. It may also include a large-scale drone or weapon carrier operating as an arsenal frontline aircraft, perhaps based on the A400M transport.[59]

The greatest attention of FCAS has been on the Franco-German manned (or optionally manned) fighter program, known as the Next Generation Weapon System (NGWS).[60] This program will produce the eventual replacement for the Eurofighter and the Dassault Rafale. According to the timeline, a flight demonstrator may take to the sky in 2025 with an introduction into service sometime between 2035 to 2040.[61] The focus on a single fighter aircraft with modular capability will be a difficult task and likely to be costly and complex. The expected timeline for the manned fighter

---

56 "Air Power Development Strategy 2016," 24.

57 Tony Garner, "The Road to FCAS," *Eurofighter.com*, Nov. 15, 2018, <https://world.eurofighter.com/articles/the-road-to-fcas>.

58 Garner, "The Road to FCAS."

59 Airbus, "Future Combat Air System – FCAS, Manned and Unmanned European System by Airbus & Dassault." *Youtube.com*, May 1, 2018, <https://www.youtube.com/watch?v=e1ekX5uwnyw>.

60 "Future Combat Air System – Anteil Next Generation Weapon System – NGWS," German Ministry of Defence, Apr. 28, 2018, 1, <https://www.bmvg.de/resource/blob/24288/6548107ac4b24da5611745b18252099a/20180427-factsheet-future-combat-air-system-anteil-next-generation-weapon-system-data.pdf>.

61 "Future Combat Air System – Anteil Next Generation Weapon System – NGWS," 2.

program is 15 to 20 years, which reflects that reality. The fighter will be a key part of the emerging information warfare architecture, providing connectivity among various platforms as well as being a decision-making center within a broad system of capabilities.[62]

While much of the public focus has been on the NGWS, the core of FCAS is arguably the network backbone, keeping with the system-of-systems concept. Units' connectivity and battlefield networking are seen as a critical part of the capability, as are developing artificial intelligence and data processing systems.[63] Furthermore, in keeping with its effort to promote interoperability, Germany is insisting that its network architecture be open and scalable in order to facilitate allies' ability to plug capabilities into the system.[64] The system would provide the foundation for all future capabilities, whether it be NGWS, the EuroMALE, updated Eurofighters, MEADS, or army helicopter units.

FCAS has a number of shortcomings, particularly looking at the entry and exit timelines for existing and future capabilities. In short, it will likely leave a decade-long gap in capabilities for the Luftwaffe. In 2016 the Bundeswehr decided that the Tornado airframes currently in service could not serve past 2025 and required a replacement. The German government has not stated a clear position on the nuclear dual-key capability, although this does not seem to be one of the driving factors behind the procurement.

Early on, the F-35 was seen as a frontrunner to replace the Tornado, particularly after comments by the Luftwaffe chief of staff in 2017 that suggested only a fifth-generation aircraft would be acceptable.[65] However, he was rebuked by the defense ministry, as his comments contradicted the

---

62 Sebastian Sprenger, "With Plans for Drone Sidekicks, Europe's Futuristic Jet Program Slowly Comes into Focus," *Defense News*, Nov. 18, 2018, <https://www.defensenews.com/air/2018/11/14/with-plans-for-drone-sidekicks-europes-futuristic-jet-program-slowly-comes-into-focus/>.

63 Ibid.

64 "Air Power Development Strategy 2016," 13; "The FCAS Programme: The Future of European Air Combat Capabilities?" *International Fighter* (2018) 4, <https://plsadaptive.s3.amazonaws.com/eco/files/event_content/j8Hp9wVXIpHDCPz0ou3XCQKksq3HPPx-1GNqgIg2P.pdf>.

65 Andrea Shalal and Sabine Siebold, "Tornado Replacement Must Be Fifth Generation: German Air Force Chief," *Reuters*, Nov. 8, 2017, <https://www.reuters.com/article/us-germany-fighter/tornado-replacement-must-be-fifth-generation-german-air-force-chief-idUSKBN1D81WR>.

ministry's view that any number of options may be able to fill the gap.[66] Moreover, significant pressure was exerted by Airbus and the French government not to procure the F-35, with the latter even threatening the withdrawal from a follow-on system.[67]

The government then issued a request for information for the Boeing F-15 and F/A-18E Super Hornet and the Lockheed Martin F-35, as well as from EADS. However, in February 2019, media outlets in Germany reported that only the Eurofighter and F/A-18E/F Super Hornet were short-listed as potential replacements for the Tornado fleet. Neither of the options are currently dual-key capable aircraft, which will require Berlin to develop this capability for either of the remaining options. Although the exclusion of the F-35 was a surprise to many observers, the decision fit with Germany's existing force structure, capability development objectives for the next 20 years, and industrial politics.

Much of the attention has been on the replacement of the dual-key capable capability, but considerations about the Tornado ECR variant likely played a significant part of the decision.[68] As noted above, these aircraft provide critical suppression of enemy air defenses (SEAD) and electronic warfare (EW) capabilities that enhance the survivability and effectiveness of the wider Eurofighter fleet, which is now expected to serve until 2040, if not 2060. There is a 10-to-15-year gap between the Tornado ECR's expected withdrawal in 2025 and the entry into service of a FCAS's manned fighter replacement program between 2035 and 2040. The prime candidate for such a platform is the EA-18G Growler, a variant of the Super Hornet that is similar in concept to the Tornado but with much updated capabilities. It can provide the existing Eurofighter fleet with area EW and SEAD that can improve their capability.

The ECR requirement also worked against the F-35, as its approach to survivability focuses on oblique EW capabilities. Such capabilities are likely

66 Andrea Shalal, "German Defense Ministry Gets Bids for Tornado Fighter Jet Replacement," *Reuters,* Apr. 28, 2018, <https://www.reuters.com/article/us-germany-airshow-fighter/german-defense-ministry-gets-bids-for-tornado-fighter-jet-replacement-idUSKBN1HV16O>.
67 Julie Carriat, "France, Germany Announce First Deals for Future Warplanes," *Reuters,* Feb.5, 2019, <https://uk.reuters.com/article/uk-germany-france-defense/france-and-germany-to-launch-first-contracts-on-future-combat-jets-idUKKCN1PU2KT>.
68 Douglas Barrie, "Berlin's Anti-Radiation-Weapon Dilemma," *IISS Military Balance Blog,* Oct. 10, 2018, <https://www.iiss.org/blogs/military-balance/2018/10/berlin-anti-radiation-weapon-dilemma>.

to be significantly more effective for it and other nearby F-35s, which do not have the same area jamming capabilities to protect other aircraft. Furthermore, the purchase of the F-35 was seen by Airbus and French partners as a threat to the broader FCAS system.[69] A similar dynamic was evident with the earlier French-U.K. partnership to develop a UAV program, which allegedly has been downgraded in importance, in part due to differing aims.[70] Whereas the British wanted primarily a reconnaissance platform, the French insisted on maintaining a focus on combat capability.[71] As one analyst has suggested, the primary reason for this divergence was the U.K.'s possession of the F-35, which would be primarily tasked with operating in contested airspace. A German purchase of F-35s would have the same effect as the United Kingdom's existing plans to utilize the capability.

Given the stated objectives of FCAS, the Eurofighter option would fit within the Bundeswehr's existing plans. The four partner nations are already working on defining an upgrade path for their existing aircraft, including greater connectivity and avionics improvements, which should increase the aircraft's effectiveness. The Luftwaffe would purchase a number of Eurofighters to replace the retiring Tornados, which would eventually be supported by a number of low observable UCAV currently under development by the French, discussed above.

The problem with this approach is that it does not immediately resolve the dual-key capability question, as neither the Eurofighter nor the F/A-18E/F is cleared to carry the B61 nuclear weapon. This will require significant engineering and development, if Berlin desires to continue deploying such capability. Given recent political debate over ending Germany's nuclear role, this may be an acceptable outcome for policy makers. Furthermore, the survivability of the Luftwaffe will likely drop significantly,

---

69 Hans Binnendijk and Jim Townsend, "German F-35 Decision Sacrifices NATO Capability for Franco-German Industrial Cooperation," *Defense News*, Feb. 8, 2019, <https://www.defensenews.com/opinion/commentary/2019/02/08/german-f-35-decision-sacrifices-nato-capability-for-franco-german-industrial-cooperation>.

70 Christina Mackenzie, "Conference Call Confusion: Is the Joint French, British Fighter Program 'Terminated'?" *Defense News*, Feb. 28, 2019, <https://www.defensenews.com/industry/2019/02/28/conference-call-confusion-is-the-joint-french-british-fighter-program-terminated/>.

71 Pierre Tran, "Britain Flip-Flops toward ISR Drone, but France Keeps Eye on Combat Capability." *Defense News*, May 11, 2018, <https://www.defensenews.com/unmanned/2018/05/11/britain-flip-flops-toward-isr-drone-but-france-keeps-eye-on-combat-capability/>.

as it will lose its primary EW and SEAD platform, the Tornado ECR. While the penetrating UCAV may allow the Luftwaffe to identify and prosecute targets in defended airspace, this will likely be a limited capability, with constrained situational awareness due to the lack of a nearby manned platform or effective EW capability to defend it.

Conclusion

As its history makes evident, the German public and political authorities are much more intent on focusing on national security, particularly with the rise of Russian intimidation in its eastern neighborhood. This has been reflected in its approach to the future of air warfare: the Luftwaffe at present has a credible roadmap within FCAS. Its application of a system of systems and its concept of operations are reasonably similar to other countries' efforts, such as the United States. By the mid-2040s, it envisions a broad, distributed network of systems, ranging from manned fighters such as the NGWS and legacy aircraft like the Eurofighter, drawing in data from a wide variety of manned and unmanned sensor platforms, while prosecuting targets from a large array of ground- and air-based capabilities. The latter may include surface-to-air or surface-to-surface systems, armed UAVs, and even perhaps arsenal aircraft operating safely behind the main area of conflict.

Unfortunately, there are a number of shortcomings to this approach. Immediately it is evident that the timetable will likely leave a significant capability gap between the retirement of the Tornado ECR around 2025 and the NGWS entry into service 15 years later. Some of the lost capability may potentially be made up by the introduction of a low observable UCAV, but this will not be a sufficient replacement for the existing Tornado fleet.

Beyond these tactical and operational problems, broader challenges remain. First, successfully executing this plan requires sustained funding and political support for the next two decades. Germany's recent history of politics, foreign relations with Russia, and lackluster defense spending raises serious questions about the country's ability to maintain long-term commitment.

Second, industrial considerations can play a critical factor. As is evident with the Eurofighter, A400M, and Tiger, inefficient industrial relationships have had a significant effect on the availability of key capabilities. Already there are indications that the NGWS is settling on the nature of inflexible workshare arrangement that has been a significant cause of spares delays. This may cripple the Luftwaffe's ability to maintain the critical edge. Also, if any of these systems arrives over budget, funding may be diverted from other critical capabilities within the system of systems, decreasing its overall combat capability. Finally, there is a possibility that industrial arrangements may contribute to the decoupling within NATO, even between European countries. The Tornado replacement program, and the emerging industrial arrangement of the FCAS program, has made that evident.

Learning from past history may well assist the Luftwaffe in exploiting the future of air warfare. Careful consideration of these challenges will ensure that it and Germany as a whole meet their objectives to provide the country and its alliance partners with a secure stable and prosperous regional environment in the future.

References

Airbus. "Future Combat Air System – FCAS, Manned and Unmanned European System by Airbus & Dassault." *Youtube.com,* May 1, 2018. <https://www.youtube.com/watch?v=e1ekX5uwnyw>.

"Air Power Development Strategy 2016." German Ministry of Defence, 2015. <https://www.epicos.com/sites/default/files/airborne_priorities.pdf>.

Albright, M. "The Right Balance Will Secure NATO's Future." *Financial Times,* Dec. 7, 1998.

Anrig, C. *The Quest for Relevant Air Power: Continental European Responses to the Air Power Challenges of the Post–Cold War Era.* Maxwell: Air University Press. 2011. <https://www.airuniversity.af.edu/Portals/10/AUPress/Books/B_0125_ANRIG_QUEST_RELEVANT_POWER.pdf>.

Banks, M. "From Cyber to Military Mobility: EU Members Endorse New Defense Objectives." *Defense News,* Nov. 20, 2018. <https://www.defensenews.com/global/europe/2018/11/20/from-cyber-to-military-mobility-eu-members-endorse-new-defense-objectives/>.

Barrie, D. "Berlin's Anti-Radiation-Weapon Dilemma." *IISS Military Balance Blog,* Oct. 10, 2018. <https://www.iiss.org/blogs/military-balance/2018/10/berlin-anti-radiation-weapon-dilemma>.

Barrie, D. "New Franco-German Combat-Aircraft Programme: A Reminder of UK's Uncertain Position." *IISS Military Balance Blog*, Jul. 14, 2017. <https://www.iiss.org/blogs/military-balance/2017/07/new-franco-german-combat-aircraft>.

Binnendijk, H. and J. Townsend. "German F-35 Decision Sacrifices NATO Capability for Franco-German Industrial Cooperation " *Defense News*, Feb. 8, 2019. <https://www.defensenews.com/opinion/commentary/2019/02/08/german-f-35-decision-sacrifices-nato-capability-for-franco-german-industrial-cooperation>.

Carriat, J. "France, Germany Announce First Deals for Future Warplanes." *Reuters,* Feb. 5, 2019. <https://uk.reuters.com/article/uk-germany-france-defense/france-and-germany-to-launch-first-contracts-on-future-combat-jets-idUK-KCN1PU2KT>.

Chivvis, C. and T. Rid. "The Roots of Germany's Russia Policy." *Survival*, Vol. 51, No. 2, 2009.

"Deployable Air Command and Control Centre (DACCC)." North Atlantic Treaty Organization.   <https://ac.nato.int/about-aircom/deployable-air-command-and-control-centre> (Accessed Sep. 12, 2019).

"EU's New Defense Union Must Not Undermine NATO, Warns Stoltenberg." *DW,* Feb. 13, 2018. <https://www.dw.com/en/eus-new-defense-union-must-not-undermine-nato-warns-stoltenberg/a-42570471>.

'The FCAS Programme: The Future of European Air Combat Capabilities?" *International Fighter,* 2018. <https://plsadaptive.s3.amazonaws.com/eco/files/event_content/j8Hp9wVXIpHDCPz0ou3XCQKksq3HPPx1GNqgIg2P.pdf>.

Fiorenza, N. "Germany Raises Equipment Readiness." *Janes Defense Weekly,* Mar. 12, 2019. <https://www.janes.com/article/87162/germany-raises-equipment-readiness>.

"Future Combat Air System – Anteil Next Generation Weapon System – NGWS." German Ministry of Defence, Apr. 28, 2018. <https://www.bmvg.de/resource/blob/24288/6548107ac4b24da5611745b18252099a/20180427-factsheet-future-combat-air-system-anteil-next-generation-weapon-system-data.pdf>.

Garner, T. "The Road to FCAS." *Eurofighter.com*, Nov. 15, 2018. <https://world.eurofighter.com/articles/the-road-to-fcas>.

Gebauer, M. "Luftwaffe hat nur vier kampfbereite 'Eurofighter'." *Der Spiegel,* May 2, 2018.   <http://www.spiegel.de/politik/deutschland/bundeswehr-luftwaffe-hat-nur-vier-kampfbereite-eurofighter-a-1205641.html>.

Glatz, R. and M. Zapfe. "NATO's Framework Nations Concept." CSS Analyses in Security Policy, No. 2018, Center for Security Studies (CSS), ETH Zurich, Dec. 2017. <http://www.css.ethz.ch/content/dam/ethz/special-interest/gess/cis/center-for-securities-studies/pdfs/CSSAnalyse218-EN.pdf>.

Gubisch, M. "France and Germany Sign First FCAS Development Contracts." *Flight Global*, Feb. 7, 2019. <https://www.flightglobal.com/news/articles/france-and-germany-sign-first-fcas-development-contr-455591/>.

Hansen, H. "Germany Should Boost Defense Spending: Defense Minister Tells Parliament." *Reuters*, June 8, 2018. <https://www.reuters.com/article/us-germany-security/germany-should-boost-defense-spending-defense-minister-tells-parliament-idUSKCN1J41R6>.

"Haushaltsausschuss des Deutschen Bundestages billigt Bundeswehrprojekte." *Bundesministerium der Verteidigung*, German Ministry of Defence, June 9, 2009. <https://www.bundeswehr.de/resource/resource/MzEzNTM4MmUzMzMyMmUzMTM1MzMyZTM2MzEzMDMwMzAzMDMwMzAzMDY3NmE2ODc1NjEzNTZkNzAyMDIwMjAyMDIw/MIP85>.

International Institute for Strategic Studies. *Strategic Survey 2015*. Routledge 2015.

Keller, P. "Germany in NATO: A Status Quo Ally." *Survival*, Vol. 54, No. 2, 2012.

Kington, T. "Taking Sides: Italian Defense Industry Rep Attacks Franco-German Fighter Deal." *Defense News*, Feb. 15, 2019. <https://www.defensenews.com/global/europe/2019/02/15/taking-sides-italian-defense-industry-rep-attacks-franco-german-fighter-deal/>.

"Konzeption der Bundeswehr." German Ministry of Defence, Jul. 20, 2018. <https://www.bmvg.de/resource/blob/26544/9ceddf6df2f48ca87aa0e3ce2826348d/20180731-konzeption-der-bundeswehr-data.pdf>.

Larsen, J. "Future Options for NATO Nuclear Policy." Atlantic Council, Aug. 30, 2011.   <https://www.files.ethz.ch/isn/132730/083011_ACUS_FutureNATONuclear.pdf>.

Lorell, M. and J. Pita. "A Review of Selected International Aircraft Spares Pooling Programs." RAND Corporation, 2016. <https://www.rand.org/content/dam/rand/pubs/research_reports/RR900/RR999/RAND_RR999.pdf>.

"Luftwaffe kann nur einen von acht A400M einsetzen." *Der Spiegel*, Feb. 11, 2017. <http://www.spiegel.de/politik/deutschland/airbus-a400m-luftwaffe-kann-nur-einen-der-acht-transporter-einsetzen-a-1134022.html>.

"Luftwaffe Preparing for Future Challenges." *Joint Air Power Competence Centre Journal*, No. 20, July 2015. <https://www.japcc.org/interview-lieutenant-general-mullner/>.

"LUNA NG Tactical Unmanned Aircraft System." EMT, No. 08, 2012. <https://www.emt-penzberg.de/uploads/media/LUNA-NG-en_02.pdf>.Mackenzie, C. "Conference Call Confusion: Is the Joint French, British Fighter Program 'Terminated'?" *Defense News*, Feb. 28, 2019. <https://www.defensenews.com/industry/2019/02/28/conference-call-confusion-is-the-joint-french-british-fighter-program-terminated/>.

Majumdar, D. "Germany Is Buying 4 Northrop Grumman MQ-4C Triton UAVs." *National Interest*, Apr. 8, 2018. <https://nationalinterest.org/blog/the-buzz/germanys-buying-4-northrop-grumman-mq-4c-triton-uavs-25262 >.

"Medium Extended Air Defense System." Lockheed Martin, 2016. <https://www.lockheedmartin.com/content/dam/lockheed-martin/mfc/pc/meads/mfc-meads.pdf>.

Metha, A. "Mattis Orders Fighter Jet Readiness to Jump to 80 Percent — in One Year." *Defense News*, Oct. 9, 2018. <https://www.defensenews.com/air/2018/10/09/mattis-orders-fighter-jet-readiness-to-jump-to-80-percent-in-one-year/>.

Miller, R. "Germany's Basic Law and the Use of Force." *Indiana Journal of Global Legal Studies*, Vol. 17, No. 2, 2010. <https://www.repository.law.indiana.edu/cgi/viewcontent.cgi?article=1419&context=ijgls>.

Neuneck, G. "European and German Perspectives," in T. Nichols, D. Stuart, and J.D. McCausland (eds.), *Tactical Nuclear Weapons in NATO.* Carlisle: Strategic Studies Institute, 2012.

Nicolai, L.M. "Aircraft Update Programmes: The Economical Alternative?" *Defense Technical Information Center Compilation Report,* June 28, 2010. <https://apps.dtic.mil/dtic/tr/fulltext/u2/p010301.pdf>.

"Operating in Contested Environments." *Defense Advanced Research Projects Agency*, Mar. 30, 2015. <https://www.darpa.mil/news-events/2015-03-30>.

"Permanent Structured Cooperation (PESCO) Updated List of PESCO Projects – Overview." Council of the European Union, Nov. 19, 2018. <https://www.consilium.europa.eu/media/37028/table-pesco-projects.pdf>.

Pieper, O. "ASSTA 3.1: Der Modernste Tornado der Luftwaffe." German Airforce Association, 2018. <http://www.idlw.de/assta-3-1-der-modernste-tornado-der-luftwaffe>.

Pocock, C. "France and Germany to Develop New Fighter Without UK?" *AIN Online*, Aug. 2, 2017. <https://www.ainonline.com/aviation-news/defense/2017-08-02/france-and-germany-develop-new-fighter-without-uk>.

Roberts, D. "Buying Old F-18 Jets Is a Bad Idea. It Just Makes Our Existing Problem Worse." *National Post*, Nov. 20, 2017. <https://nationalpost.com/opinion/dennis-roberts-buying-old-f-18-jets-is-a-bad-idea-it-just-makes-our-existing-problem-worse>.

Rutten, M. "From St. Malo to Nice: European Defence: Core Documents." Joint Declaration issued at the British-French Summit, St. Malo, Dec. 3–4, 1998, Paris: European Union Institute for Security Studies, 2001, 8–9.

Shalal, A. "German Defense Ministry Gets Bids for Tornado Fighter Jet Replacement." *Reuters*, Apr. 24, 2018. <https://www.reuters.com/article/us-germany-airshow-fighter/german-defense-ministry-gets-bids-for-tornado-fighter-jet-replacement-idUSKBN1HV16O>.

Shalal, A. and S. Siebold. "Tornado Replacement Must Be Fifth Generation: German Air Force Chief." *Reuters*, Nov. 8, 2017. <https://www.reuters.com/article/us-germany-fighter/tornado-replacement-must-be-fifth-generation-german-air-force-chief-idUSKBN1D81WR>.

Sprenger, S. "With Plans for Drone Sidekicks, Europe's Futuristic Jet Program Slowly Comes into Focus." *Defense News*, Nov. 14, 2018. <https://www.defensenews.com/air/2018/11/14/with-plans-for-drone-sidekicks-europes-futuristic-jet-program-slowly-comes-into-focus/>.

Stuart, D. "Introduction of European Policies and Opinions Relating to Tactical Nuclear Weapons," in T. Nichols, D. Stuart, and J.D. McCausland (eds.), *Tactical Nuclear Weapons in NATO*. Carlisle: Strategic Studies Institute, 2012.

"Team Tempest, From Rolls Royce Came thrust." *European Defense Review*, Jul. 17, 2018. <https://www.edrmagazine.eu/rolls-royce-the-tempest>.

Tran, P. "Britain Flip-Flops toward ISR Drone, but France Keeps Eye on Combat Capability." *Defense News*, Jul. 17, 2018. <https://www.defensenews.com/unmanned/2018/05/11/britain-flip-flops-toward-isr-drone-but-france-keeps-eye-on-combat-capability/>.

"Wales Summit Declaration." North Atlantic Treaty Organization, Sep, 5, 2014. <https://www.nato.int/cps/en/natohq/official_texts_112964.htm>.

"White Paper 2016: On German Security Policy and The Future of The Bundeswehr." German Federal Government (2016). <https://www.bundeswehr.de/resource/resource/MzEzNTM4MmUzMzMyMmUzMTM1MzMyZTM-2MzIzMDMwMzAzMDMwMzAzMDY5NzE3MzM1Njc2NDYyMzMyMDIw-MjAyMDIw/2016%20White%20Paper.pdf>.

Wiegold, T. "Aufgestockter Verteidigungshaushalt soll große Rüstungsprojekte ermöglichen." *Augen Geradeaus!* Nov. 5, 2018. <https://augengeradeaus.net/2018/11/aufgestockter-verteidigungshaushalt-soll-grosse-ruestungspro-jekte-ermoeglichen>.

8 INDIA
Missions, Technologies, and Constraints

Amit Gupta

Introduction

Asia is remerging as the arena for conflicts among the great powers and, consequently, one has to examine how regional air forces are configuring themselves to face future threats in the region. One important air force in this context will be the Indian Air Force (IAF), which has plans to be a significant actor in Asian security. The IAF would like to be at 42 squadrons (around 800 aircraft) by the mid-2020s and, eventually, perhaps go up to 45 squadrons. The rationale for such a buildup is obvious: a challenge on two fronts; a nuclear mission; a large maritime domain that has to be protected in conjunction with the efforts of the Navy; the ability to participate in multinational expeditionary efforts; and the development of intelligence, surveillance, and reconnaissance (ISR) capabilities to combat internal challenges such as insurgencies and terrorism.

These requirements, however, have to be balanced against the needs of the other services as well as the continued demand to limit defense expenditures in order to fund development projects in India. Further, there is the continued pressure exerted by the domestic arms industries to "Make in India," which means that a share of the weapons acquired will have to be domestically produced (and regardless of the outcome of the 2019 elections, the demand to build weapons domestically will remain). Given these facts, what will the IAF's future force structure look like and what missions will it realistically be able to carry out?

The Mission

The Indian government would like the IAF to carry out the following missions over the coming decades:

- Deter the Chinese and Pakistani militaries from carrying out strikes against India
- Carry out nuclear strikes as part of India's nuclear doctrine of assured destruction
- Provide an air defense capability
- Help maintain the sovereignty of India's maritime domain
- Provide ISR against both terrorists and insurgents and, if needed, be used for tactical strikes against terror groups
- Undertake humanitarian relief both domestically and internationally and participate in multinational security efforts.

Conventional Deterrence

The primary missions of the IAF are to deter conventional attacks by the Chinese and Pakistani air forces and to provide close air support to the Indian Army if it pursues its doctrine of Cold Start.[1] In the case of Pakistan, this would entail maintaining air superiority along the India-Pakistan border and, when needed, carrying out interdiction efforts against targets deeper in Pakistani territory. Against China, the objective would be even more modest in that it would seek to deter the People's Liberation Army Air Force (PLAAF) along the Himalayan border between the two countries.

The IAF is seeking to build a force structure that can deal with this two-front threat. Specifically, this means acquiring, in sufficient numbers, an effective frontline aircraft to carry out air combat missions. In the coming years the Sukhoi Su-30 MKI will be the backbone of the IAF, with the service now having close to the 272 planes that Hindustan Aeronautics Limited (HAL) was supposed to provide. Alongside the Sukhois will be a force of aging Jaguars, the venerable Mirage 2000, the new Rafale, and, hopefully, sufficient numbers of the Tejas light combat aircraft (additionally, the MiG-21 and MiG-27 will continue in service until the early 2020s).

---

1   For a discussion of Cold Start see Walter C. Ladwig, "A Cold Start for Hot Wars? The Indian Army's New Limited War Doctrine," *International Security*, Vol. 32, No. 3 (2007–2008).

The Jaguars, procured in the 1980s, have undergone a recent moderniza-tion, and the IAF is satisfied that the plane can carry out the roles of ground attack and interdiction in support of the Army. The Rafale, however, will be the cutting edge of India's strike force as its avionics, radar, and Meteor missile are going to be game changers in the skies.[2]

India has also reopened the proposal to induct 114 more combat aircraft into the Air Force. The usual aircraft companies have submitted proposals to sell aircraft—including, reportedly, a bid by Sukhoi to sell its Su-35 fighter.[3] Complicating the issue is the fact that India has pulled out of the Indo-Russian venture to build a fifth-generation fighter and even Mos-cow seems to be ramping up production of an earlier generation of Sukhoi since the fifth-generation fighter has run into delays.[4] Buying more planes in the open market, therefore, becomes a priority. India has also sought to augment the fleet by cannabalizing old Jaguars from Britain, France, and Oman and has recently decided to purchase mothballed MiG-29s from Rus-sia.[5] So, what should India buy and how quickly? This will be determined by the nature of the threat on India's borders coupled with the need to make the airborne nuclear deterrent more robust.

China

In the Chinese context, the IAF and Indian strategic analysts are thinking in terms of a conventional air war. In fact, Chinese military modernization is moving to the next generation of warfare as it is bringing stealth aircraft, hypersonic missiles, and anti-satellite weaponry into its arsenal. Current Indian planning is not significantly accounting for these shifts in Chinese

2 Ajit K. Dubey, "Meteor Missile Deal Set to Win Back India's Aerial Supremacy against Rivals," *Daily Mail* (UK), Dec. 18, 2017, <https://www.dailymail.co.uk/indiahome/indian-ews/article-5192397/Meteor-missile-deal-win-Indias-aerial-supremacy.html>.
3 "Saab Proposes to Make 96 Gripen Jets in India to Win Air Force Deal," *Economic Times*, New Delhi, Feb. 19, 2019, <https://economictimes.indiatimes.com/news/defence/saab-proposes-to-make-96-gripen-jets-in-india-to-win-air-force-deal/articleshow/68065031.cms>.
4 Rajat Pandit, "To Avoid Sukhoi 'mistake', India to Go for Russian 5th-Generation Fighter Only on Complete-Tech Transfer Pact," *Economic Times*, Jul. 11, 2018, <https://economictimes.indiatimes.com/news/defence/after-sukhoi-mistake-india-to-go-for-russian-5th-generation-fighter-only-on-full-tech-transfer-pact/articleshow/57551801.cms>.
5 Rajat Pandit, "IAF 'Harvesting Organs' of Globally Retired Jets," *Times of India*, Jul. 23, 2018, <https://timesofindia.indiatimes.com/india/iaf-harvesting-organs-of-globally-retired-jets/articleshow/65096225.cms>.

acquisitions and, instead, depends on a twentieth-century force structure to deal with the Chinese conventional threat along the border. As Air Marshall Arvind Subramaniam has written about the Chinese threat:

> The IAF's combat edge in the air has narrowed down significantly with one Rand report pegging the number of fourth-generation aircraft with the PLAAF as close to 700 ... Of immediate worry for the IAF is the ability of the latest variant of the H-6 bomber, the H-6K, to carry six DH-10 cruise missiles, which have a range of around 1,500 km. These bombers have a combat radius of 1,800 km, which means that they do not need to get airborne from airfields in Tibet and can launch their cruise missiles on critical Indian military targets from well outside any kind of air defence umbrella that the IAF can put in place over the next decade. Though the Chinese J-20 and JFv31 Stealth fighters are some time away from operationalisation, the IAF's acquisition of the Rafale and induction of the indigenously-built Tejas would take time to translate into operational capability.[6]

He continues,

> Coupled with the high probability of widespread and debilitating surface-to-surface missile strikes on IAF combat capability that could severely hamper sortie generation rates, the PLAAF's distinctly superior network paralysing capabilities could seriously impact the IAF's command and control systems. If one moves sequentially, the next comparison leads towards the PLAAF's significantly superior air defence capability led by a multi-layered and highly potent missile umbrella that is rated very highly even by US standards. Comprising the newly acquired S-400, the older S-300 and its improved and reverse engineered version named the HQ-9—that the IAF is particularly concerned about—and the shorter-range HQ-12, the IAF's offensive air operations over the Tibetan plateau would be hotly contested even over the Tactical Battle Area, which is envisaged

---

6   Arjun Subramaniam, "Closing the Gap: A Doctrinal & Capability Appraisal of the IAF & the PLAAF," Observation Research Foundation, May 2, 2018. <https://www.orfonline.org/expert-speak/closing-gap-doctrinal-capability-appraisal-iaf-plaaf/>.

to be closer to the LAC [line of actual control] and hundreds of kms away from PLAAF missile locations.[7]

Subramaniam's analysis makes the important point that China is seeking to create a twenty-first-century air warfare capability and moving quite rapidly toward achieving it. In fact, one of the major implications of China's 2025 technology development program is that it will significantly enhance the military capabilities of the PLAAF.[8] India, on the other hand, is lagging behind because of its lengthy procurement and domestic production processes. Although there is an understanding of how the modern battlefield is changing, India's domestically designed missiles, network-centric warfare systems, and satellite capabilities are taking far too long to be developed and incorporated into service. What will work in favor of the IAF is the fact that the People's Liberation Army (PLA) and the PLAAF primarily focus on the eastern maritime regions of Taiwan, the South China Sea, and the East China Sea, where they face American and Japanese navies that are both technologically advanced and formidable opponents.

Although time may be on India's side, it will be of no avail unless there is a corresponding shift in the thinking on the modern-day air battle as well as the systems that go with it. India has recently carried out an anti-satellite test and is in the process of developing an advanced version of the BrahMos that is supposed to be a hypersonic missile.[9] This will require, however, that India allocate adequate resources and push for the rapid development of such systems. Yet, if India persists in its lackadaisical approach to the indigenous development of weapons systems, it could be decades before it gets the necessary weaponry. It would not be able to sustain a conflict against a better armed and more modern Chinese military.

Pakistan

In the context of India's traditional security challenge with Pakistan, the IAF has the ability to deter the Pakistan Air Force (PAF), but a more coercive

---

7   Ibid.
8   James McBride, "Is 'Made in China 2025' a Threat to Global Trade?", Council on Foreign Relations Backgrounder, Aug. 2, 2018, <https://www.cfr.org/backgrounder/made-china-2025-threat-global-trade>.
9   For a discussion of India's development of hypersonic weapons see Amit Gupta, "The Missile Age: Is India Ready?" *Geopolitics Magazine*, June 2018.

role may require acquiring new systems to implement doctrine, engage in long-range strikes, and improve readiness. Indian analysts tend to be confident about the IAF's ability to maintain air superiority over the PAF. As one analyst has written, "The PAF today is no match for the IAF, as it is beset by both technological and numerical inferiority. The PAF today has very limited stand-off Precision Guided Munitions (PGM) delivery capability. Strategically and tactically, PAF is biased primarily toward Air Defence and limited battlefield air strike (BAS) support to the Pakistan Army. Thus, today's PAF does not pose a major challenge to the IAF."[10] The IAF views itself as fully capable of achieving air superiority, something that has been enhanced with the acquisition of the S-400 missile defense system. Thus, the IAF could both maintain air superiority and use it to provide the kind of close air support that the Indian Army would require if it were to carry out its putative doctrine of Cold Start (although the IAF itself is not keen to play the provider of close air support in Cold Start since it would work against its primary mission of strategic bombing).[11] The February 2019 confrontation between the IAF and the PAF would suggest that the IAF will have to improve its capabilities in some areas if it is to maintain air superiority in the near to medium term.

In response to the killing of 40 paramilitary officers by a Jaish-e-Mohammed (JeM) group suicide bomber on February 12, 2019, on February 26 the IAF crossed the line of control and bombed a Jaish madrassa near Balakot in Pakistan. The Indian response was carefully crafted because the government wanted to make clear that it was not attacking the Pakistani government, the Pakistani military, or Pakistani civilians. Instead, it was attacking terrorists. The success of the strike is disputed because the IAF says it used penetration bombs to kill the JeM operatives at the madrassa (hence there was little damage to the buildings) while the Pakistani government claims that the IAF bombed a forest.[12]

---

10 Vivek Kapur, "Challenges for the Indian Air Force 2032," *Journal of Defence Studies*, Vol. 7, No. 1 (2013), 83. <https://idsa.in/system/files/jds_7_1_VivekKapur.pdf>.

11 See Jaganath Sankaran, "The Enduring Power of Bad Ideas: 'Cold Start' and Battlefield Nuclear Weapons in South Asia," *Arms Control Today*, Nov. 2014. < https://www.armscontrol.org/ACT/2014_11/Features/Cold-Start-and-Battlefield-Nuclear-Weapons-in-South-Asia>.

12 Abu Arqam Naqash and Sanjeev Miglani, "India Launches Air Strike in Pakistan; Islamabad Denies Militant Camp Hit," *Reuters*, Feb. 25, 2019, <https://www.reuters.com/article/us-india-kashmir-pakistan/india-launches-air-strike-in-pakistan-islamabad-denies-militant-camp-hit-idUSKCN1QF07B>.

On February 27, the PAF launched an attack with 24 aircraft that was checked by Indian aircraft. In the ensuing dogfight, however, the IAF lost a MiG-21 and New Delhi claimed that it had shot down a Pakistani F-16 but the plane fell on the Pakistani side of the border so the kill could not be confirmed.[13] With such conflicting claims, it is difficult to judge the effectiveness of either the IAF or the PAF, but some lessons can be learned from the air battle. First, the IAF was able to go about 60 kilometers into Pakistan and carry out a strike, thus giving some credence to its claim that it could penetrate Pakistani airspace to inflict damage. Second, however, it seems that the IAF was surprised by how quickly the PAF responded to the Indian intrusion and that India launched 12 aircraft to challenge a much larger Pakistani formation—the Indian formation included the venerable MiG-21 because a squadron of the aircraft was based closest to the border and it had the quickest reaction time.[14] The outcome of the brief conflict would suggest, therefore, that India may have to boost its readiness at the border and keep more planes in forward positions to prepare for a surprise attack.

Nuclear Mission

India's nuclear doctrine stresses no first use and is particularly concerned about preventing the unauthorized and accidental launch of nuclear weapons. It also reinforces civilian control over the military in decision-making. The nuclear doctrine is, therefore, as Shyam Saran, the former head of the Indian National Security Advisory Board, stated, "The very nature of nuclear deterrence as practiced by a civilian democracy dictates that decisions relating to the nature and scope of the arsenal, its deployment and use, be anchored in the larger architecture of democratic governance. It is the civilian political leadership that must make judgments about domestic political, social and economic priorities as well as the imperatives imposed by a changing regional and global geopolitical environment."[15]

---

13 "Pakistan Denies Indian Claims It Used US F-16 Jets to Down Warplane," *Guardian*, Mar. 3, 2019, <https://www.theguardian.com/world/2019/mar/03/pakistan-denies-indian-claims-it-used-us-f-16-jets-to-down-warplane>.
14 Interview with a senior Indian military officer, New Delhi, Mar. 8, 2019.
15 Shyam Saran, "Is India's Nuclear Deterrent Credible?," public lecture, India Habitat Centre, New Delhi, Apr. 24, 2013, <https://www.armscontrolwonk.com/files/2013/05/Final-Is-Indias-Nuclear-Deterrent-Credible-rev1-2-1-3.pdf>.

The desire to retain civilian control and prevent accidental or unauthorized use has essentially led India to keep its nuclear arsenal demated as a cheap and easy way to maintain the security of the weapons. But the demating of the weapons, coupled with a doctrine of massive retaliation, has constrained what India can do to deter a lower range of threats or even a rogue strike or an unauthorized launch from Pakistan. Indian strategic analysts recognize this limitation and, therefore, have called for rewriting the doctrine to provide the government with a flexible response based on the type of an attack launched on India.[16] There is also the question of just how effective are India's land-based and sea-based missile deterrents.

Questions have been raised about the effectiveness of the land-based missile force. Those concerns range from the proven quality of these systems to the ability to deploy them quickly when an attack has to be countered. As for the sea-based deterrent, the conservative estimate is that it will take another 10 to 20 years to make India's proposed nuclear submarine force operational and armed with nuclear weapons.[17] The question is, however, will the Indian government be willing to allow nuclear weapons to be deployed aboard nuclear vessels since it would make them, as W.P.S. Sidhu argues, autonomous of civilian control.[18] In the light of these technological and doctrinal considerations, the role of the IAF becomes particularly central to the nuclear mission.

Because aircraft can be recalled it lessens the danger of an accidental or unauthorized launch. The IAF has been particularly careful to only train a small number of units to carry out this mission. Further, aircraft can be shifted to other targets if the battlefield scenario changes—something that cannot be done with missiles. If the Indian nuclear doctrine is changed to do limited strikes, then aircraft are the easiest way to attack tactical or theater targets. If the doctrine remains one of assured and massive retaliation, then aircraft again play a central role in the mission.

---

16 Kumar Sundaram and M.V. Ramana, "India and the Policy of No First Use of Nuclear Weapons," *Journal for Peace and Nuclear Disarmament*, DOI: 10.1080/25751654.2018.1438737.

17 Vinay Kaura, "India Will Continue Boosting Its Nuclear and Missile Capabilities to Counter China—and Beijing Will Keep Responding," *South China Morning Post*, Feb. 8, 2019. <https://www.scmp.com/comment/insight-opinion/asia/article/2185271/india-will-continue-boosting-its-nuclear-and-missile>.

18 W.P.S. Sidhu, "Updating India's Nuclear Doctrine," Brookings Institution, May 1, 2014, <https://www.brookings.edu/opinions/updating-indias-nuclear-doctrine>.

Given India's limited capabilities regarding command, control, communications, computers, and intelligence, any nuclear exchange with Pakistan is likely to be a single salvo with the aim to get as many targets as possible. If so, then the Indian military will have to attack countervalue targets (cities) as opposed to counterforce targets (military assets) because it is not clear if the Indian military would have the intelligence and surveillance capabilities to conduct multiple layers of attacks. Thus, it is likely that India would concentrate on attacking Pakistan's three major cities—Karachi, Lahore, and Islamabad—because this would create unacceptable political, economic, administrative, and even military devastation. In that case, the IAF would be able to destroy two of these cities because they are easy to attack. Lahore is 20 kilometers from the border, and Karachi is a coastal city.

Until recently, manned nuclear strikes by the IAF were seen as one-way missions because it was unlikely that the aircraft, which had not been hardened, would survive a nuclear blast. With the induction of the Rafale and the Su-30, such a suicide mission should no longer be the case as the pilots are likely to be able to return to Indian territory. If indeed the BrahMos is nuclear capable, and if India has miniaturized its warheads to fit it, the missile also gives the IAF a stand-off capability because it has a range between 400 and 700 kilometers.[19]

Against China, the nuclear mission at best can be one of using tactical weapons in the Himalayan theater because India does not have the aircraft to reach the Chinese heartland. One option is to use tactical nuclear weapons to block the passes in the Himalayas and thus prevent Chinese advances into Indian territory. Yet it is not in India's interests to escalate a conventional skirmish to the nuclear level because China not only has a numerical advantage in nuclear weapons, but it also has more credible delivery systems.

Internal Security and Terrorism

India has generally not used air power in its attempts to counter insurgencies even though in 1960 a DC-3 transport was shot down by Naga rebels

---

19 Akshaya Kumar Sahoo, "BrahMos Missile Successfully Test Fired from Sukhoi Fighter Jet for First Time," *Asian Age*, Nov. 23, 2017. <https://www.drdo.gov.in/drdo/pub/npc/2017/november/din-23november2017.pdf>.

and in 1966, unconfirmed reports claim, combat aircraft were used to strafe Mizo rebels. However, the Indian government has preferred not to use air power against insurgents for several reasons: insurgents are viewed, at least nominally, as being citizens of India and, therefore, are not to be bombed; it is also feared that the use of air power leads to collateral damage and that the use of air power would lead insurgents to escalate the conflict by obtaining anti-aircraft weapons.[20] Faced with a Maoist insurgency, the Indian government has now approved the use of unmanned aerial vehicles (UAVs) to carry out surveillance operations against the insurgents.[21] Although surveillance UAVs are now used, it is less likely that the Indian government will permit the use of air power to wage war on insurgents within the country's borders. It is also likely that the mission will be given to the Air Force rather than a paramilitary unit, which are not adequately trained to handle UAV systems. It would also cost less to give the mission to the Air Force.

Air power will be increasingly required to respond to terrorist attacks from across the Pakistani border. The Mumbai terror attack of 2008 could have been dealt with more effectively if India had a UAV capability that would have maintained maritime security over its maritime spaces and ports. The terrorists were able to enter Mumbai harbor undetected in 2008 and dock close to one of their primary targets, the Taj Mahal Palace hotel. India has yet to build the UAV capability to provide domestic security, but the Indian government has been seeking UAVs from the United States, including unmanned combat aerial vehicles (UCAVs), to deal with this threat. The Indian government has signed an agreement with the United States to purchase 22 Guardian drones. However, recent reports suggest that it will be cut to 10 because of costs (other reports suggest that the IAF would like to buy up to 100 UAVs and UCAVs, because they are seen as not

---

20 See, for example, Rajesh Rajagopalan, "Force and Compromise: India's Counter-Insurgency Grand Strategy," *Journal of South Asian Studies*, Vol. 30, No. 1 (Apr. 2007), 87.
21 "Strategic UAV Base Shifted to Maoist Hotspot in Bastar," *Economic Times*, Oct. 28, 2017. <https://economictimes.indiatimes.com/news/politics-and-nation/strategic-uav-base-shifted-to-maoist-hotspot-in-bastar/articleshow/60992553.cms>.

only strengthening the deterrent but also growing the strategic partnership with the United States).[22]

Terrorism and Pakistan: The Role of Air Power

India has long accused Pakistan of both sending terrorists across the border and allowing its territory to be used by such groups to carry out attacks on neighboring countries. India has faced the problem of Pakistan-supported insurgent groups in Kashmir, but the attack by a group of 10 terrorists on Mumbai in 2008 was more worrisome. India's response was tempered because of what the American strategist Paul Nitze called the advantage provided by nuclear weapons. Nitze argued that once a country acquired nuclear weapons, it could pursue conflict at lower levels secure in the knowledge that its nuclear weapons would deter a broader conflict.[23]

After the 2001 JeM attack on the Indian parliament, the IAF drew up plans for cross-border strikes, but the Indian political leadership, worried about the potential for escalation, rejected such an option. This changed following the JeM terror attack February 12, 2019, that killed 40 Indian police officers in Pulwama. In response, Prime Minister Narendra Modi decided to send the IAF across the border and Pakistan retaliated, as mentioned earlier, with its own conventional strike. Indian commentators say that now that the taboo against cross-border strikes has been removed, the IAF will be used again because it allows for a quick response against terror attacks.

As a result of the February 2019 air strikes and the Pakistani response, both countries exercised caution: having made their political and military points, they reduced tensions along the border. Western observers have long worried about the escalatory dangers of conflict in South Asia, but both India and Pakistan were careful about how they fought the skirmish. They had limited objectives and both countries were quick to tell the other to use restraint and not cause an escalation that would lead to a nuclear war.

---

22 Huma Siddiqui, "Modi Government to Fast-Track Acquisition of Lethal Predator Drones for Indian Navy and Indian Army," *Financial Express* (New Delhi), Jan. 1, 2019. <https://www.financialexpress.com/defence/modi-government-to-fast-track-acquisition-of-lethal-predator-drones-for-indian-navy-and-indian-army/1430508>.
23 Paul Nitze, "Atoms, Strategy, and Policy," *Foreign Affairs*, Vol. 34, No. 2 (Jan. 1956), 195.

Multinational Missions

In the near to medium term, one of the IAF's most successful missions, and the one that will increase India's soft power, is its ability to contribute to multinational humanitarian missions. India has a long history of partici-pating in humanitarian missions, as seen by its participation in tsunami relief efforts in 2004 and, more recently, being the first responder in the 2015 Nepal earthquake. The country now has the airlift capability to engage in multinational and extra-regional relief efforts since it bought 10 C-17 Globemaster transports from the United States (the IAF, by contract, has the option to purchase another six but the closing of the production line has meant that India was allowed by the Trump administration to purchase the last remaining C-17).

As India becomes a bigger player in international affairs it will increas-ingly have to act to protect the global commons and carry out humanitarian assistance. The support role, however, is not well understood in India, where there is a common assumption that there are too many transports so it is now time to focus on fighters. But countries such as Australia and Qatar have recognized the value of such heavy lifts for making their pres-ence felt in coalition efforts. The Australian government initially sanctioned the purchase of four C-17s but, having recognized the value of the plane, bought four more. Its rationale was that if they provided logistical support in a crisis, they made their presence felt, reaped political benefits, but did not suffer casualties, which would have been badly received by Australian public opinion.

India, likewise, is cautious about committing troops abroad, espe-cially in combat situations, but would be less concerned about playing a support role by using its heavy-lift aircraft to transport supplies. The C-17 production line was closed by Boeing, although several countries would buy the aircraft if it were still available. It might make sense for India, therefore, to ask the United States to transfer that production line to HAL. HAL would retain scientific employment, would build up experience in manufacturing large, wide-bodied aircraft, and could sell these aircraft far less controversially around the world with American assistance (as opposed to combat aircraft like the F-16, now called the F-21 for sale to India, since that would mean the plane went to countries such as Pakistan).

The Weapons Acquisition Process and the IAF

India's weapons acquisition process is lengthy, cumbersome, and plagued with the constant charges of lack of transparency, kickbacks, corruption (witness the current Rafale controversy). There is a common belief that the longer one holds out on a deal, the more concessions are made by the seller. Instead, there are lengthy delays in procuring the weapons systems, as well as cost overruns. Thus, it took India 20 years to complete the negotiations for the Hawk trainer, leading to the joke at British Aerospace that it took one-fifth of the first 100 years of flight to sell the aircraft to India. Similarly, although the demand for the medium multi-role combat aircraft first emerged around 2005, the deal was only brought to fruition in 2015 with the purchase of the Rafale. Even then, the ambitious plan to buy 26 planes off the shelf and then manufacture over 100 at HAL was scrapped. Instead, the Indians agreed to a more modest and direct purchase of 36 aircraft. The reasons to stop the domestic production of the Rafale symbolize the inefficiencies in the Indian arms industry.

Media reports have revealed that when Dassault began negotiating with HAL for the domestic production of the aircraft, HAL quoted man-hours for production that were three times greater than those at Dassault. Moreover, the costs of the plane escalated as HAL admitted that it lacked the machine tools, technology, and trained manpower to build the plane. The IAF further complicated the domestic production process by demanding that Dassault guarantee the quality of the aircraft produced by HAL, something that the French company refused to do because it did not have control over the production process. The IAF's desire to have externally imposed quality control stemmed from its problems with HAL's building of the Sukhoi-30 MKI—particularly with the plane's engine. This was despite the fact that HAL was actually assembling the aircraft rather than producing them from scratch. As one analysis states, "HAL assembles the Sukhoi Su-30 MKI fighter jet from knocked-down kits imported from Russia. As there was no technology-transfer agreement in the Sukhoi purchase deal, HAL can't make the jets on its own."[24]

---

24 See, "Forget Rafale, HAL Cannot Even Produce and Deliver Su-30 MKI's on Time to the IAF," *Defence News*, Oct. 2, 2018, <http://www.defencenews.in/article/Forget-Rafale,-HAL-cannot-even-produce-and-deliver-Su-30-MKIs-on-time-to-the-IAF-580869>.

The problem with indigenously designing, developing, and producing aircraft is that India has, at best, had a mixed record in doing so. In the 1950s it sought to develop a piston engine trainer, a sub-sonic trainer (Kiran), and a supersonic combat aircraft (Marut). All three planes were produced but with a range of problems associated with both the process and the eventual product. The process was marked by lengthy delays in design and production, and the actual products did not meet the expectations or requirements of the IAF. The Marut, for example, never reached supersonic speeds because Indian engineers were unable to produce the needed powerplant and the Indian government, being in this instance pennywise and pound foolish, refused to import an engine. More recently, the Tejas, the Sitara jet trainer, and the HTT-40 basic trainer have all been marked by developmental delays and incorporated a large number of imported components, making these aircraft expensive and vulnerable to embargoes by foreign suppliers. Thus, the development of the Tejas was hindered by an American embargo of technical assistance after the Indian nuclear tests of 1998. While domestic production has not met the IAF's combat requirements, its threat environment has evolved into a two-front confrontation against China and Pakistan.

Budgetary Constraints

The desired modernization plan of the IAF would allow it to carry out conventional deterrence, the nuclear mission, and some degree of interoperability (which is discussed below). But the ability to achieve such capabilities is determined by the continued tension in India between the competing demands for guns or butter. As India is a democracy, the main emphasis of its governmental efforts is to build human capacity in the country by providing subsidized food, subsidized health care, and subsidized primary, secondary, and tertiary education. All this has been achieved, and admittedly several other countries in Asia have done this better, by keeping a check on excessive military expenditure.

When India had a planning commission, it was willing to commit up to 3 percent of gross domestic product (GDP) on defense. That was at a time when the Indian economy was socialist and had low growth rates. Buying weaponry then was based on the availability of hard currency or the willingness of suppliers to provide certain types of weaponry.[25] Thus in the 1960s, following the debacle of the 1962 war, India invested heavily in defense spending, but when the Bihar famine occurred toward the end of the decade, the government cut defense expenditure to stave off a serious crisis.

More recently, for at least the past decade and adhered to by different political parties, the Indian government has spent between 1.6 and 1.8 percent of GDP on defense.[26] This was possible partly because the Indian economy had grown so rapidly that the country spent more on defense while using a smaller portion of GDP. But another reason is Indian demographic pressures and the resulting need for a commitment to economic development. With the current Indian median age around 27, between 700,000 to 1 million youth join the work force every month. Providing employment for these people is the priority for any successful Indian government, and given that the Indian state is still the largest employer in the country this determines where Indian state investment should go.

The IAF, therefore, has been restricted in its attempts to modernize by the pressures exerted by developmental forces in the country. Thus, the number of Rafales was cut from 126 to 36 aircraft and the number of Guardian drones may be halved because of the lack of finances. Indian strategic analysts feel that the IAF's demand for 114 combat aircraft may also stall because of budgetary pressures.[27]

A Long-Term Production and Procurement Plan for the IAF

A long-term production and procurement plan for the IAF will primarily have to address the problem of a shortfall in the number of sanctioned squadrons and the growing obsolescence of the fleet. In contrast, both Pakistan and, especially, China are modernizing their air fleets, and this

---

25 Amit Gupta, *Building an Arsenal: The Evolution of Regional Studies Force Structures* (Westport, CT: Praeger, 1997).

26 Rajat Pandit, "Budget 2018: Govt Hikes Defence Budget by 7.81%, But It's Just 1.58% of GDP & Lowest Since 1962," *Times of India*, Feb. 1, 2018, 28–29. <https://timesofindia.indiatimes.com/india/budget-2018-govt-hikes-defence-budget-by-7-81/articleshow/62740525.cms>.

27 Interview with Abhijit Iyer-Mitra, Senior Fellow, Institute for Peace and Conflict Studies, Sep. 15, 2018.

threatens India's capability to deter a one-front or two-front war. The budgetary pressures will also remain in place as India's demography will continue to shape the economic policies of the country. What needs to be recognized, therefore, is that the IAF will have four long-term suppliers— Russia, France, United States, and HAL—and successful procurement and force development will require working with all four partners.

Russia

Each of the four partnerships brings certain advantages to India but there are also constraints that need to be understood by India's strategic planners. Despite talk of diversifying India's arms supplies, Russia remains the primary supplier of weaponry to the IAF. It continues to fly the aging MiG-21, as well as the MiG-27 and MiG-29, and approximately 270 Sukhoi-30 MKIs. In a modern-day combat environment technology matters, but so do numbers, and the Russian link has always allowed India to build up the numerical strength of the force. Russia has also been willing to transfer systems that other nations have been unwilling to provide and to resupply India in the event of a war. Thus, India got nuclear submarines and reactor technology from Russia, and Moscow has never cut off the delivery of spare parts to New Delhi in a war—contrast this with the United States cutting off arms supplies to both India and Pakistan in the 1965 and 1971 wars. More recently, Russia has partnered with India to build the BrahMos missile, thereby putting India in the second tier of nations that are developing hypersonic missiles.[28]

Although the arms sales relationship with Russia gives India several advantages, it is not the same as the one with the erstwhile Soviet Union. Russia has been unable to modernize its economy and, instead, depends on the same exports it did in the Soviet days—and arms sales remain a key part of those exports. Further, unlike the Soviet Union, which saw the utility of having India as one its main friends in the international system, Vladimir Putin's Russia sees India through the lens of commercial transactions. Modern Russia considers itself part of the West and feels that links with the nations of the nonwestern world are not conducive to improving

---

28 Manu Pubby, "New Brahmos Missile Will Take Down Enemy 'Force Multipliers'," *Economic Times*, Feb. 21, 2019, <https://economictimes.indiatimes.com/news/defence/new-brahmos-missile-will-take-down-enemy-force-multipliers/articleshow/68094383.cms>.

its status in the international system. When the concept of the BRICS (Brazil, Russia, India, China, and South Africa) was introduced, Russians went to international conferences saying that they should not be compared with the other nations in the grouping since theirs was a "first world" nation. Given the rise of China and India, this was an outdated western ethnocentric approach.

There is also the continuing problem of spares from Russia. During the Soviet days, India had a hard time acquiring spare parts. Russia has still not improved on that part of arms sales, as acquiring spares remains a major problem for any recipient of Russian weaponry. One can buy Russian, but the problems of maintenance and upkeep become a major headache. So, what then should India buy from Russia?

The simple thing would be to sign a deal to buy another four to six squadrons of the Su-30 MKI because HAL already has a mature production line in place and the IAF has nearly two decades of experience of flying and maintaining the aircraft. As things stand, the IAF is not interested in acquiring more Su-30 MKIs because the plane would be built in India, which would lead to further clashes with HAL over the quality of domestically manufactured engines for the aircraft. Given the multiple problems associated with quality control at HAL, this option—the cheapest and most effective—is not available to the Indian government although HAL is pushing to build 18 more Sukhoi to meet the shortfall in IAF squadrons.[29] India could potentially buy another plane from Russia and three planes could be acquired by the IAF—the MiG-35, the Su-34, and the Su-57.

The MiG-35 is a development of the MiG-29, which has been in service with the IAF since the 1980s. The plane has been modernized to include an active electronic scanned array radar, which is one of the priorities of the IAF. It is claimed to be able to fire both Russian and western weaponry (the Swedish Gripen is also able to do this and given India's collection of both Russian and western munitions this would be an asset). However, the plane is viewed as being an upgrade of the MiG-29 with no significant improvement in quality. The plane was, in fact, at one time offered as an alternative to the Tejas and, such is the bizarre nature of Indian weapons

---

29 Ajai Shukla, "India Wants 18 More Sukhoi 30 MKIs," *Rediff News*, Feb. 8, 2019, <https://www.rediff.com/news/report/india-wants-18-more-sukhoi-30mki-fighters/20190208.htm>.

procurement, is being considered yet again as part of the 114 aircraft purchase for which India issued a request for information in 2018.

The Su-34 Fullback is a dedicated attack version of the Su-27 family. It has forged a reputation in the Syrian conflict as an effective tactical bomber.[30] The problem is that the IAF has no interest in the aircraft and does not see it as fulfilling the current missions of the Air Force, especially since the Jaguars have been modernized and will continue in this role for some time. This leaves the Su-57.

Under normal circumstances, the Su-57 would be an ideal follow-on aircraft for the IAF because it has the potential to be a fifth-generation fighter and will be significantly less expensive than the Lockheed F-35. But that is where the advantages of the plane end. The IAF and Indian design engineers feel cheated by Russia because it did not take Indian design requirements into consideration and locked the design of the plane as a single seater rather than the twin-seater that the IAF wanted. Further, disputes broke out because Russia refused to transfer the technologies India wanted, leading to New Delhi pulling out of the program. Unless Russia is willing to accede to Indian demands, and there is no evidence to suggest it will, the Su-57 may be another example of failed Indo-Russian collaboration.[31] How then can the air power relationship with Russia be continued and expanded? The answer lies in the BrahMos and S-400 antimissile systems.

The BrahMos has given India a high-speed cruise missile. The next generation, the BrahMos Mark II, may be a genuine hypersonic missile. In an era when missiles are becoming an integral part of air power, this will provide a technological advantage to India as it seeks to deter Pakistan and China. Similarly, more purchases of the S-400 would not only continue an investment in Russia but would also give the IAF an easy way to make up for its lack of fighter squadrons by providing a substantially improved ground-based air defense capability. The plan, therefore, must be to show imagination and move away from the traditional emphasis on aircraft to a push for futuristic missiles.

---

30 Tom Cooper, "Here's the Key to Understanding the Russian Air Force's Actions in Syria," *War Is Boring*, June 6, 2016, <https://warisboring.com/heres-the-key-to-understanding-the-russian-air-forces-actions-in-syria>.

31 Rahul Bedi and Reuben F. Johnson, "India Withdraws from FGFA Project, Leaving Russia to Go It Alone," *IHS Jane's Defence Weekly*, Apr. 20, 2018.

Another area where India-Russia purchases must continue is in the realm of helicopters. India has an ambitious plan for the domestic manufacture of rotary winged aircraft. The deal to buy 60 and indigenously produce 140 Kamov helicopters is an important one because it not only leads to employment at HAL but also provides the lift that is badly needed in India's Himalayan region.[32] India's Cheetah helicopters are aging and the Kamov becomes a timely replacement. The deal also keeps Russia within the Indian arms market and provides Moscow with an incentive to sell advanced weaponry to India.

France

The IAF's link with France goes back to the 1950s when it acquired the Ouragan and Mystère fighters from that country. It was with the Mirage 2000, however, that the full value of French aircraft became apparent. The plane performed admirably in the Kargil war and the IAF was able to use its electronic countermeasure capabilities to blunt the use of shoulder-fired surface-to-air missiles by Pakistan. At that time, France reportedly gave India permission to refit the Mirages to deliver nuclear weaponry. The Rafale deal provides similar advantages.

The Rafale builds up the capability of the IAF, especially when mated with the Meteor missile, which has a proven long-distance kill ability. Despite all the talk about missiles as nuclear delivery systems, manned aircraft remain the most dependable way of delivering nuclear weapons—especially because they can be recalled in case the political leadership changes its mind—and the Rafale adds to that strategic delivery capability. The plane is also hardened so it could survive the shockwaves of a nuclear blast and allow the pilot to return to base after performing the mission.

What makes the French connection especially attractive is that it allows for both quality and flexibility. So what to do with this partnership in the future? First, the Rafale deal needs to be doubled to give the IAF a significant number of aircraft that are lighter than the Sukhoi. Second, the Indian government should talk to Dassault about future partnerships

---

32 "Production of Kamov Helicopters for India to Be Done in Four Stages: Russian Official," *Economic Times*, New Delhi, Jul. 14, 2018, <https://economictimes.indiatimes.com/news/defence/production-of-kamov-helicopters-for-india-to-be-done-in-four-stages-russian-official/articleshow/61891706.cms>.

to jointly develop an aircraft. This could include the joint development of India's advanced medium combat aircraft program, which could be rolled into Dassault's plans to build a follow-on to the Rafale. Alternatively, the IAF should consider a joint venture with France to design and develop an armed drone. While India has asked the United States for an armed version of the Predator, it may make technological sense to work with France since it is more willing to transfer technology and will work to Indian specifications. France is working with Airbus and the Italian company Leonardo to manufacture a medium-altitude long-endurance armed drone, which might be a project for the Indian government to invest in.

United States

The link with the United States provides both advantages and disadvantages to India. The advantages lie in acquiring world-class technology and, because of foreign military sales, the accusation of taking bribes is removed. No one in the Indian opposition has ever questioned, as has happened in the case of the Rafale, the legality of the C-17 and C-130 purchases for the IAF. American weaponry also comes with a first-rate link to supplier companies for maintenance of systems and the supply of spare parts. This is a welcome change from the headaches the IAF has faced in maintaining Russian weaponry and acquiring a substantial reserve of spare parts. Nonetheless, the old suspicions remain in India about the United States being a reliable supplier because Indian generals worry about the cut-off of spare parts in a war. India is also concerned about the conditions that the United States imposes—such as the attempt to prevent India from purchasing the Russian S-400—that are seen as constraining Indian autonomy in the acquisition of weapons systems.

Despite the formal dialogue between the U.S. and Indian foreign and defense ministers and the recent renaming of the U.S. Indo-Pacific Command to recognize the Indian subcontinent, and the claims by U.S. president Donald Trump that he gets along well with Prime Minister Modi, what exactly has been achieved in the U.S.-India relationship, particularly in the field of weapons acquisition and production? The answer is not much. No major U.S. weapons system is being built in India. The IAF has to be careful about acquiring the F-16—which it may eventually have to buy because of domestic political pressure—because the Pakistan Air Force is familiar with the performance parameters of the plane. So, realistically speaking, what should the IAF seek from the United States?[33]

The relationship is crucial for India since the United States will be a major partner in Asia to counterbalance China. At the same time, India is a long way from being a major strategic partner for the United States in the way that Britain, Israel, or even Australia is. This may change as the two countries sign agreements in the future that will allow the transfer of sensitive technologies and interfaces such as the Link 16 communications system. Such a positive change, however, will take some time as the Indian decision-making process moves at a glacial pace.

The U.S.-India arms relationship, therefore, should be based, for now, on the purchase of support systems that enhance the capabilities of the IAF but, at the same time, do not place the Air Force in a vulnerable position in a future crisis. Thus, it would make sense to buy more C-130s from Lockheed because the aircraft has performed admirably for the IAF and increased India's medium-range lift capability. The Indian government should ask, if possible, for setting up the production line of the C-17 Globemaster in India. Several nations around the world—Australia, Qatar, and Kuwait to name a few—were keen to buy more C-17s, which would also be less technologically demanding for it to build (in contrast to a fourth- or fifth-generation fighter aircraft for which it does not have the trained personnel or machine tools). The IAF should also be seeking reconnaissance drones from the United States because they would strengthen border security and be used for domestic surveillance against terrorists and insurgents. This is one area where the United States would be a willing partner and would make a valuable contribution to Indian society.

Hindustan Aeronautics Limited

HAL has a major advantage in the arms acquisition and production process because there is a consensus among politicians and security analysts that India not only requires an indigenous arms production capability but one that produces increasingly complex weapons systems. It was with this in mind that Prime Minister Jawaharlal Nehru, in the 1950s, sanctioned the development of the HF-24 Marut supersonic fighter, the Kiran trainer, and

---

33 In order to sell the plane to India, Lockheed has renamed the plane as the F-21 to make it more palatable to the Indian defense establishment.

the HT-2 primary trainer. Nehru believed that if India could learn to build combat aircraft it could eventually make jetliners. His dream was never realized because HAL's track record in this area has been disappointing.

Although HAL has been able to license-produce aircraft such as the MiG-21 and the Su-30 MKI, despite facing quality complaints from the IAF, it has had less success in indigenously produced aircraft. The Marut, while a good design, never attained supersonic speeds. The IJT Sitara was scrapped. The medium transport aircraft is making slow progress, as is the indigenous airborne early warning system. The light combat aircraft has been marred by lengthy delays and the Indian Navy has rejected the naval version of the aircraft. The Chinese point out that in a modern-day combat environment the Tejas is too small to carry the weaponry, fuel, and sensors that are required to succeed in twenty-first century warfare.[34] Nor has HAL enhanced its capabilities sufficiently to be able to build the next generation of fighter aircraft.

When Dassault asked HAL how many man-hours it would take to build the Rafale, HAL gave a labor cost that was almost three times as expensive compared to producing the plane in France.[35] The IAF was concerned about quality control so it asked Dassault to accept liability for the shortcomings in HAL's production of the aircraft.[36] Dassault naturally refused since it had no control over the production process. The dilemma for the IAF is, therefore, how to give HAL projects so that it continues to survive and flourish but at the same time the IAF's warfighting mission is not compromised.

The IAF has to take a realistic approach and not keep asking HAL to build the perfect twenty-first century plane with all the bells and whistles that imported systems have. All that does is delay projects and escalate costs. The IAF should accept that the Tejas is a replacement for the MiG-21 and therefore not a frontline fighter but one that increases the quantitative edge of the Air Force (as the Russians say, quantity has its own quality).

---

34 Interview with Dr. Xiaoming Zhang, Professor, USAF Air War College, Maxwell AFB, Alabama. Dec. 17, 2018.
35 Sudhi Ranjan Sen, "HAL's Jets Costlier than Foreign Ones, Says Defense Ministry Audit," *Hindustan Times*, Oct. 19, 2018, 1.
36 Gautam Datt, "HAL's Poor Track Record Overshadows IAF Plans to Buy 126 Dassault Rafale Jets," *India Today*, Apr. 5, 2013, <https://www.indiatoday.in/india/north/story/rafale-jets-hal-hindustan-aeronautics-limited-iaf-dassault-aviation-158290-2013-04-05>.

Given the high accident rate of the IAF, a trainer version of the Tejas, which could double as an attack aircraft in a conflict, would increase the numbers of planes produced by HAL and keep the aircraft manufacturer in business.

Last, in an era of privatization and market reforms in India, one needs to keep in mind that HAL is not the only game in town. Private companies such as Larsen and Toubro, Reliance, Adani, and Kirloskar should be supported as they enter the aviation business for, unlike the public sector HAL, they are easier to hold accountable on quality control and the timely completion of projects.

India's future air power will depend on the ability to balance these four suppliers and continue to provide projects to all of them. What also needs to be done is to reform the acquisition process so that it does not take 15 to 20 years to acquire a weapons system. This would lead to major delays in developing forces and to escalating costs. It is ridiculous to expect that a cost estimate made in 2019 will be valid in 2035. The IAF needs to work with these suppliers to develop weapons for a futuristic battlefield as opposed to buying big ticket items meant for fighting past wars. To sum up, if the IAF wants to be a modern combat force that can project power extra-regionally, it has to buy quickly and buy smart.

India wants to retain its domestic aeronautical sector. No Indian political party will allow HAL to be closed down since it is, however flawed, a symbol of technological achievement and a generator of scientific employment. The solution lies in holding HAL's feet to the fire and forcing it to ramp up production of the Tejas. HAL has been producing eight Tejas a year, which means that it will take two to three years to get one squadron of aircraft for the IAF. Such a slow production rate will not permit the development of a capable Tejas force for the IAF.

The government also needs to take a firm stand with the IAF, which has set up unrealistic expectations for the Tejas. The IAF has sought a finished first-rate product from HAL, and this has been further complicated by the Air Force's desire to bring about changes while production and development are ongoing. The government needs to ask HAL to manufacture as many planes as possible to build up numbers in the second-rank squadrons that are not based along the borders—and the Air Force should be made to accept these planes. After sufficient squadrons are in place, the earlier squadrons can be sent back to HAL for modernization. What will emerge in the IAF is a high-low mix of aircraft building toward a 42-squadron force, which would allow for training pilots and providing sufficient flying hours to combat pilots.

Conclusion

The United States Air Force is now going for a mix of F-22s and F-35s, F-15s and F-16s, and is going to procure a propeller-driven light attack aircraft, at the low end of the technology spectrum, to provide close air support and carry out residual strikes after the high-technology systems have wreaked havoc on the enemy's air defenses. The Tejas, similarly, should be seen as the low-technology end of the spectrum and be used accordingly by the IAF. That would remove the pressure for the Tejas to be a frontline combat aircraft but would permit the IAF to have sufficient squadrons. The Tejas could also be used as an attack trainer to allow young pilots to grasp the characteristics of supersonic flight.

UCAVs have become a key part of the force structure of modern air forces. They are useful for surveillance and reconnaissance, for loitering for long periods over sensitive areas, and for carrying out pinpoint precision strikes against insurgents and terrorists. One only has to look at the success of the U.S. Central Intelligence Agency's drone program to see how effective the modern UCAV is. For India, UCAVs would be useful for patrolling the maritime spaces as well as the long border with Pakistan. The IAF will be getting the armed Harpy drone from Israel and Israel has talked about transferring a drone production line to India. But India should be looking at the United States, the world leader in UCAV technology, to jointly develop a set of drones that can do long-range surveillance as well as tactical missile strikes. This would be an easy goal to pursue in the dialogue with the U.S. defense secretary since there are several U.S. companies that could help India co-develop such a UCAV.

The planned acquisition of the S-400 would certainly boost the air defense capabilities of the IAF as it seeks to shoot down aircraft and missiles at greater distances. That would enhance India's anti-access/area denial capabilities. The purchase also has political implications because it will be a test for shaping the future course of the India-U.S. relationship since India has taken the stance that it will not comply with U.S. sanctions against Russia. Not abiding by U.S. sanctions will be a test to see how much autonomy India can retain in the new relationship that it is attempting to create with the United States.

In conclusion, the key to modernizing the IAF is to buy quickly and to buy smart. The last time that happened was after the 1962 war, when India developed a comprehensive defense modernization plan that saw a modern military emerge from the ashes of defeat. India now faces challenges on two fronts as well as internal threats that require the use of air power. How effectively these challenges can be met will be determined by how quickly and efficiently procurement is carried out.

References

Bedi, R. and R.L. Johnson. "India Withdraws from FGFA Project, Leaving Russia to Go It Alone," *IHS Janes Defence Weekly*, Apr. 20, 2018.

Cooper, T. "Here's the Key to Understanding the Russian Air Force's Actions in Syria," *War Is Boring*, June 6, 2016. <https://warisboring.com/heres-the-key-to-understanding-the-russian-air-forces-actions-in-syria>.

Datt, G. "HAL's Poor Track Record Overshadows IAF Plans to Buy 126 Dassault Rafale Jets," *India Today*, Apr. 5, 2013. <https://www.indiatoday.in/india/north/story/rafale-jets-hal-hindustan-aeronautics-limited-iaf-dassault-aviation-158290-2013-04-05>.

Dubey, A.K. "Meteor Missile Deal Set to Win Back India's Aerial Supremacy against Rivals," *Daily Mail* (UK), Dec. 18, 2017. <https://www.dailymail.co.uk/indiahome/indianews/article-5192397/Meteor-missile-deal-win-Indias-aerial-supremacy.html>.

"Forget Rafale, HAL Cannot Even Produce and Deliver Su-30 MKI's on Time to the IAF," *Defence News*, Oct. 2, 2018. <http://www.defencenews.in/article/Forget-Rafale,-HAL-cannot-even-produce-and-deliver-Su-30-MKIs-on-time-to-the-IAF-580869>.

Gupta, A. *Building an Arsenal: The Evolution of Regional Studies Force Structures.* Westport, CT: Praeger, 1997.

Gupta, A. "The Missile Age: Is India Ready?" *Geopolitics Magazine*, 2018.

Kapur, V. "Challenges for the Indian Air Force 2032," *Journal of Defence Studies*, Vol. 7, No. 1 (2013), 83. <https://idsa.in/system/files/jds_7_1_VivekKapur.pdf>.

Kaura, V. "India Will Continue Boosting Its Nuclear and Missile Capabilities to Counter China—and Beijing Will Keep Responding," *South China Morning Post*, Feb. 8, 2019. <https://www.scmp.com/comment/insight-opinion/asia/article/2185271/india-will-continue-boosting-its-nuclear-and-missile>.

Ladwig, W.C. "A Cold Start for Hot Wars? The Indian Army's New Limited War Doctrine," *International Security*, Vol. 32, No. 3 (2007–2008).

McBride, J. "Is 'Made in China 2025' a Threat to Global Trade?" Council on Foreign Relations Backgrounder, Aug. 2, 2018. <https://www.cfr.org/backgrounder/made-china-2025-threat-global-trade>.

Naqash, A.A. and S. Miglani. "India Launches Air Strike in Pakistan; Islamabad Denies Militant Camp Hit," *Reuters*, Feb. 25, 2019. <https://www.reuters.com/article/us-india-kashmir-pakistan/india-launches-air-strike-in-pakistan-islamabad-denies-militant-camp-hit-idUSKCN1QF07B>.

Nitze, P. "Atoms, Strategy, and Policy," *Foreign Affairs*, Vol. 34, No. 2 (Jan. 1956), 195.

"Pakistan Denies Indian Claims It Used US F-16 Jets to Down Warplane," *Guardian*, Mar. 3, 2019. <https://www.theguardian.com/world/2019/mar/03/pakistan-denies-indian-claims-it-used-us-f-16-jets-to-down-warplane>.

Pandit, R. "To Avoid Sukhoi 'Mistake', India to Go for Russian 5th-Generation Fighter Only on Complete-Tech Transfer Pact," *Economic Times*, Jul. 11, 2018. <https://economictimes.indiatimes.com/news/defence/after-sukhoi-mistake-india-to-go-for-russian-5th-generation-fighter-only-on-full-tech-transfer-pact/articleshow/57551801.cms>.

Pandit, R. "IAF 'Harvesting Organs' of Globally Retired Jets," *Times of India*, Jul. 23, 2018. <https://timesofindia.indiatimes.com/india/iaf-harvesting-organs-of-globally-retired-jets/articleshow/65096225.cms>.

Pandit, R. "Budget 2018: Govt Hikes Defence Budget by 7.81%, but It's Just 1.58% of GDP & Lowest since 1962," *Times of India*, Feb. 1, 2018. <https://timesofindia.indiatimes.com/india/budget-2018-govt-hikes-defence-budget-by-7-81/articleshow/62740525.cms>.

"Production of Kamov Helicopters for India to Be Done in Four Stages: Russian Official," *Economic Times*, New Delhi, Jul. 14, 2018. <https://economictimes.indiatimes.com/news/defence/production-of-kamov-helicopters-for-india-to-be-done-in-four-stages-russian-official/articleshow/61891706.cms>.

Pubby, M. "New Brahmos Missile Will Take Down Enemy 'Force Multipliers'," *Economic Times*, Feb. 21, 2019. <https://economictimes.indiatimes.com/news/defence/new-brahmos-missile-will-take-down-enemy-force-multipliers/articleshow/68094383.cms>.

Rajagopalan, R. "Force and Compromise: India's Counter-Insurgency Grand Strategy," *Journal of South Asian Studies*, Vol. 30, No. 1 (2007), 87.

"Saab Proposes to Make 96 Gripen Jets in India to Win Air Force Deal," *Economic Times*, New Delhi, Feb. 19, 2019. <https://economictimes.indiatimes.com/news/defence/saab-proposes-to-make-96-gripen-jets-in-india-to-win-air-force-deal/articleshow/68065031.cms>.

Sahoo, A.K. "BrahMos Missile Successfully Test Fired from Sukhoi Fighter Jet for First Time," *Asian Age*, Nov. 23, 2017. <https://www.drdo.gov.in/drdo/pub/npc/2017/november/din-23november2017.pdf>.

Sankaran, J. "The Enduring Power of Bad Ideas: 'Cold Start' and Battle-field Nuclear Weapons in South Asia," *Arms Control Today*, Nov. 2014. <https://www.armscontrol.org/ACT/2014_11/Features/Cold-Start-and-Battlefield-Nuclear-Weapons-in-South-Asia>.

Saran, S. "Is India's Nuclear Deterrent Credible?" Public lecture, India Habi-tat Centre, New Delhi, Apr. 24, 2013. <https://www.armscontrolwonk.com/files/2013/05/Final-Is-Indias-Nuclear-Deterrent-Credible-rev1-2-1-3.pdf>.

Sen, S.R. "HAL's Jets Costlier than Foreign Ones, Says Defense Ministry Audit," *Hindustan Times*, Oct. 19, 2018, 1.

Shukla, A. "India Wants 18 More Sukhoi 30 MKIs," *Rediff News*, Feb. 8, 2019. <https://www.rediff.com/news/report/india-wants-18-more-sukhoi-30mki-fighters/20190208.htm>.

Siddiqui, H. "Modi Government to Fast-Track Acquisition of Lethal Predator Drones for Indian Navy and Indian Army," *Financial Express* (New Delhi), Jan. 1, 2019. <https://www.financialexpress.com/defence/modi-government-to-fast-track-acquisition-of-lethal-predator-drones-for-indian-navy-and-indian-army/1430508>.

Sidhu, W.P.S. "Updating India's Nuclear Doctrine." Brookings Institution, May 1, 2014. <https://www.brookings.edu/opinions/updating-indias-nuclear-doctrine>.

"Strategic UAV Base Shifted to Maoist Hotspot in Bastar," *Economic Times*, Oct. 28, 2017. <https://economictimes.indiatimes.com/news/politics-and-nation/strategic-uav-base-shifted-to-maoist-hotspot-in-bastar/articleshow/60992553.cms>.

Subramaniam, A. "Closing the Gap: A Doctrinal & Capability Appraisal of the IAF & the PLAAF," Observation Research Foundation, May 2, 2019. <https://www.orfonline.org/expert-speak/closing-gap-doctrinal-capability-appraisal-iaf-plaaf>.

Sundaram, K. and M.V. Ramana. "India and the Policy of No First Use of Nuclear Weapons," *Journal for Peace and Nuclear Disarmament*. DOI: 10.1080/25751654.2018.1438737.

9 AUSTRALIA
A Uniquely Strategic Environment

Pete Wooding

Introduction

The Royal Australian Air Force (RAAF) is on the cusp of celebrating its centenary of existence. In this time, military aviation in Australia has developed from a nascent independent capability struggling for relevance, recognition, and resources to become an integral aspect of the nation's military power in meeting national security needs and supporting national interests. This is no different from the anguished growth of many air arms over the last century and similarly Australia's journey reflects the opportunities and challenges that its adherents have sought to address in establishing the role of air power as a key enabler in modern warfare. In this sense, any discussion of the future of air power from a national perspective must duly recognize the uniqueness of each nation's circumstances in the ongoing journey, for there is no future without a past.

Aviation has long featured as a central element of Australia's development as a nation and, consequently, in its national security. The utility of aviation has reflected a progressively greater importance as technological developments and enhanced capabilities have made the role of national air power, in its broadest sense comprising civil and military considerations, of growing relevance to Australia's circumstances. In the main, this has reflected the country's unique geographic conditions, both as a continental landmass with a small population and the effects of the "tyranny of distance" in relation to its friends and neighbors and, in the early part of the twentieth century, maintaining its connection with Great Britain and the Commonwealth. These considerations have been further shaped by the enduring maritime nature of Australia's strategic environment, which reflects an island nation dependent upon secure sea lines of communication through a region of growing strategic importance, both in terms of economic power and areas of potential conflict. In the contemporary context, this is clearly highlighted by the growing economic importance of the Asian region and the tensions involving the rise of an increasingly powerful China and its regional aspirations, as these issues directly impact Australia's primary area of strategic interest, the Indo-Pacific.

As Australia seeks to adapt to an increasingly uncertain and complex international environment, the RAAF is capitalizing upon its rich history to develop as a capable and credible fifth-generation air force. In assessing Australian air power from a military perspective, this chapter reflects on three distinct aspects shaping the way ahead. First, the development of air power from an historical perspective, which reflects Air Marshal Fisher's contention that "the Air Force of today is the product of many and varied forces in the past."[1] This context is an important feature for understanding Australia's aviation foundations as an air-minded nation and the strategic culture that underpins the focus in going forward. The second feature of importance is the evolving geostrategic context and the implications for air power, especially as this relates to the continuities of Australia's strategic outcomes and policy perspectives. This is particularly important in light of the changing dynamic of the Indo-Pacific region. This was a key element of the 2016 Defence White Paper (DWP16) and underscores Australia's focus for national security, which also overlaps with key security concerns of the United States, Australia's principal alliance partner.[2] And finally, there is the need to address air power's strategic direction in meeting the Australian government's national security needs. This encompasses aspects of focus and doctrine, and how this is shaping the recapitalization of capability, especially in relation to the joint operating space. And while much of this discussion reflects the Air Force's development, it is important to recognize that it also broadly applies to the integration of and developments across air elements of the Australian Army and the Royal Australian Navy.

---

1   Alan Stephens, *Going Solo: The Royal Australian Air Force 1946–1971* (Canberra: Australian Government Publishing Service, 1995), v.
2   Australian Government: Department of Defence. *2016 Defence White Paper* (Canberra: Commonwealth of Australia, 2016).

The Development of Australian Military Air Power

Setting the context is essential to having a meaningful understanding of the future of air power from an Australian perspective. For Australia, a key part of this perspective is the notion of an air-minded nation. Aviation has long been an important feature in Australia's development as a nation: from its earliest days, the allure of flying and flyers captured the fledging nation's imagination. Australian aviators were recognized as dashing pioneers whose daring exploits held the promise to overcome the challenges of a far-flung empire. Beyond the adventure, however, successive governments regarded aviation as a topic of national importance when addressing the immensity of the Australian continent and its international isolation. This perspective was implicit in the Director General of Civil Aviation A. B. Corbett's assertion in 1943 that: "A nation which refuses to use flying in its national life must necessarily today be a backward and defenceless nation."[3] This sentiment would come to shape Australia's subsequent aviation development, including that of the world's second oldest air force.

Australian military air power exhibits an uneven development that reflected changing circumstances, perceptions, and expectations. In this, there have arguably been three distinct periods of Australian air power development. First, the period from the early twentieth century up to World War II was an uncertain period of establishment, neglect, and rebuilding in relation to the role of military aviation for Australia's security needs. This was in part attributable to the entrenched political sway of the Navy and Army, despite the air power lessons of 1914–1918, which was compounded by inter-service rivalry in a period of financial crisis and strictures. The second phase encompassed the period of World War II to the latter part of the Cold War, which marked the maturing of air power as a significant feature of contemporary warfighting. Various conflicts during this period, especially the role of air power during the Pacific Campaign, sought to underscore the significance and utility of military aviation to Australia's national security and interests. This was particularly relevant post-Vietnam with the drive toward self-reliance in the defense of Australia and a maritime strategy that was primarily dependent on integrated air and naval

---

3 John Gunn, *Contested Skies: Trans-Australian Airlines, Australian Airlines, 1946–1992* (St. Lucia: University of Queensland Press, 1999), 11.

capabilities. Finally, the post–Cold War period has seen the recapitalization of Australia's air power to meet the needs for a next-generation air force that is matched to the complexity of a world order that appears to be in transition. For Australia, the capability edge now resides in fifth-generation capabilities that include stealth, precision engagement, and electronic attack, as well as the ability to exploit the space and cyber domains. These drivers have the capacity to enhance the RAAF's flexibility as the force of choice in terms of rapid response, small footprint, and the ability to deliver strategic effects encompassing lethal and non-lethal options.

The potential role of aviation in Australia's defense was an early topic of discussion and debate. Impressed by the exploits of Bleriot, in 1909 the Commonwealth government offered a £5,000 prize to the Australian inventor of an airplane suitable for military purposes.[4] While little came of the competition, it nonetheless marked the start of Australia's involvement with the military aspects of aviation, which was further progressed by the decision of the 1911 Imperial Conference that military aviation should become a focus of the armed forces across the British Empire. Australia became the first Commonwealth country to pursue this policy, which resulted in the government's decision in 1912 to establish the Australian Flying Corps (AFC) as part of the Australian Army and a Central Flying School.[5] However, it was the Great War that would be the watershed for Australian military aviation as the importance of the third dimension in industrial-age warfare became apparent. Notably, Australian air elements performed with distinction in both the Middle East and France under Britain's Royal Flying Corps.

Despite the demonstrated efficacy of air power in modern war, Australia's ability to build upon its experience was largely neglected during the interwar period. Following the cessation of hostilities, the AFC was disbanded in 1919 before being reestablished as the independent RAAF in 1921. Despite the government's commitment to retaining an aviation element within the defense forces, there was no clear plan to develop it as an integral part of Australia's defense strategy. The lack of a cohesive approach

---

4 Steve Campbell-Wright, "Fledglings: Australia's Military Aviation and Preparations for War," *The First Fateful Shot: Port Phillip Bay, August 1914* (Melbourne: Military History and Heritage, 2014).

5 Royal Australian Air Force, *The Australian Experience of Air Power* (Canberra: Air Power Development Centre, 2013), 14.

resulted in the air force largely withering to become a training organization with little priority assigned to combat capability. This situation was
compounded by the Great Depression that resulted in severe cost cutting,
which was further impacted by the greater political influence wielded by
the Navy and Army as the established services. In a climate of severe financial constraint, the very existence of the RAAF was left seriously in doubt.
However, the threatening rise of nationalism and militarism in Germany,
Italy, and Japan in the 1930s led to the establishment of rapid rearmament
programs among western nations, including Australia, which secured the
RAAF's status as an independent service. While the RAAF underwent a
period of rapid expansion to improve its operational capabilities it still
suffered from structural, capability, and doctrinal deficiencies from the prolonged period of neglect. Despite this, Australia's air power had survived
as a core feature of its defense force albeit still in the broader context of
imperial strategy.

World War II was to finally bring home to Australia the salient lessons of air power and in a much more direct way than previously. This
would come to solidify the critical importance of air power in Australia's
strategic culture in future decades. This was made all the more poignant
given that Australia's security had long been steeped in the blind faith
placed in imperial defense and the primacy of the fleet for security, which
were deeply entrenched in the prevailing mindset of the pre-war Singapore strategy. However, the rapid German blitzkrieg victories in Europe
and Japan's bombing of Pearl Harbor underscored that air power had come
of age as a weapon of war. And while this appeared finally to vindicate
air power's proponents, it nonetheless still came as a rude awakening for
many military traditionalists as to war's evolving character. For Australians, this awakening was brought home with the Japanese onslaught in the
Pacific and the dire threat to Australia.

Australia's vulnerability to air attack, and possible invasion, was
underscored by the bombing of Darwin. More than 100 air raids were
undertaken by the Japanese against Darwin and its shipping, with the
raid of February 19, 1942 being the largest single attack ever conducted
against Australia.[6] This lesson in modern air power was further reinforced

---

6 "75 Years Ago – 1942," VeteransSA, Jan. 13, 2017, <https://veteranssa.sa.gov.au/
story/75-years-ago-1942/>.

by actions at the Battle of the Coral Sea and the Battle of Midway, which marked strategic victories for the Allies whereby carrier aviation averted the threat to the Australian mainland and the supply lines between the United States and Australia.[7] Critically, the victory at Midway also marked the turning point in the war in the Pacific. Further evidence of the efficacy of air power was given by McArthur's island-hopping strategy in the South West Pacific campaign in which air power enabled sea power projection that in turn enabled operational maneuver of land forces through the island chains.[8] Finally, as if further evidence were needed, the efficacy of air power, especially American air power, was clearly displayed with the strategic bombing campaign against Japan that would culminate with the dropping of the atomic bomb. For Australia, the lessons were stark and had a marked impact on the nation's psyche—air power had finally appeared to realize the promises of its earlier proponents and the rise of America as a principal security partner was now clear. These lessons would be reinforced throughout the Cold War.

The Cold War period marked a significant period of advancement for air power in general and no less so for Australia in particular. Unlike the uncertainty and neglect of the interwar years, World War II proved to be a watershed in the maturing of Australia's understanding and appreciation of what air power offered the nation's unique geostrategic circumstances, which had been dramatically underscored by the threat of invasion. For Australia, the Cold War emphasis was on assisting the containment of the global spread of communism through multiple concurrent small-scale expeditionary commitments within an alliance framework. In its commitments[9], Australian air power played a limited but valued role in supporting allied efforts through the contribution of personnel and air elements that comprised both combat and support aircraft. Central to this approach was a broader planning focus for the ongoing modernization of RAAF air power around high-end capabilities that were primarily shaped by the prospect

---

7 Historically, the battle is significant in air power terms as the first action that aircraft carriers engaged each other with neither side's ships sighted or firing directly upon the other.

8 Peter Dean, "MacArthur's War: Strategy, Command and Plans for the 1943 Offensives," in *Australia 1943: The Liberation of New Guinea*, ed. Peter J. Dean (Melbourne: CUP, 2014), 53.

9 These were typified by air power contributions to missions that ranged from the Berlin airlift to conflicts in Korea, Malaya, and Vietnam, as well as support to various humanitarian assistance and UN-mandated missions.

of a crisis involving a regional power, most notably Indonesia during the tensions of the 1960s and 1970s under the concept of Forward Defence.

In the post-Vietnam period, and shaped by the Guam Doctrine, air power was further developed under the defense of Australia construct that emphasized self-reliance. A key element of these strategies was the pursuit of a capability edge within a balanced force structure that provided an air power advantage over regional adversaries, and which was underpinned by engagement with allies, principally the United States. This concept would remain a point of continuity as a central developmental pillar in the years ahead to good effect, although this advantage began to erode as regional air forces also modernized force structures.[10] By the end of the Cold War, the effect was that the world's second oldest air force had firmly entrenched the critical role of air power in Australia's national mindset and defense strategic thinking.

The post–Cold War strategic environment underwent a dramatic change in challenges and threats. The most immediate event being the First Gulf War, in which coalition air power played a seminal role as epitomized by the images of precision warfare and stealth. The lessons of air power's renewed efficacy in the conflict were not lost upon the RAAF, which was only able to play a limited role due to issues of interoperability and preparedness. Given demonstrated shortcomings in contributing to the coalition air effort,[11] the key lesson was that it needed to ensure that involvement in future global security operations required the RAAF to be a balanced and ready force able to undertake a wide range of operations encompassing the spectrum from coalitions to domestic activities. To achieve this change in outlook necessitated the development of credible force options, especially for coalition operations with interoperability as a central tenet. Notably, a focus on the defense of Australia as the sole driver of force development could no longer suffice in a world of globalized challenges and threats, although the Australian-led 1999 East Timor intervention underscored the ongoing importance of self-reliance in regional

---

10 Indicative of the decline in relative competitive edge has been the introduction within the region of advanced combat platforms such as the MiG-29, Su-27, F-15 and, looking forward, the likelihood of the F35.

11 This was typified by the inadequacy of the self-protection systems for the F111 strike aircraft to meet the anticipated threat environment. Rather than platforms, the main contribution became a small element of intelligence staff, primarily photo-interpreters.

security matters. This approach came to fundamentally shape the development of Australia's air power in the subsequent decades, most notably with the acquisition of next-generation capabilities such as the E-7 Wedgetail airborne early warning and control (AEWC), KC-30 tanker, EA-18 Growler electronic attack aircraft, and C-17 strategic airlift platforms, as well as the impending introduction to service of the F-35 fifth-generation fighter and the sophisticated MQ-4 Triton and MQ-9 Reaper unmanned aerial vehicles (UAVs). These acquisitions have significantly reinvigorated Australia's ability to provide flexible air power options through the means to contribute to and sustain high-end outputs to a multitude of post-9/11 commitments, most notably but not exclusively in the fight against extremism in the Middle East area of operations. That platforms such as the E7 have come to be recognized as the capability of choice among allies is testimony that this experience has reinforced Australia's reputation in the early part of the twenty-first century as one of the most modern and capable small air forces in the world.

Australia's Evolving Strategic Future

Strategic context shapes destiny in guiding the approach to defense needs, development, and prioritization. And it is clear that over the last century, Australian defense policy has constantly evolved to meet changing strategic conditions and emerging challenges in terms of both global and regional issues. In terms of continuity, there are three fundamental factors that will continue to impact Australia's approach to strategic thinking and policy, all of which will continue to influence Australia's consideration of air power's efficacy in contributing to national security and national interests in uncertain times.

Strategic Geography

First, is the notion of strategic geography. A nation's geography is arguably a primary factor in driving national security considerations and, hence, force structure. In Australia's case, the physical reality is that the nation is the world's largest island, one that is resource rich with a low population density, and that has one of the world's largest coastlines and exclusive economic zones. Also, its history is that of a principally western nation—culturally and demographically, much like New Zealand—that is positioned on the periphery of a vast Asian archipelagic and maritime region that is of growing economic and security importance. The physical nature of Australia's regional geography also ranges from Antarctica in the south to China to the north, and with India, the Middle East, and the African continent to the west. Collectively, this region now accounts for approximately half the world's population and a significant, and growing, proportion of global gross domestic product (GDP) and trade.

The importance of Australia's strategic geography has long been a fundamental feature of the nation's security thinking. However, as noted by Paul Dibb, it was not until the 1976 Defence Review that the uniqueness of Australia's geographic circumstances reached prominence, as reflected in the assertion that "Australian geography had a compelling influence on Australian security."[12] This would become a key theme that would recur in subsequent Defence White Papers (DWPs).[13] In the post-Vietnam period, this had the effect of focusing Australia's security strategy around the direct defense of Australia and its northern maritime approaches through protection of the air-sea gap. From this primary focus came the need to provide security in the immediate neighborhood and the wider Asia-Pacific region as well as to contribute to Australian interests globally. This nesting of priorities was generally perceived as a pragmatic reflection of Australia's capacity to balance its national power with its national interests. In this, the DWP16 represents a more nuanced perspective on the importance of strategic geography for defense planning.

The DWP16 clearly articulates the importance of the Indo-Pacific region to Australia's future national security and national interests.[14] Of note, the paper assesses that half of the world's economic output will be focused in the Indo-Pacific by 2050.[15] In this context, and as a maritime nation, Australia sits adjacent major shipping routes that link the Middle East and Africa with the Pacific, which Australia depends upon for external trade and sustainability, most notably for energy imports. The importance

---

12 Paul Dibb, "DWP 16: The Return of Geography," *The Strategist*, Mar. 1, 2016, <https://www.aspistrategist.org.au/dwp-2016-the-return-of-geography/>.
13 Ibid., 1.
14 Australian Government: Department of Defence, *2016 Defence White Paper* (Canberra: Commonwealth of Australia, 2016), 14.
15 Ibid.

of this focus is all the more critical when noting the effects of globalization and major power rivalry, which are creating what Nick Bisley describes as "a China-centric integrated Asian strategic system" that will "have maritime and continental dimensions."[16] This evolving state of regional affairs is increasingly challenging the past seven decades of global order and the past notion of a "fortress Australia" that was distanced from external threats that could directly threaten the nation.

From a security perspective, the changing nature of Australia's regional circumstances and its strategic geography have important implications for Australia's air power capabilities. This is particularly evident in terms of being able to create and sustain strategic effects over significant distances in a maritime region of growing military power and security challenges, and potentially one of relative decline in U.S. influence and power. In this context, the recognized importance of air power is reflected over the last decade in capability acquisitions such as the C-17, KC-30A, E-7 Wedgetail, P-8 Poseidon, MQ-4C Triton, EA-18 Growler, and the F-35 Lightening II. What these platforms produce, especially in terms of a capability edge, is an integrated ability to monitor, shape, and respond across a vast and geographically diverse operating space, both in terms of independent operations and coalition activities. It also reflects a focus on high-end capabilities suited to mid- to high-intensity conflict that can enable Australia to punch above its weight.

Independent Middle Power

This brings us to the second aspect, which is Australia's identification of itself as an independent middle power, which has security implications in terms of both influence and prosperity.[17] While there is no clear definition of what a middle power is, Robert Potter postulates that pragmatically the concept recognizes that a state may be contextually powerful within its

---

16 Nick Bisley, *Integrated Asia: Australia's Dangerous New Strategic Geography* (The Centre of Gravity Series, Canberra: Strategic and Defence Studies Centre, ANU, 2017), 2.
17 Zoe Halsted, Nicholas Taylor, Gillian Davenport, Breanna Gabbert, Adam Bell, and Shixi Guo, *In Brief: Foreign Policy White Paper* (Canberra: Australian Institute of International Affairs, 2018).

geographic circumstances.[18] For Australia, this perspective is reflected in Australia's 2017 Foreign Policy White Paper (FPWP17) that identifies that the "Indo-Pacific encompasses our most important economic partners and its dynamism supports economic growth in Australia" and that "priorities also reflect the global scope of our national interests and foreign policy."[19] Notably, this perspective reflects Australia's long-held world view that its security is inextricably linked to international order and stability, which it has sought to influence. In this sense, the pursuit of middle power status is underpinned by Australia's robust liberal democratic institutions and economy, as well as its security alliance with the United States, which have enabled it to enjoy a disproportionate level of international status and influence. This is reflected in the Lowry Institute's 2018 Asia Power Index, which ranks Australia as an overperforming middle power in the region, its status commensurate with that of Singapore and South Korea.[20]

Critical to achieving stability and security in the burgeoning Indo-Pacific is Australia's commitment to a normative international rules-based order. Over the past 70 years, the international system has largely enjoyed peace and prosperity founded upon the international institutions established post–World War II and underpinned by U.S. global leadership. In this context, Australia has sought to exercise its middle power influence to strongly support the U.S.'s international role as the preeminent western power; central to Australia's policy position has been the unique security relationship between the two nations. While this relationship has been mutually beneficial, Australia has also recognized the emerging shift in global power balance that directly impacts the international system, a trend that is clearly articulated in the Lowry Institute's observation that "three of the world's four largest economies are in Asia, and the fourth, the United States, is a Pacific power. By 2025, two-thirds of the world's population will live in Asia, compared with just over a tenth in the West."[21] For Australia,

---

18 Robert Potter, "Stuck in the Middle – 'Middle Power' Defined." *Australian Institute of International Affairs*, June 25, 2015, <http://www.internationalaffairs.org.au/australianoutlook/stuck-in-the-middle-middle-power-defined-2/>.
19 Australian Government: Department of Foreign Affairs, *2017 Foreign Policy White Paper* (Canberra: Australian Government, 2017), 3.
20 Lowy Institute, "Asia Power Index 2018," <https://power.lowyinstitute.org/downloads/LowyInstitute_AsiaPowerIndex_2018-Summary_Report.pdf>.
21 Ibid.

this shift in global power has underscored the growing dilemma of balancing its principal security relationship with the United States and the importance of its growing economic relationship with China.

Noting the growing tensions between the United States and China, there exist potential opportunities for influential middle powers, such as Australia, to shape regional affairs through their defense posture and enhanced regional defense engagement. In this, there is a very clear role for Australia to exploit its air power in a manner that aligns with the reality of its strategic geography. In particular, air power uniquely provides the capacity for Australia to demonstrate the ability to deliver strategic effects across a maritime and archipelagic environment that is flexible, responsive, and precise and which befits a middle power status in supporting regional security initiatives and promoting stability through closer security cooperation. Indicative of this stance is the evolving nature of the Five Power Defence Arrangement (United Kingdom, Australia, New Zealand, Malaysia, and Singapore) and Australia's promotion of major regional multinational exercises, such as Bersama Lima, that are indicative of the strong focus on air power in promoting national intent and reinforcing regional commitment.

Great and Powerful Friends

And third, is Australia's long-held tradition of aligning with benevolent great power friends. Underpinning its approach to defense, Australia has always sought a close relationship with a great power as a guarantor of its security, as noted by Michael Wesley: "Australian strategic policy making has always been strongly invested in regional unipolarity: it has been most confident and stable when its great power ally held primacy in the Indo-Pacific."[22] Over the last century, this facet of strategic policy has seen the transfer in the primary security relationship from the United Kingdom to the United States, which was formalized in 1953 with the Australia, New Zealand, United States (ANZUS) Treaty. From the early 1950s to the Vietnam War, these relationships were based around the concept of forward defence, which primarily sought to prevent the spread of communism across Asia. Post-Vietnam, and following the announcement of the Guam

---

22 Michael Wesley, "Australian Grand Strategy and the 2016 Defence White Paper," *Security Challenges* 12 (2016), 19–30.

Doctrine in 1969, Australian strategic thinking evolved toward the concept of defense of Australia, which emphasized a more self-reliant approach to defense policy and notably a primary focus on protecting the country's maritime areas of responsibility. This posture was articulated in the 1976 DWP, which stated: "for practical purposes, the requirements and scope for Australia's defence activity are limited essentially to the areas closer to home."[23] Despite the focus on self-reliance, this has been underpinned by the enduring reliance on great power alignment, which has seen the relationship with the United States significantly strengthen over the decades, especially in terms of interdependence.

The DWP16 clearly reinforced the strategic necessity to enhance the Australia-US relationship as a centerpiece of the nation's security strategy. Notably, it identifies that the likelihood of a direct attack on Australia remains remote, increased cyber activity notwithstanding, and that future planning must recognize "the regional and global nature of Australia's strategic interests and the different sets of challenges created by the behaviours of countries and non-state actors."[24] In this, alliance relationships, most notably with the United States, remain central to Australia's long-term security. This focus is reflected in Australia's standing in accessing and acquiring advanced capabilities and enabling support from the United States, high levels of engagement on defense policy and cooperation, and the political will to contribute to American-led military operations and endeavors. This focus reflects common values and broad alignment of strategic interests. In terms of air power, this is reflected in Australia's acquisition of advanced capabilities such as the F-35 fighter, the EA-18 Growler, and the MQ-4 Triton and MQ-9 UAVs. Such acquisitions, which are complemented by personnel exchanges, provide significant benefits in terms of both global support and capability sustainment, as well as enhancing interoperability with Australia's principal ally and security guarantor.

While there are enduring features that shape Australia's approach to pursuing its national interests, the world is also constantly evolving, which presents unique challenges and threats. Despite 70 years of relative peace

---

23 Australian Government: Department of Defence, *Australian Defence* (Canberra: Australian Government Publishing Service, 1975), 6.
24 Australian Government: Department of Defence, *2016 Defence White Paper* (Canberra: Commonwealth of Australia, 2016), 15.

and prosperity since the end of World War II, there are emergent challenges that now threaten to disrupt the extant rules-based global order that has largely been founded on U.S. global leadership and the promotion of western values and interests. Australia recognizes the pressures of change, especially in terms of the rise of China and its potential impact on the global power balance in general and the implications for the Indo-Pacific region in particular. In seeking to manage a world that is arguably more complex and volatile, especially in terms of power challenges, asymmetric threats, and disruptive technologies, the Australian government's defense strategy is built around three strategic defense interests (SDIs) and associated strategic defense objectives (SDOs), which will guide force structure and force posture.[25] These reflect both continuity and change.

The government's defense strategy sets out the expectations of Defence in contributing to peacetime stability and, if necessary, the conduct of decisive operations in times of conflict. In this, the most basic defense interest is a secure, resilient Australia with the goal of deterring, denying, and defeating any attempt by a hostile country or non-state actor to attack, threaten, or coerce Australia.[26] This is to be achieved by providing Defence with the capability and resources required to respond independently and decisively to military threats, including incursions into Australia's northern maritime approaches. The second defense interest is in a secure nearer region, encompassing maritime Southeast Asia and the South Pacific with a focus on the security of maritime Southeast Asia and supporting the stability of Australia's near neighbors.[27] In Southeast Asia, the focus is on strengthening engagement, including helping to build the effectiveness of regional capabilities and operations to address shared security challenges. Critically, the Australian government has articulated the need to continue its commitment to strengthened regional security architectures that support transparency and cooperation; and, where necessary, for the Australian Defence Force (ADF) to have increased capabilities to make contributions to any regional security operations. The third defense interest is to support a stable Indo-Pacific region and the rules-based global order.[28]

---

25 Ibid., 68.
26 Ibid.
27 Ibid., 69.
28 Ibid., 70.

This is to be achieved through the provision of meaningful contributions to global responses to address threats to the rules-based global order that may threaten Australia and its interests. Central to this approach, and as previously identified, is the need to work closely with the United States, and other international partners, to play a role in coalition operations wherever Australia's interests are engaged. Collectively, these elements shape the nexus between Australian security and interests, and the contribution of Australian air power in a disruptive world.

Future Air Power: An Australian Approach

A century of air power in meeting the challenges of peace and war has underscored Australia's enduring utility as an increasingly key enabler of national security. In this sense, Australia's ongoing approach to exploiting the third dimension will continue to focus on ensuring it remains an essential element of Australia's national military power that supports current strategic guidance for a future force that will be more capable, agile, and potent.[29] Notably, the future force must be more capable of conducting independent operations in Australia's immediate region and enabling enhanced contributions to coalition operations. This will be achieved through a future force structure that is joint by effect and integrated by design, which will enable the ADF to apply "more force more rapidly and more effectively" when required across a multi-domain battlespace.[30]

In developing the future force, the Australian approach recognizes that the nature of air power will remain an enduring feature of future warfare, even though its character will continue to evolve. In this, air power will continue to be defined by flexibility and precision in its employment across the battlespace, but the effects at the strategic to tactical level will adapt in response to the confluence of the future operating environment and emergent technologies. Notably, air power will play a key role in Australia's ability to effectively pursue an effects-based approach to national security that exploits its inherent attributes of flexibility and responsiveness to a broad range of security challenges and contexts. In this, the Air Force's contribution to Australia's military strategy will continue to be defined

---

29 Ibid., 87.
30 Ibid., 84.

by the key effects of shaping, deterring, and countering. This will emphasize the key tenets of seeking a capability edge, maintaining a balanced
force structure, and enhancing interoperability. And finally, air power will
continue to follow the established principle that planning adheres to centralized command and decentralized execution if it is to optimize flexibility
and adaptability, which accords with the philosophy of mission command.

In looking to the future, the Air Force's raison d'être will remain the
generation of air power effects to secure Australia and its interests all the
while noting the increased complexity of the future operating environment. This environment will increasingly reflect reemergent state-based
security challenges, as characterized by regional tensions over the South
China Sea and North Korean nuclearization, as well as ongoing nonstate
threats encompassing terrorism and intrastate conflicts. Collectively these
challenges will be manifest in both traditional threats from conventional
forces as well as asymmetric threats, including the growing disruption
involving cyber activities. In responding to a more complex environment,
the Air Force will continue to conduct air operations from air bases both
domestically and through host nations by engaging with other services,
allies, coalitions, and security partners. This reflects air power's inherent
flexibility and adaptability that enables employment across a range of military operations, from disaster relief to high-end conventional warfighting
involving the application of precise kinetic and, increasingly, non-kinetic
effects. And critical to generating air power, the Air Force will continue
to adopt the long view for creating a force that is balanced in its capability mix and ensures that it can best meet the broadest range of Australia's
enduring and emerging strategic requirements. While it will adopt and
field niche capabilities, it will not develop as a niche force as this would
undermine its utility.

In taking the long view, changing force structure is an involved process
in both cost and lead time for the acquisition and exploitation of capability.
For Australia, this perspective is strongly underpinned by a risk-managed
approach to change that emphasizes a measured development process
while accommodating the fielding of new capability in response to immediate needs. In this, the RAAF is currently embarking upon an extensive
program to establish itself as a fifth-generation air force, which reflects the
realities of the last two decades of the ADF's operational experience from

East Timor to the Middle East. Indicative of the success of this approach has been Operation Okra, the ADF's contribution to the coalition mission to counter the threat posed by the terrorist group the Islamic State of Iraq and Syria. A key element of this contribution was the Air Task Group that comprised the KC-30A tanker, E-7A Wedgetail AEWC aircraft, and F/A-18 Hornet fighters. Despite both the KC-30A and E-7A being deployed prior to achieving initial operational capability, the platforms nonetheless came to be considered coalition air assets of choice for their respective missions. This level of adaptability and responsiveness to serving national security commitments will remain a central feature of the Air Force's future development.

In contributing to the national effort, Air Force operations will become more adept at exploiting an information advantage for delivering measured and integrated effects to shape and control the battlespace. This outcome is underscored by three key concepts:

- First, the establishment of a smart, high-capacity network that seeks to establish the Air Force as a fully networked entity that can exploit a capability edge in the information age. In particular, the way information is collected, collated, shared, and exploited will be underpinned by a decentralized and distributed information architecture that is scalable and dynamic and characterized by inherent redundancy and graceful degradation. This will enable air power to be applied as a seamless whole rather than as discrete elements operating in isolation, which will be a key element in managing operational risk.

- Second, the integration of more pervasive and interleaved surveillance and reconnaissance capabilities utilizing high-resolution, all-weather sensors to facilitate enhanced situational awareness to underpin decision superiority. This will enable real-time flexibility and adaptiveness for the responsive delivery of precision effects using fifth-generation air power capabilities.

- Third, enhanced decision support tools to ensure that air power's inherent flexibility utilizing finite resources can be optimally employed to rapidly deliver effects from the tactical to the strategic. In the future, the utilization of emerging technologies such as artificial intelligence (AI) will be critical to optimizing air power effects.

Collectively, these developments will ensure that the RAAF continues to develop an all-important operational advantage and facilitating interoperability with key partners, most notably the United States. Australian air power doctrine recognizes that within this future construct the essential nature of exploiting the third dimension remains fundamentally unchanged. Reflecting on a century of dramatic changes in technology, opportunities, and evolving threats, the application of air power will continue to focus upon the four core roles of control of the air, strike, air mobility, and intelligence, surveillance, and reconnaissance (ISR).[31] Notably, these roles are not unique to Australia, but represent the enduring functions that balanced air forces have always provided since military aviation's initial days. Of note, the four core air power roles are underpinned by three enabling air power roles, namely: command and control (C2); force protection; and force generation and sustainment.[32] Collectively, these aspects enable air power to continue to achieve mass effect with precision in a contested battlespace through the interconnectedness of accurate and timely situational understanding, seamless decision-making, and flexible and adaptable platforms. And in developing the future force, it will be increasingly critical for air power to be able to contribute to effects across five interdependent operating domains: air, land, maritime, space, and cyber. This focus will enable Australian air power to deliver a key asymmetric advantage in achieving dominant effects in the joint fight.

Future Drivers Shaping Australian Air Power

While air power roles are enduring, the manner in which air power effects are achieved typically changes in response to the evolving character of warfare, especially in relation to technological change. In this, air power has always reflected an affinity with and dependence upon the adoption of advanced systems. However, advances in technology do not necessarily translate into advances in capability unless they are adequately aligned

31 Air Power Development Centre, *AAP-1000D: The Air Power Manual* (Canberra: Air Power Development Centre, 2013), 46.
32 Ibid., 79.

with the enabling air power roles. The way air forces contextualize and incorporate advances in technology is, therefore, only relevant to strategic needs and how this meets the future employment of air power for national interests. In the Australian context, and reflective of the nation's middle power focus, the RAAF's future-looking document Beyond the Planned Air Force has identified nine key technological disruptors: autonomous systems, uninhabited systems, stealthy systems, advanced weapon modes, C3 and sensor systems, quantum technology, space systems, training and education systems, and the human interface.[33]

Autonomous systems represent the emergence of machine learning to facilitate the execution of self-directed tasks. This has a direct bearing on the decision-making function associated with the delivery of air power effects in general and the role of man-in-the-loop decision-making considerations, especially in relation to unmanned aerial systems. Aligned with the technological challenges associated with the development of advanced learning systems, especially where this impacts the autonomous delivery of effects, are the legal and ethical dimensions of developing, accrediting, and employing trusted systems. In this, the incorporation of the command and control of such systems into an integrated model of decision-making system-of-systems represents both great opportunity and significant challenges for air power with respect to the notion of decentralized execution. This will demand fail-safe considerations, especially where this may involve the delivery of lethal or non-reversible effects.

Uninhabited systems have been an area of significant growth over the last decade. While UAVs have existed for several decades, it has only been in recent times that they have grown in sophistication and utility, as evidenced by their employment in the fight against terrorism. The upsurge in the use of unmanned systems has been a result of the increased sophistication of the systems and a growing trust in the reliability and utility of the platforms. There has also been greater recognition of the political imperative to minimize the risk to own forces, primarily aircrew, and the potential for reduced system costs. From an operating perspective, unmanned vehicles have addressed one of the limitations of air power, which is that of persistence over the battlespace. Notably, advanced systems, such as the

---

33 Air Power Development Centre, *Beyond the Planned Air Force: Thoughts on Future Drivers and Disruptors* (Canberra: Air Power Development Centre, 2017), 19–33.

MQ-4C Triton that is being acquired by Australia for the maritime surveillance role, will provide an operating endurance that can far exceed that of manned platforms. And while uninhabited systems are now a critically important feature of the ISR realm, the biggest future developments are increasingly focused on uninhabited combat aerial vehicles (UCAVs).

While UCAVs are currently operational, existing systems represent a nascent capability with a restricted mission set. Their specialized and contentious nature also mean that such systems remain limited to a restricted number of military forces although this is likely to expand as their utility, availability, and affordability grows. Therefore, it can be expected that over time UCAVs will develop into robust capabilities that will increasingly complement and potentially dominate manned air systems in such traditional high-risk roles as attack and air control. The benefits of high levels of performance in terms of endurance, altitude, and maneuverability that overcome human limitations make such options increasingly attractive to planners and politicians, especially in non-permissive environments. For nations, such as Australia with its strategic geography and desire for maintaining a regional capability edge, UCAVs make for a potentially attractive security investment. Aligned with the acquisition of such advanced systems is also the need to develop new employment concepts to optimize their utility, such as "swarms" of expendable platforms to overwhelm local air defenses.

Stealth has been one of the defining features of the current generation of air power, especially in the public mindset. It has primarily come to be associated with the minimization of radar and infrared signatures for combat platforms. This obviously has significant benefits in terms of platform survivability and conferring an operating advantage: therefore, it will continue to be an area of important research and development. What will be significant into the future, however, will be a broader application of stealth technology and techniques across the spectrum of air power capabilities. This will apply not only to noncombat platforms but also to supporting entities ranging from operating sites to other operational domains. Additionally, the focus will become not just one of denying detection but will include aspects of distraction and disguise, that is the ability to mimic other platforms to confer an advantage.

Two emerging technological developments for advanced weapons systems that have significant application to future air power are hypersonics and directed energy weapons (DEW). These systems confer advantages in terms of enhanced standoff ranges and speed of engagement; however, they also present targeting challenges and in the case of DEW the ability to precisely deliver effective levels of lethal energy. The development of hypersonic missiles, that is those capable of speeds above Mach 5, provides improved survivability of the weapon and the launch platform, as well as complicating the ability for an adversary to respond to a high-speed inbound threat. These advantages make hypersonic weapons ideally suited to highly contested, non-permissive targets, which is reflective of the emerging threat environment in Australia's primary area of strategic interest, the Indo-Pacific, especially China's focus on an anti-access / area denial (A2/AD) strategy for contesting the South China Sea. For Australia, the acquisition of hypersonic missiles combined with fifth-generation platforms such as the F-35 provides a capability edge against both land and maritime targets.

Compared to hypersonic systems, DEWs are at a more embryonic stage of operational development. DEWs are systems that utilize high-energy lasers or high-power microwaves to disrupt and damage an adversary's capabilities through both lethal and non-lethal effects. These systems have potential application on both airborne platforms as potential attack mechanisms, but also, and possibly more significantly, as ground-based counter-air weapon systems. Although deliverable power and cooling considerations are significant limitations currently, as are targeting aspects against moving targets, these challenges are not necessarily insurmountable. The potential benefits of operationalizing such capabilities are significant noting the low cost-per-shot of DEWs and instantaneous nature of striking a target given the ability to engage a target at the "speed of light." The potential non-lethal element of DEWs, especially microwave systems, also provides an important consideration regards collateral damage and the potential for reversible effects.

In a network construct, modern platforms—manned and unmanned—are increasingly key elements as collection assets. This is particularly true of fifth-generation combat platforms such as the F-35A, EA-18G, and P-8A that are capable of capturing prodigious volumes of data; however, it now also potentially extends to using noncombat platforms as collectors. This trend will only increase in the future as a truly distributed and interdependent information environment is further developed, which will be key to fully exploiting future air power capabilities. To fully exploit this information-rich environment will demand real-time systems that can sense, assimilate, interpret, and disseminate actionable intelligence, especially for time-sensitive targets, faster and more reliably than in the past. This will be reliant on more powerful information systems that will include AI for decision superiority and quantum technologies for enhanced sensor and analytical capabilities. The ability to rapidly disseminate bandwidth-intensive information packages across a distributed network will also require higher-capacity communication systems that will need to be based around emerging technologies, including optical systems and laser bearers. A key aspect of this development will be consideration of interoperability, which for Australian platforms operating in a coalition environment means its primary partner, the United States.

Finally, future air power will be increasingly reliant on space and cyber capabilities. Exploiting these domains will be dependent on the Air Force continuing to build expertise and capability both in terms of alliance linkages and organically. In terms of space capability, this has already commenced through Defence's involvement in the American wideband global satcom (WGS) communications constellation, relocation of a U.S. C-band radar to Australia to support space situational awareness, and exploring the acquisition of a radar satellite. Similarly, for cyber, the latest DWP has acknowledged the bolstering of the cyber security capacity while it has also been separately acknowledged that Australia has an offensive cyber capability. While these represent significant steps, the need to develop ways of operating with the most advanced space and cyber capabilities must also recognize that while dependence creates operating advantages and opportunities it also produces potential vulnerabilities. Space and cyber reliance creates increasingly contested environments with decreasing barriers to entry for potential adversaries. While it is difficult to accurately predict the systemic effects of attack on the ADF, it is critical to develop the capacity to operate with degraded space and cyber capabilities.

Implementing the Future

The Air Force Strategy 2017–2027 is the keystone document articulating RAAF's transformation into a fifth-generation air force over the next decade.[34] The strategy provides a framework to align Australia's future air power development with a period of growing strategic uncertainty. Notably, it provides a focus on warfare in the information age, which recognizes that the exponential growth in the volume of data that underpin contemporary society has fundamentally altered its character.[35] In this context, the strategy focuses primarily on those key areas that must evolve to ensure the Air Force's continued development as a modern fighting force capable of undertaking operations as part of a genuinely joint ADF. In looking to the future, the strategy focuses on five core functional areas:

1. Joint warfighting capability.
2. People capability.
3. Communication and information systems.
4. Infrastructure.
5. International engagement.[36]

The backbone for the strategy's implementation is Plan Jericho.

Plan Jericho provides the foundation for transforming the Air Force into "a future force that is agile and adaptive, fully immersed in the information age, and truly joint."[37] To meet this goal, the Air Force is currently undergoing the largest technological upgrade in the force's history, which will ensure that it is a modern, fully integrated element of military power that is able to deliver air and space power effects in the information age. To achieve this vision, Plan Jericho aims to drive a culture of innovation and improvement in the way it acquires and sustains new capabilities for a complex and uncertain future. This focuses on a holistic approach integrating elements of operations, personnel, and capability.

The plan recognizes that the Air Force has always operated leading-edge technology, which must now be best exploited to deliver

---

34 Air Power Development Centre, *Air Force Strategy 2017–2027* (Canberra: Air Power Development Centre, 2017).
35 Ibid., 8.
36 Ibid.
37 Royal Australian Air Force, *Plan Jericho* (Canberra: Air Power Development Centre, 2015), 2.

decisive air power in the information age. In this context, the Air Force will transition to an even more technologically advanced force to maintain a capability edge and maintain relevancy in the changing international security environment. Key to this goal is the ability of the Air Force to integrate fifth-generation air power into a joint ADF and broader national capabilities to ensure that it can support all phases of operations and domains. In this construct, the heart of future systems will be the ability to facilitate superior knowledge in a form that makes it relevant, timely, and actionable. This will ensure that tomorrow's air force can "integrate Air Force component capabilities that will operate in a Joint environment under a joint operational level C2 construct enabled by integrated national, strategic, operational and tactical level systems."[38]

Critical to realizing this transformation is prioritization of capability acquisition and associated defense funding. In terms of prioritization, DWP16 clearly underscores that the government will provide significant investment in new defense capabilities to meet evolving security challenges. Notably, the paper identifies that: "Central to the development of this Defence White Paper has been the Government's direction to align defense strategy, capability and resources" and that "the Government's long-term funding commitment provides a new 10-year Defense budget model to 2025–26, over which period an additional $29.9 billion will be provided to Defence. Under this new budget model, the defense budget will grow to $42.4 billion in 2020–21, reaching two percent of Australia's gross domestic product (GDP) based on current projections."[39] Aligned with the funding commitment is the fully costed, 10-year Integrated Investment Program that provides the framework for managing risk for all capability-related defense investment, which is a new initiative.

Overall, the new funding approach is primarily focused on Defence's capital acquisitions. In terms of the Defence's six capability streams, the breakdown for the 10-year division of investment to FY 2025–26 capability streams are:

- Air and Sea Lift (6 percent)
- ISR, EW, Space and Cyber (9 percent)

---

38 Ibid., 12.
39 Australian Government: Department of Defence, *2016 Defence White Paper*. (Canberra: Commonwealth of Australia, 2016), 24.

- Maritime and Anti-submarine Warfare (25 percent)
- Land Combat and Amphibious Warfare (18 percent)
- Strike and Air Combat (17 percent)
- Key enablers (25 percent).[40]

Within this construct, the funding for air power capabilities more specifically identifies that the EA-18G Growler, P-8A Poseidon, and F-35A Lightening II are in the top 10 acquisition projects by total approved project expenditure and account for 44.1 percent of the allocated funding.[41] Further, for the 2018–19FY project expenditure, air capabilities account for 56 percent of the budgeted spend among the top 10 acquisition projects.[42] This funding distribution reflects that, in looking to the future, air capabilities represent a significant proportion of defense investment in contributing to national security.

Conclusion

Australia's future air power is in the ascent as it builds upon a century of rich history. As an air-minded nation, Australia has embraced aviation to overcome the tyranny of distance as well as effectively exploiting the third dimension in meeting its national security and national interests, especially in terms of its unique geostrategic circumstances. Notably, it has learned the lessons of crises and conflicts to establish Australia's reputation in the early part of the twenty-first century as one of the most modern and capable small air forces in the world—one that is demonstrably well matched to its middle power focus in times of significant change, both regionally and globally.

In looking to the future, the challenge will be to ensure that Australia's air power focus and development remain relevant to achieving Australia's national objectives through its military strategy. And key to this focus will be ensuring that air power can achieve a key asymmetric advantage in projecting dominant effects in supporting the joint fight. This will be characterized by a more capable, agile, and potent force structure that enables

---

40 Ibid., 85.
41 Australian Strategic Policy Institute, *The Cost of Defence: ASPI Defence Budget Brief 2018–2019. Briefing Note.* (Canberra: Australian Strategic Policy Institute, 2018), 12.
42 Ibid.

effective contributions to independent operations as well as coalition activities. To achieve this, the future focus for Australia's air power will continue to build upon enduring aspects while adapting to the changing character of warfare. Doctrinally, the RAAF will maintain its focus on the four key air power roles of control of the air, strike, air mobility, and ISR. In this context, capability will continue to be shaped by three core tenets: the focus on maintaining a capability edge, the importance of enhancing interoperability with key allies, and the need to continue a balanced approach to developing force structure. In going forward, however, what will be significant will be the application of air power in an age increasingly characterized by the pervasive nature of the information domain in future conflict, which reflects aspects of opportunity, dependence, and vulnerability.

The pathway to change has already commenced under Plan Jericho, which aims to position the RAAF as a fifth-generation air force in an information-dominated future. Notably, this focus is reflected in the need to develop as a fully integrated and network-enabled force that is capable of fighting and winning in the information age, one that is fully aligned with its key allies and partners, most notably the United States. Investment in new capabilities encompassing airborne warning and control, electronic attack, and advanced ISR—both manned and unmanned—highlight the significant enhancements now underway in achieving a capability edge when operating in an information-rich battlespace. And while the platforms provide significant enhancements in shaping the battlespace through the delivery of kinetic and non-kinetic effects, these platforms form one element of an integrated and networked enabled infrastructure. Collectively, and with an eye on integrating emerging technologies, Australia's air power will remain credible, agile, and relevant for the uncertain times ahead.

References

Air Power Development Centre. *AAP-1000D: The Air Power Manual*. Canberra: Air Power Development Centre, 2013.

Air Power Development Centre. *Air Force Strategy 2017–2027*. Canberra: Air Power Development Centre, 2017.

Air Power Development Centre. *Beyond the Planned Air Force: Thoughts on Future Drivers and Disruptors*. Canberra: Air Power Development Centre, 2017.

Australian Government: Department of Defence. *Australian Defence*. Canberra: Australian Government Publishing Service, 1976.

Australian Government: Department of Defence. *2016 Defence White Paper*. Canberra: Commonwealth of Australia, 2016.

Australian Government: Department of Foreign Affairs. *2017 Foreign Policy White Paper*. Canberra: Australian Government, 2017.

Australian Strategic Policy Institute. *The Cost of Defence: ASPI Defence Budget Brief 2018–2019*. Canberra: Australian Strategic Policy Institute, 2018.

Bisley, N. *Integrated Asia: Australia's Dangerous New Strategic Geography*. The Centre of Gravity Series, Canberra: Strategic and Defence Studies Centre, ANU, 2017.

Campbell-Wright, S. "Fledglings: Australia's Military Aviation and Preparations for War," *The First Fateful Shot: Port Phillip Bay, August 1914*. Melbourne: Military History and Heritage, Victoria, 2014.

Clarke, S. Strategy, Air Strike and Small Nations. RAAF Base Fairbairn: Air Power Studies Centre, 1999.

Dean, P.J. "MacArthur's War: Strategy, Command and Plans for the 1943 Offensives." *Australia 1943: The Liberation of New Guinea*, edited by Peter J. Dean, Melbourne: CUP, 2014.

Dibb, P. "DWP 16: The Return of Geography." *The Strategist* (2016). <https://www.aspistrategist.org.au/dwp-2016-the-return-of-geography/>.

Gray, C.S. *Airpower for Strategic Effect*. Montgomery: Air University Press, 2011.

Gunn, J. *Contested Skies: Trans-Australian Airlines, Australian Airlines, 1946–1992*. St. Lucia: University of Queensland Press, 1999.

Halsted, Z., N. Taylor, G. Davenport, B. Gabbert, A. Bell and S. Guo. *In Brief: Foreign Policy White Paper*. Canberra: Australian Institute of International Affairs, 2018.

Kainikara, S. *Essays on Air Power*. Canberra: Air Power Development Centre, 2012.

Lambert, A. and A.C. Williamson. *The Dynamics of Air Power*. Bracknell: MOD, Joint Services Command and Staff College, 1996.

Lowy Institute. "Asia Power Index 2018." *Lowy Institute*. <https://power.lowyinstitute.org/downloads/LowyInstitute_AsiaPowerIndex_2018-Summary_Report.pdf>.

Olsen, J.A. *Global Air Power*. Dulles: Potomac Books, Inc., 2011.

Peach, S. *Perspectives on Air Power: Air Power in its Wider Context*. Norwich: The Stationary Office, 1998.

Potter, R. "Stuck in the Middle – 'Middle Power' Defined." *Australian Institute of International Affairs* (2015). <http://www.internationalaffairs.org.au/australianoutlook/stuck-in-the-middle-middle-power-defined-2/>.

Royal Australian Air Force. *Plan Jericho*. Canberra: Air Power Development Centre, 2015.

Royal Australian Air Force. *The Australian Experience of Air Power*. Canberra: Air Power Development Centre, 2013.

Stephens, A. *Power Plus Attitude*. Brisbane: Watson, Ferguson and Co., 1992.

Stephens, A. *Going Solo: The Royal Australian Air Force 1946–1971.* Canberra: Australian Government Publishing Service, 1995.

VeteransSA. "75 Years Ago – 1942." January 13, 2017. <https://veteranssa.sa.gov.au/story/75-years-ago-1942/>.

Wesley, M. "Australian Grand Strategy and the 2016 Defence White Paper." *Security Challenges* (2016), 19–30.

10 PAKISTAN Overcoming Strategic Constraints

Sharad Joshi

Introduction

An important element of Pakistan's defense strategy is the role of air power. As with most modern nation-states, expansion of a country's air weapons—whether through aircraft, missiles, or space capabilities—is important to its development as a military power. This chapter examines air power in Pakistani security strategy through multiple perspectives and questions. What is Pakistani military strategy and where does air power fit into this strategy? What are the characteristics of Pakistani air power capabilities? How successful have these capabilities been for Pakistani security requirements?

Pakistani security strategy focuses on different threat perceptions, ranging from state actors to nonstate actors such as terrorist groups, based on which it has expanded its arsenal. But there are important constraints and concerns stemming from Islamabad's expansion of its air power. According to one scholar, "the weapons production and arms acquisition process of regional powers constrain their actual capabilities and subsequently determine their behavior in the emerging international system," leading to "incomplete force structures in these powers."[1] Such assessments are applicable to Pakistan also, where domestic and international constraints

---

1 Amit Gupta, *Building an Arsenal: The Evolution of Regional Power Force Structures* (Westport, CT: Praeger, 1997), 1–2.

are present. Pakistani defense strategy is thus driven by a need to overcome these constraints in its pursuit of a revisionist agenda in South Asia.

Examining Pakistan's air power capabilities in the context of its defense strategies is crucial for at least a few reasons. Aerial capabilities are an important indicator of defense expansion strategies. Therefore, Islamabad's air power policies provide some guidance on its overall foreign and defense policies. This is relevant to multiple aspects of Pakistani security policy and objectives, including its rivalry with India, which involves nuclear weapons, counterterrorism, and counterinsurgency objectives against groups such as the Pakistani Taliban (or the Tehrik-i-Taliban Pakistan). The air power element is also a salient feature of Pakistan's relations with extra-regional major powers, especially the United States and China, and is an important symbol and indicator of the ties with the two global rivals. Pakistan is also interested in an export role for aerial platforms, especially those developed in collaboration with China, which would be a source of foreign exchange and signify higher status as a supplier of weaponry. Finally, this survey of Pakistani air power issues and strategies, which relies primarily on academic writings, policy assessments, and media reporting, is an important case study through which to examine how regional powers consider air power and what it represents for them.

Pakistani Perceptions of Its Threat Environment and the Role of the Air Power

Pakistan's security thinking and military strategy have been driven primarily by its deep-rooted rivalry and animosity with India, ever since the two countries gained independence in 1947. The antagonism is based especially on Islamabad's desire to wrest control of the state of Jammu and Kashmir from India. Pakistan and India have fought several wars: in 1947–1948 over Kashmir following an invasion by Pakistan-backed tribesmen into the state; in 1965, after Pakistan pushed mujahideen to foment an insurgency in Kashmir; and in 1971, which led to the creation of Bangladesh from East Pakistan. The two countries also fought a limited war in the Kargil region of Kashmir in the summer of 1999, following the occupation of positions on the Indian side of the line of control (LOC) in northern Kashmir by Pakistani army troops and affiliated terrorist proxies. In addition, Pakistan has executed a strategy of state sponsorship of terrorism in Kashmir and other parts of India since the 1980s, which continues. Both sides have also sought control of the Siachen glacier located beyond the LOC in the Himalayas, with the Indian army occupying the glacier since 1984.[2]

Air power has played an important role in India-Pakistan conflicts, especially in support of the army, and is likely to be even more important in future. In 1947–1948, the Indian Air Force (IAF) airlifted troops for the defense of Jammu and Kashmir, along with bombing missions.[3] The Pakistan Air Force (PAF) had lesser capabilities. The 1965 war occurred after Pakistan pushed mujahideen into Kashmir to foment an insurgency under Operation Gibraltar, and followed it up with a military invasion through Operation Grand Slam to try to cut off India from the Kashmir Valley, not anticipating India's escalation of the war on other fronts, including Punjab and Rajasthan.[4] The PAF—smaller, but much better equipped, prepared, and more agile—posed a tough challenge to the IAF, which suffered major aircraft losses on the ground after the PAF's preemptive strike on the first day of the conflict. But the IAF was able to push back Pakistan's subsequent invasion.[5] In the 1971 war over East Pakistan/Bangladesh, the IAF was able to defeat the PAF more decisively, especially in the eastern sector, while the PAF was unable to relieve pressure in the east through an ineffective preemptive attack on Indian airbases in the west.[6] In the 1999 Kargil war, the PAF could play very little of an operational role, given that the war was fought on the Indian side of the LOC, and the IAF's success in preventing PAF helicopters from resupplying troops on the Indian side, even

---

2 Sharad Joshi, "The Story of a Glacial Trust Deficit," StratPost.com, Dec. 23, 2008, <https://www.stratpost.com/the-story-of-a-glacial-trust-deficit>.

3 Arjun Subramaniam, "Doctrinal Evolution in the Indian Air Force," in Harsh V. Pant (ed.), *Handbook of Indian Defence Policy: Themes, Structures, and Doctrines* (New York, NY: Routledge, 2016), 222.

4 C. Christine Fair, *Fighting to the End: The Pakistan Army's Way of War* (New York, NY: Oxford University Press, 2014), 82.

5 Jasjit Singh, "IAF's Ground Reality," *Sunday Tribune*, May 6, 2007; Arjun Subramaniam, "Doctrinal Evolution in the Indian Air Force," 223; George K. Tanham and Marcy Agmon, *The Indian Air Force: Trends and Prospects* (Santa Monica, CA: RAND, 1995), 25–32. Tanham and Agmon also discuss some of the problems of the IAF's strategy and doctrine in the 1965 war.

6 Ahmad Faruqui, "Failure in Command: Lessons from Pakistan's Indian Wars, 1947–1999," *Defense Analysis*, Vol. 17, No. 1 (2001), 35.

as IAF fighter aircraft, especially the Mirage 2000s, effectively targeted and destroyed Pakistani posts, while under instructions not to cross the LOC.[7]

Since the Kargil war, and under its 2012 doctrine, the IAF has aspired to a greater role to strengthen conventional deterrence and escalation dominance and to achieve strategic effects through its operations.[8] After the Kargil war, the most serious confrontation between the air forces of Pakistan and India occurred in February 2019 following a suicide terrorist attack on an Indian security convoy in Kashmir by the Pakistan-backed Jaish-e-Mohammed (JeM). The IAF launched an airstrike against a JeM camp in Balakot, Pakistan, using Mirage 2000s, and in the aftermath there was little certainty over whether the strike had destroyed the complex as claimed by India. Nonetheless, this was the first time that Indian fighter aircraft had carried out an attack across the border since 1971. In its response, Pakistan attempted to attack Indian military targets across the border in Kashmir, without much success. One of the Indian MiG-21 Bison aircraft chasing back the Pakistani fighters was shot down in Pakistani territory, even as India claimed to have brought down an F-16 aircraft, which remained unconfirmed. What is clear, however, is that the IAF was able to penetrate into Pakistani territory in the Khyber Pakhtunkhwa province, where Balakot is located, to carry out a demonstrative and punitive attack, which was a major change from traditional Indian caution in the use of air power in response to Pakistan-backed terrorism.[9] Historically, the PAF has acted in support of the army (as has the IAF), and the February 2019 skirmish was possibly the first instance of air power taking the lead in an attack.

It should be reiterated that air power is only one military tool in pursuit of certain political objectives, and therefore evaluation of air power should assess whether it enables the state to prevail in wars, either on its own or in coordination with other services.[10] In that regard, in its most important focus, the PAF has not succeeded in military confrontations with

---

7 Benjamin S. Lambeth, "Airpower in India's 1999 Kargil War," *Journal of Strategic Studies*, Vol. 35, No. 3 (June 2012), 289–316.

8 Christina Goulter and Harsh Pant, "Realignment and Indian Airpower Doctrine: Challengers in an Evolving Strategic Context," *Journal of Indo-Pacific Affairs*, Fall 2018.

9 Amit Gupta, "The February Flashpoint: Lessons Learned," *Geopolitics*, Mar. 2019, 18–21.

10 Viktoriya Fedorchak, "Air Warfare," in David J. Galbreath and John R. Deni (eds.), *Routledge Handbook of Defence Studies* (New York, NY: Routledge, 2018), 186.

India. This is despite the fact that throughout the rivalry between Pakistan and India, Islamabad has never had to deal with a long-term simultaneous two-front war scenario that India has to prepare for, which has only been exacerbated with the intensification of the Sino-Pakistani military and nuclear nexus.

Pakistan seeks to prevent a conventional Indian attack along the lines of the 1965 and 1971 wars and Indian responses in the two wars represent two related aspects of Pakistani threat perceptions. The demise of Pakistan's eastern wing in the 1971 war influenced contemporary threat perceptions of successive generations of defense leaders and analysts in the country, in the belief that India might once again attack Pakistani territorial sovereignty. These views have been bolstered by India's conventional military superiority. One view is that in an Indo-Pakistan war, the IAF would achieve superiority, or even air supremacy, over the PAF, as the latter's conventional military capabilities are only capable of dealing with India's military in limited border clashes defensively.[11] Although the IAF is certainly superior to the PAF in terms of the number and quality of aircraft, the decline in IAF squadrons in recent years, and India's need to prepare for a simultaneous two-front scenario against the PAF and the Chinese Peoples' Liberation Army Air Force, has decreased the relative IAF-PAF gap.[12] Pakistani analysts (and others) also believe that India's long delays with the light combat aircraft project and the acquisition of medium multirole combat aircraft (MMRCA) will further decrease the IAF's squadron strength in the coming years.[13]

Concerns over Indian conventional military superiority and the history of defeats by India have become an important narrative for Pakistan despite the fact that Pakistan is considered the revisionist power in South Asia based on its objectives and strategies in the region, especially with

---

11 Jonah Blank, Richard S. Girven, Arzan Tarapore, Julia A. Thompson, and Arthur Chan, *Vector Check: Prospects for U.S. and Pakistan Air Power Engagement* (Santa Monica, CA: RAND, 2018), 14–15.

12 Saurabh Joshi, "IAF Admits Unprepared for 2-Front Defense," StratPost.com, Mar. 12, 2016, <https://www.stratpost.com/iaf-admits-unprepared-for-2-front-defense>; Sushant Singh, "IAF Worry: in Next Two Years, Only 26 Fighter Squadrons, Short by 16," *Indian Express*, Jan. 24, 2019.

13 Zia Ul Haque Shamsi, "Indian Air Force Modernization Plan of 2020: Challenges for Regional Air Forces," Australian Defence College, Oct. 2012, <http://www.defence.gov.au/ADC/Publications/Commanders/2012/04_IAF%202020%20AWC.pdf>.

respect to India, and that Indo-Pak crisis and confrontations are generally provoked by such revisionist policies.[14] Additionally, it should also be noted that in 2010 Pakistan army chief General Ashfaq Parvez Kayani said that, "we plan on adversaries' capabilities, not intentions."[15] For Pakistan, one concern is that the IAF plans to possess capabilities that would allow it to inflict strategic impact on the country.[16] This could perhaps include operations such as the February 2019 strike in Balakot, which, although under dispute regarding its effects, did demonstrate emerging resolve by India to use punitive airpower. India's impending acquisition of the S-400 anti-aircraft missile system has also caused some worry in Pakistan, which would be the main target of this system, as it would allow New Delhi to be more confident in carrying out punitive air strikes on Pakistani territory in response to a potential terrorist attack by an Islamabad-backed group. In a crisis like the 2019 terrorist attack in Pulwama and the subsequent IAF airstrike in Balakot, the S-400 would allow India to destroy PAF fighters and its airborne early warning and control (AEWC) systems.[17]

Thus, Islamabad is ostensibly concerned about a Cold Start–like strategy by India that involves quick, shallow, conventional military attacks into Pakistani territory across the border as punishment for Pakistani support for terrorist groups—a strategy that emerged especially after the 2001–2002 Indian strategy of coercive diplomacy under Operation Parakram following the terrorist attack on the Indian parliament. That strategy demonstrated some gaps in Indian military doctrine, especially in scenarios that envisioned limited attacks below full-scale conventional war, which could provoke a nuclear response from Pakistan.[18] In a major conventional war, one of the strategies considered by the Pakistani military includes the "Riposte" strategy under which a strike corps would

---

14 On Pakistani revisionism, see Rajesh Basrur, "India and Pakistan: Persistent Rivalry," in Sumit Ganguly, Andrew Scobell and Joseph Chinyong Liow (eds.), *The Routledge Handbook of Asian Security Studies* (New York, NY: Routledge, 2018), 153–63.

15 "Kayani Spells Out Threat Posed by Indian Doctrine," *Dawn*, Feb. 4, 2010.

16 Shamsi, "Indian Air Force Modernization Plan of 2020: Challenges for Regional Air Forces," 20.

17 Ejaz Haider, "S-400: Pakistan Faces a Major Asymmetric Threat," *Newsweek Pakistan*, Oct. 11, 2018, <https://newsweekpakistan.com/s-400-pakistan-faces-major-asymmetric-threat>.

18 Rajeswari Pillai Rajagopalan, "Pakistan's Nasr Missile: 'Cold Water' over India's 'Cold Start'?" *Diplomat*, Jan. 31, 2019; Ali Ahmed, "Towards a Proactive Military Strategy: 'Cold Start and Stop'," *Strategic Analysis*, May 2011, Vol. 35, No. 3, 404–05.

advance deep into India while other corps would repel Indian advances.[19] The belief is that the initial momentum and the international community's subsequent intervention would lead to a ceasefire and subsequently Pakistan would exchange the territory it captured for concessions from India.[20]

In addition, Pakistani threat perceptions have been noted by its National Command Authority (NCA), the primary nuclear weapons administration entity, which specified three main threats in December 2017:

1. Increasing conventional weapons imbalance with India.
2. Nuclearization in the Indian Ocean region.
3. India's focus on ballistic missile defense (BMD).[21]

Islamabad's analysts and decision-makers also believe, without any reliable evidence, that India seeks to subvert Pakistan by supporting rebellions in Balochistan and Sindh.[22] In recent years, Pakistan has combated separatist insurgents in the province of Balochistan, most notably those belonging to the Balochistan Liberation Army. It has also been hit by terrorist and insurgent violence, by jihadi groups including the Pakistani Taliban, al-Qaeda, the Islamic State group, and Lashkar-e-Jhangvi. The Pakistani Taliban insurgency has posed a dilemma for the Pakistani military on at least two levels: its doctrine was, and continues to be, geared toward the perceived conventional military threat from India; and how to fight jihadist militants, who were part of the broader family of Sunni jihadist terrorist groups that the military has closely supported in India and Afghanistan.

But despite the immediate security threat since 9/11 from a variety of internal jihadist terrorist and insurgent groups, Pakistan's defense strategy remains focused on the perceived India threat. This is reflected in the PAF's doctrines, capabilities, and aspirations, even though the public rationale for acquiring more fighter aircraft is to combat the Pakistani Taliban and other terrorists more effectively. Thus, for instance, in context of Islamabad's quest for 10 more F-16 C/D Block 52 aircraft from the United States in 2016, the government asserted that it needed the multirole fighters and

19 Francisco Aguilar, Randy Bell, Natalie Black, Sayce Falk, Sasha Rogers, and Aki Peritz, "An Introduction to Pakistan's Military," Harvard Kennedy Center, Belfer Center for Science and International Affairs, Jul. 2011, 10, <https://bit.ly/2AtiW3e>.
20 Aguilar et al., "An Introduction to Pakistan's Military," 10.
21 Baqir Sajjad Syed, "Pakistan to retain full spectrum deterrence policy," *Dawn*, Dec. 22, 2017.
22 See, for example, Munir Akram, "India's Pakistan Strategy," *Dawn*, Apr. 3, 2016.

their precision-strike capability in its campaign against terrorists to reduce civilian collateral damage.[23] (These aircraft were in addition to the eight that it was already seeking.)

Pakistan's threat perceptions also include concerns about U.S. intentions in Pakistan and U.S. drone activity in the country. In the last decade and a half, U.S. drone strikes have attacked terrorist targets in the tribal areas of Pakistan. Islamabad's public protests against drone strikes belie the fact that successive governments, including the Musharraf regime, extended tacit cooperation and permission for drone strikes.[24] It is crucial to note that by 2017–2018, defense ties between the United States and Pakistan deteriorated steadily after years of suspicion between the two sides.[25] The May 2011 U.S. Navy Seals operation to kill al-Qaeda leader Osama bin Laden in Abbottabad, Pakistan, provoked broader suspicions in Pakistan about U.S. intentions, especially concerning Pakistan's nuclear weapons. It is therefore unsurprising that in December 2017, the head of the PAF, Air Chief Marshal Sohail Aman, ordered the shooting down of any drones, even American ones, that intruded into Pakistani airspace.[26] A persistent public concern for Pakistan in the post-9/11 period has been that the United States aims to roll back Pakistan's nuclear capability with the justification that Islamabad's nuclear weapons are insecure in the face of terrorist groups, and the legacy of the A.Q. Khan nuclear proliferation network. And in 2018, as the United States suspended most military aid to Pakistan, Air Chief Marshal Aman said that the PAF was "fully prepared to defend all the aerial frontiers of the country."[27]

Outside Pakistan's territory, Pakistani military decision-makers have historically spoken of a need for "strategic depth" through a sphere of influence in Afghanistan. It is not clear how strategic depth would be

---

23 "Pakistan Air Force Needs to Replace 190 Planes by 2020," *Dawn*, Mar. 15, 2016; Khuldune Shahid, "Pay for Your F-16s, US Tells Pakistan," *Friday Times*, May 6, 2016.

24 C. Christine Fair, "Studying Drones: The Low Quality Information Environment of Pakistan's Tribal Areas," in Kerstin Fisk and Jennifer M. Ramos (eds.), *Preventive Force: Drones, Targeted Killing, and the Transformation of Contemporary Warfare* (New York, NY: New York University Press, 2016).

25 Ahmad Faruqui, "Imran's Foreign and Defense Policy: An Initial Assessment," *Daily Times*, Aug. 1, 2018.

26 Omer Farooq Khan, "Pakistan Air Force Chief: Shoot Down US Drones," *Times of India*, Dec. 7, 2017.

27 Mujib Mashal and Salman Masood, "Cutting Off Pakistan, U.S. Takes Gamble in Complex Afghan War," *New York Times*, Jan. 5, 2018.

operationalized militarily and in terms of force posture across the Durand Line. But some instances have provided hints. For example, during the summer 1999 Kargil war, out of fear that India would attack Pakistan's nuclear assets, Islamabad contacted the ruling Taliban in Afghanistan about moving some nuclear weapons to Afghanistan for their protection.[28] Although this plan was not executed, stationing warheads, delivery systems, and other air assets to protect its second-strike capability might be one possible military use of the strategic depth strategy. More generally, Pakistan has sought decisive influence in Afghanistan, through proxies such as the Afghan Taliban and the Haqqani Network, to ensure that India does not pose a challenge on its western borders. The weak and fragile Afghan military is not capable of challenging the Pakistani military, including the PAF, or even posing a limited defense.

Pakistan has frequently had an uneasy relationship with Shia-majority Iran, with which it shares a border and which also has a long border with Afghanistan. Iran has also periodically accused Pakistan of supporting Sunni jihadist groups such as Jaish al-Adl in carrying out attacks in Iran. In the past, both Iran and Pakistan have supported opposing groups in Afghanistan, with Iran backing the Northern Alliance and Pakistan supporting the Taliban.[29] Although there are no major border disputes between Islamabad and Tehran, issues such as the Saudi-Iranian rivalry, Sunni-Shia sectarian divide in Pakistan, and some degree of competition for influence in Afghanistan ensures that Iran is part of Pakistan's threat assessment. It is not clear to what extent Pakistani military doctrine, including its air power strategy, focuses on scenarios involving Iran, but it might be noted that in 2017 a PAF JF-17 aircraft shot down an Iranian drone allegedly spying in Pakistan's Balochistan province.[30]

---

28 Molly Moore and Kamran Khan, "Pakistan Moves Nuclear Weapons," *Washington Post*, Nov. 11, 2001, <https://wapo.st/2todE57>.

29 Lisa Curtis, "The Reorientation of Pakistan's Foreign Policy Toward Its Region," *Contemporary South Asia*, Vol. 20, No. 2, June 2012, 262–63; Maysam Behravesh, "What Does Iran Want in Afghanistan?" Al Jazeera.com, Feb. 4, 2019, <https://www.aljazeera.com/indepth/opinion/190204092658549.html>.

30 Naveed Siddiqui, "Iranian Drone Shot Down by PAF, Confirms FO," *Dawn*, June 21, 2017.

Pakistani Air Power Structure and Capabilities

According to the PAF official website in January 2019, its mission is to "maintain peace with honour, should deterrence fail [to] ensure sovereignty of Pakistani airspace," while its vision is "to be among the most respected Air Force of the world."[31] In conventional military operations, the primary task of the PAF is to achieve theater superiority and provide air support to the army.[32] Apart from aerial defense of the country, the PAF is tasked with retaliatory attacks, nuclear strikes, protection of sea assets, maritime attacks, support for ground operations, and search and rescue missions.[33] These are the more traditional responsibilities. The PAF and Pakistan's air power are also tasked with bolstering Islamabad's strategy of state sponsorship of terrorism against India, by seeking to deny New Delhi a conventional military response.

Pakistani defense strategy has historically been dominated by the Army, which has also controlled the country's politics since independence. Thus, the Air Force and the Navy have largely been subordinate services to the Army. (Operationally, for instance, the PAF was informed of the Army's 1999 Kargil infiltration only after the troops were discovered on the Indian side of the LOC.[34]) Pakistani air capabilities comprise its aircraft in the Air Force, and also naval and army aerial capabilities. In addition, its air power also comprises its missile forces, including ballistic and cruise missiles. Offensive capabilities can be grouped into fighter aircraft and missiles, both for conventional and nuclear roles.

The PAF, which comprises about 70,000 personnel, is divided into three regional commands: northern, southern, and central. As of 2019, it has 425 combat ready aircraft.[35] PAF's fighter aircraft include three squadrons of F-7PGs, the export version of the Chengdu J-7, including the trainer version, the FT-7PG; one squadron of F-16 A/B MLU; one squadron of F-16 A/B ADF; and one squadron of Mirage IIID. Fighter aircraft with ground attack capabilities include two squadrons of JF-17 Thunder, two squadrons

---

31 PAF website, accessed Jan. 29, 2019, <http://paf.gov.pk>.
32 Aguilar et al., "An Introduction to Pakistan's Military," 31.
33 Blank et al., *Vector Check: Prospects for U.S. and Pakistan Air Power Engagement*, 31.
34 John H. Gill, "Military Operations in the Kargil Conflict," in Peter R. Lavoy (ed.), *The Causes and Consequences of the Kargil Conflict* (Cambridge, UK: Cambridge University Press, 2009), 95.
35 "Chapter Six: Asia," *The Military Balance* (London: International Institute for Strategic Studies, 2019), 119:1, 298.

of JF-17 Thunder Block II, one squadron of F-16 C/D Block 52, and three squadrons of Mirage 5 (5PA).[36] As of 2019, Pakistan has 33 F-16As, 24 F-16Bs (including some that received mid-life upgrades), 12 F-16Cs, and 6 F-16Ds, apart from 49 JF-17 Thunder Block Is, and 36 JF-17 Thunder Block IIs.[37] Other estimates also say that the PAF has about 110 JF-17 fighters, which were first inducted in 2007, and are multirole, single-engine aircraft, with a supposed operational range of 1,200 kilometers.[38]

PAF aircraft also include one squadron of Mirage 5PA2/5PA3 armed with AM-39 Exocet anti-ship missiles. For AEWC it has one squadron with Saab 2000 and Saab 2000 Erieye, and one squadron with ZDK-03, a variant of the Shaanxi Y-8.[39] The PAF has one squadron of IL-78 Midas Tankers; one squadron of C-130B/E Hercules, CASA CN235M-220, and Lockheed L-100-20 aircraft; a VIP squadron with Boeing 707, Cessna 560XL, CN235M-220, Fokker F-27, Dassault Falcon 20E, and Gulfstream GIV-SP; and a communications squadron with Embraer Phenom-100 and Harbin Y-12 aircraft.[40] Its helicopter fleet consists of one squadron of Mi-171Sh and six squadrons of Aerospatiale's SA316 Alouette III, and its electronic warfare capabilities include one squadron of Dassault's Falcon 20F aircraft.[41]

The Navy and Army also have some limited aviation capabilities. For the Army, air defense, reconnaissance, and attack capabilities are part of its planning. In 2017, it concluded a deal for eight Airbus H-125M Fennec armed helicopters, and the United States will supply the Army with 12 Bell AH-1Z attack helicopters under foreign military financing, as agreed to in 2015.[42] The PAF also maintains a JF-17 squadron based in Karachi for a maritime strike role and, given the Navy's limited air attack capabilities, it is likely to rely on submarine-launched cruise missiles, including the Babur-3.[43] Pakistani F-16s are equipped with advanced medium-range

36 Ibid.
37 Ibid., 299.
38 Liu Zhen, "Did Pakistan Use Its Chinese JF-17 Jets to Shoot Down Indian Planes?" *South China Morning Post*, Mar. 1, 2019; Franz-Stefan Gady, "Confirmed: Pakistan Air Force Now Operates 70 JF-17 Fighter Jets," *Diplomat*, Dec. 13, 2016; Naveed Siddiqui, "JF-17B Fighter Jet Takes Maiden Test Flight," *Dawn*, Apr. 28, 2017.
39 "Chapter Six: Asia," *The Military Balance*, 2019, 119:1, 298.
40 Ibid.
41 Ibid.
42 Brian Cloughley, "Pakistan's Armed Forces," *Asian Military Review*, May 7, 2018, <https://asianmilitaryreview.com/2018/05/pakistans-armed-forces>.
43 Ibid.

air-to-air missiles (AMRAAMs), which fulfilled PAF's desire for beyond visual-range missiles, and which complement its AEWC planes, bolstering its capabilities against India.[44]

The PAF aims to acquire 150 JF-17s with Block I, Block II, and Block III versions of the plane, with domestic production of 58 percent of the airframe, as it replaces the aging F-7 and Mirage III/5 fighters.[45] Other reports say that the PAF is likely to induct 250-300 JF-17s, to replace older aircraft, making it the mainstay of the PAF.[46] Some of the Block II and Block III variants will be equipped with in-flight refueling, which would imply longer-range maritime strike capabilities and operations planned by the PAF, especially in the Arabian Sea.[47] This should be viewed in context of future joint operations involving the PAF and the Pakistan Navy; as an example, in March 2018, the two services conducted a joint-firing exercise of anti-ship missiles from a JF-17 aircraft and a naval vessel.[48] According to a January 2019 IAF intelligence assessment, the PAF has asked China to speed up the supply of up to 62 more JF-17 Block II and Block III planes as soon as possible, including one batch by summer 2019.[49] Although JF-17 fighters are not on par with Rafale, Eurofighter, and Gripen, according to various experts, it does seem to fulfill Pakistan's requirements more cheaply.[50]

According to Pakistani estimates, about 190 fighter aircraft will be phased out of the PAF by 2020, which would then require replacements that could potentially include French aircraft and Russian Sukhoi-35s, a fifth-generation plane that Russia has reportedly offered to sell to Pakistan, although in other reports Moscow denied such a possibility.[51] Pakistani officials say that India's quest for major air force acquisitions (including the proposed 126 fighter aircraft) and its desire for fifth-generation aircraft

44 Jamal Hussain, "Pakistani Air Power," in John Andreas Olsen (ed.), *Routledge Handbook of Air Power* (New York, NY: Routledge, 2018), 346.
45 Gady, "Confirmed: Pakistan Air Force Now Operates 70 JF-17 Fighter Jets."
46 Siddiqui, "JF-17B Fighter Jet Takes Maiden Test Flight."
47 Franz-Stefan Gady, "Pakistan Stands Up New Fighter Squadron," *Diplomat*, Mar. 8, 2018.
48 "Navy Test-Fires Anti-Ship Missile," *Nation*, Mar. 5, 2018.
49 Manish Shukla, "Pakistan Eyes 62 JF-17 Jets from China as Rafale Dogfight Continues in India," *Zee News*, Jan. 7, 2019.
50 Saurabh Joshi, "Video: PAF JF-17 at #PAS 15," StratPost.com, Jul. 5, 2015, <https://www.stratpost.com/video-paf-jf-17-at-pas15/>.
51 "Pakistan Air Force Needs to Replace 190 Planes by 2020," *Dawn*, Mar. 15, 2016; Dipanjan Roy, "Russia-Pakistan Ties to Remain 'Quite Limited'," *Times of India*, Jul. 12, 2018.

provide a justification for Islamabad to look for similar air assets.[52] But they ignore that India's proposed expansion of air capabilities is driven primarily by threat perceptions of China's air capabilities, including along the eastern and northeastern sectors of the border between the two countries. India's expansion includes its proposed acquisition of the Russian S-400 air defense missile system and Rafale fighter jets that will be deployed along the border, as well as existing assets such as the Sukhoi Su-30 MKI fighters and the Akash missile system.[53]

News reports in 2018 stated that Pakistan was seeking a new lead-in fighter-trainer, although until recently defense planners had discounted their need on the grounds of expense and lack of requirement.[54] In 2009, there were reports that Pakistan would receive China's J-10 single-engine, multirole aircraft, but the deal and delivery have not been confirmed since.[55] This proposal might have been overtaken by consideration of other, more advanced aircraft purchases, such as the J-20 stealth fighter, currently under development by China.[56]

Pakistan has managed to refurbish and upgrade several variants of its decades-old Mirage aircraft (5-EF, III-DP, III Rose-I), which have performed many roles, including reconnaissance flights against India, bombing runs during the 1971 war with India, and shooting down Soviet and Afghan aircraft during the anti-Soviet campaign in Afghanistan.[57] Due to the expense involved in overhauling large numbers of the aircraft abroad, Pakistan set up the Mirage Rebuild Factory at the Kamra aeronautical complex near Islamabad, along with purchasing decommissioned Mirages from other countries and developing, with South African assistance, air-to-air refueling mechanisms.[58] One view is that the ideal replacement for the Mirages would be the Rafale, manufactured by France's Dassault, but given the

---

52 Ibid.

53 Sushant Singh, "China Builds Air Assets in Tibet, Indian Missile Units Head East," *Indian Express*, Dec. 20, 2018.

54 "Pakistan Air Force Is Seeking a Lead-in Fighter Trainer," Quwa.org, Dec. 12, 2018, <https://quwa.org/2018/12/12/ideas-2018-pakistan-air-force-is-seeking-a-lead-in-fighter-trainer-2>.

55 Blank et al., *Vector Check: Prospects for U.S. and Pakistan Air Power Engagement*, 37; Mehmood ul Hassan Khan, "Pakistan-China Defence Cooperation," *Defence Journal*, Mar. 31, 2014.

56 "Pakistan Eyes Export Variant of China's First Stealth Aircraft," *Times of India*, Nov. 1, 2016.

57 "Thrifty at 50: How the Pakistan Air Force Keeps Ageing Mirages Flying," *Dawn*, Apr. 29, 2018.

58 Ibid.

hefty price tag, Islamabad opted for the JF-17, co-produced with China.[59] The PAF launched the Air Force Development Plan (AFDP) some years ago, to add force multipliers including airborne electronic warfare and AWACS planes and refuelers such as IL-78s, but this program has been suspended.[60]

Increasingly, Pakistan is working on incorporating drones into its military strategy, both against state and nonstate actors, and for surveillance and carrying out attacks through armed drones. In October 2018, Beijing announced that China would sell to Islamabad 48 Wing Loong II armed drones, which are reportedly equivalent to American MQ-9 Reaper drones, and are meant for reconnaissance and attacks through air-to-surface weapons.[61] According to official Chinese media, the Pakistan Aeronautical Complex in Kamra and China's Chengdu Aircraft Industrial Company will jointly produce the drones in the coming years.[62] This might appear to have been a response to the U.S. decision to offer India 22 armed Sea Guardian drones, which would reportedly be the first transfer of an armed drone from the United States to a nonmember of the North Atlantic Treaty Organization (NATO).[63] But this is belied by the chronology, because by September 2015, Pakistan was carrying out armed drone strikes of its own on jihadist groups in the tribal areas of western Pakistan, using drones that "strongly resembled" Chinese drones.[64]

So even before the recent military drone-related interactions between New Delhi and Washington, Islamabad had already incorporated armed drones into its military operations. One the one hand, it showed that Islamabad was not averse to carrying out military action against the likes of the Pakistani Taliban, but, on the other hand, for Islamabad this was a way of pushing back against the United States and its controversial drone strikes in Pakistan.[65]

---

59 Ibid.
60 Shamsi, "Indian Air Force Modernization Plan of 2020: Challenges for Regional Air Forces," 18.
61 "China to Sell 48 High-End Military Drones to Pakistan," *Economic Times*, Oct. 10, 2018.
62 Ibid.
63 Mike Stone, "Exclusive: U.S. Offers India Armed Version of Guardian Drone – Sources," Reuters, Jul. 18, 2018, <https://reut.rs/2Ax74gy>.
64 Kiran Stacey, "Pakistan Shuns US for Chinese High-Tech Weapons," *Financial Times*, Apr. 17, 2018.
65 Ibid.

Pakistan has regularly justified seeking new aircraft for counterinsurgency purposes. Since at least 2014, Pakistani security forces have carried out sustained counterinsurgency operations against the Pakistani Taliban in North Waziristan in the Federally Administered Tribal Areas through Operation Zarb-e-Azb, which has also involved air support from the PAF, as well as more direct air attacks against militants. According to the PAF, after August 2008, air operations were launched against Taliban elements using new F-16 aircraft and C-130s and army attack helicopters, and in the first two years the operations covered thousands of sorties against over 4,000 terrorist targets.[66] By 2009, air power had acquired a greater role in counterinsurgency operations against the Pakistani Taliban, with more precise targeting using laser bombs and more accurate imagery to avoid excessive civilian casualties, while the army provided the Air Force with targeting and location data.[67] Nevertheless, according to some sources, air strikes as part of Operation Zarb-e-Azb from 2014 onward led to substantial civilian casualties.[68] Part of the reason why the PAF was pressed into service was concern over deployment of ground forces and casualties in the Waziristan tribal areas, even though it was clear that air strikes cannot be a substitute for counterinsurgency operations.[69] Air attacks in association with helicopter gunship attacks, infantry, and artillery firing were carried out by F-16s in Swat Valley and South Waziristan, using some limited technical U.S. assistance such as high-resolution, infrared sensors, and intelligence provided by secret U.S. drone flights.[70] But, worried about prevailing anti-U.S. sentiments in Pakistan, the PAF repeatedly sought to develop its own technical capabilities for target selection and imagery analysis, and according to the Pakistani air chief, "the biggest handicap we had in Bajaur was that we didn't have good imagery."[71]

Pakistan has frequently made the case for further supply of F-16 aircraft, saying that they would be used against the Pakistani Taliban more

---

66 Stephen Trimble, "F-16s Powered up Pakistan's Counter-Insurgency Strikes," *FlightGlobal*, Nov. 13, 2011, <https://bit.ly/2GGCEhr>; Brian Cloughley, "Pakistan's Armed Forces."
67 Eric Schmitt, "Pakistan Injects Precision into Air War on Taliban," *New York Times*, Jul. 29, 2009.
68 Marvin Weinbaum, "Insurgency and Violent Extremism in Pakistan," *Small Wars and Insurgencies*, 2017, Vol. 28, No. 1, 46.
69 Schmitt, "Pakistan Injects Precision into Air War on Taliban."
70 Ibid.
71 Ibid.

effectively, although Washington's concern has been that the aircraft would be used to bolster Islamabad's offensive military capabilities against New Delhi.[72] As it happens, the AMRAAMs installed on F-16s are meant to strike aerial targets, rather than the ground targets of the Pakistani Taliban, which does not have offensive air capability—which further detracts from the counterinsurgency rationale for F-16s.[73] One route toward increased aerial counterinsurgency capabilities is by modifying some of its 300 MFI-395 Super Mushak military trainer plans with precision-guidance bombs and intelligence, surveillance, and reconnaissance capabilities, which Islamabad began in 2017.[74] This could be a more cost-effective way of adding aircraft for counterinsurgency operations. It is also worth noting that Islamabad has exported the MFI-395 to several countries already, including Qatar and Nigeria.[75]

So while the PAF has an expanded set of threats, including counterinsurgency missions, the increase in capabilities that focus on counter-air and longer-range attack operations would primarily be meant for missions against India, something that would negatively affect Islamabad's ties with Washington.[76] Therefore, the Pakistan Navy's P3C Orion aircraft, which the U.S. government justified supplying to Islamabad in 2003 on the grounds that they would help counter maritime drug trafficking and al-Qaeda, would actually be operated against India, considering that Pakistan does not really have any major maritime security threats from nonstate actors.[77]

Newer F-16 variants have competed (unsuccessfully) for the 126 Indian MMRCA contract a few years ago.[78] This might have reduced the possibilities of any future supplies of F-16s to Pakistan. Pakistan plans to rely on JF-17s to expand its force, but they lack the precision targeting

---

72 Schmitt, "Pakistan Injects Precision into Air War on Taliban"; "Pakistan Could Use F-16 Jets against India: US Lawmakers to Obama," *Times of India*, Apr. 28, 2016.
73 Seema Sirohi, "India-Pakistan Crisis: F-16s and the Missile Mystery," TheWire.in, Mar. 11, 2019, <https://thewire.in/security/india-pakistan-crisis-f-16s-and-the-missile-mystery>.
74 Franz-Stefan Gady, "Pakistan to Arm Super Mushak Aircraft for Counterinsurgency Operations," *Diplomat*, May 24, 2017.
75 Ibid.
76 Blank et al., *Vector Check: Prospects for U.S. and Pakistan Air Power Engagement*, 5.
77 Ibid., 31.
78 Saurabh Joshi, "Lockheed Martin to Announce New F-16 Block 50 Variant," StratPost.com, Feb. 15, 2012, <https://bit.ly/2IgTEfd>; Saurabh Joshi, "LMT Tata to Build F-16 Wings in India," StratPost.com, Sept. 4, 2018, <https://bit.ly/2BEKqn1>.

capabilities available in F-16s.[79] Islamabad has sought the Damocles targeting pod for the JF-17, manufactured by Thales, but that deal did not work out due to Indian pressure, financial constraints, and worries over technology leakage to China.[80] The advantages of the JF-17 for Pakistan are that it is much less expensive than the F-16, involves joint production with China, has export potential, and is useful for counterinsurgency and ground support operations, although it is not clear if it can match Indian fighters in air superiority.[81] It is notable that Pakistan plans to station JF-17s close to the border with India, and weapons such as beyond visual range missiles for JF-17 further emphasize its India focus.[82] The PAF has been part of the Azm-e-Nau military exercises between 2009 and 2013 to test operational readiness and joint services operations in scenarios such as a Cold Start–like limited attack by India.[83] In these exercises (including High Mark 2010, 2016), the PAF's F-7P and Mirage III/V planes also practiced landing and taking off from motorways, which is not necessarily a major achievement, according to IAF officers.[84]

Beyond the more short-term fighter aircraft acquisitions, Pakistan seeks to develop a fifth-generation fighter, and according to the PAF it is working on this project indigenously with the aim of making it independent of International Traffic in Arms Regulations constraints due to concerns over U.S. sanctions, and not relying on "western or eastern partners" according to the PAF chief Air Chief Marshal Mujahid Anwar Khan in 2019.[85] But there is little evidence that Islamabad has made much headway with this project, especially considering that few such aircraft are in service

79 John Irish, "Pakistan Wants Air Force Upgrade for Prolonged Militant Fight," Reuters, Apr. 7, 2016.

80 Ibid.

81 Blank et al., *Vector Check: Prospects for U.S. and Pakistan Air Power Engagement*, 38.

82 Manjeet Singh Negi, "Pakistan Builds Airbase near Gujarat Border, Deploys Chinese JF-17 Fighters," *India Today*, Jul. 9, 2018, <https://bit.ly/2HQrt6A>; "Pakistan Successfully Test-fires 'Beyond Visual Range' Missile from JF-17 Thunder," *Express Tribune*, Feb. 2, 2018.

83 Muhammad Ali Baig and Hamid Iqbal, "A Comparative Study of Blitzkrieg and Cold Start Doctrine: Lessons and Countermeasures for Pakistan," *IPRI Journal*, Winter 2018, 20.

84 Saurabh Joshi, "PAF Road Test: No Big Deal," StratPost.com, Apr. 2, 2010, <https://www.stratpost.com/paf-road-test-no-big-deal>; Zafar Khan, "Cold Start Doctrine: The Conventional Challenge to South Asian Stability," *Contemporary Security Policy*, Dec. 2012, Vol. 33, No. 3, 586; Pamela Constable, "Pakistan Military Prepares for a Possible Indian Attack," *Washington Post*, Sept. 22, 2016.

85 Alan Warnes, "Pakistan Air Force Builds for the Future," *Asian Military Review*, Jan. 10, 2019, <https://asianmilitaryreview.com/2019/01/pakistan-air-force-builds-for-the-future>.

globally. The PAF might be referring to the Chengdu J-20, which has been mentioned in context of future PAF acquisitions. But this would hardly be indigenous, and, more crucially, it is unclear if Islamabad would have the financial means to support such an acquisition or even a joint production.

Finally, Pakistan also regards as a threat India's rapidly expanding satellite launch capabilities, which allow Delhi to detect Pakistani missile launches and carry out intelligence gathering, communications, and high-resolution imagery analysis.[86] Islamabad has always linked its limited space program, run by the Space and Upper Atmospheric Research Commission, to its ballistic missile program.[87] India is also focusing on anti-satellite capabilities, including Delhi's successful anti-satellite test in March 2019, which was meant to be demonstrator against China's own anti-satellite weapons.[88] But given Islamabad's alliance with Beijing, it would likely benefit from Beijing's own substantial anti-satellite weapons, and therefore it might not actually need to invest in such technologies on its own.

Air Power and Pakistan's Nuclear Weapons

Pakistan's nuclear weapons are, along with its strategy of supporting terrorist groups in India, an important mechanism for Islamabad to neutralize New Delhi's conventional military superiority, and its air capabilities are crucial to its nuclear program. Pakistan sees the expansion of its offensive air power capabilities as integral to preserving its conception of "strategic stability" or "strategic balance."[89] Islamabad has an expanding nuclear weapons stockpile of about 140–150 warheads, according to 2018 estimates by the *Bulletin of the Atomic Scientists*, which will further increase more in the coming years to about 220–250 warheads by 2025, based at least in part

---

86 Raja Mansoor, "Pakistan Is Losing the Space Race," *Diplomat*, Feb. 1, 2018.

87 Mian Zahid Hussain and Raja Qaiser Ahmed, "Space Programs of India and Pakistan: Military and Strategic Installations in Outer Space and Precarious Regional Strategic Stability," *Space Policy*, Vol. 47, Feb. 2019, 63–75.

88 Ashley J. Tellis, "India's ASAT Test: An Incomplete Success," Carnegie Endowment for International Peace, Apr. 15, 2019, <https://carnegieendowment.org/2019/04/15/india-s-asat-test-incomplete-success-pub-78884>.

89 "Pakistan Has Cost-Effective Solution to India's Latest Ballistic Missile Defence System: Report," *Economic Times*, Nov. 7, 2018.

on the number of nuclear-capable launchers available to Islamabad.[90] The nuclear element of Pakistan's air power is organized under the Strategic Forces Command (SFC), which has 12,000–15,000 personnel, and consists of medium-range ballistic missiles (Ghauri, Ghauri-2, Shaheen-2, Shaheen-3), short-range ballistic missiles (Ghaznavi, Shaheen-1, Abdali, and Nasr), and nuclear-capable cruise missiles (the Babur series, and Ra'ad), with some of the cruise and ballistic missiles in the testing phase (Shaheen-2, Shaheen-3, and Ra'ad).[91] Pakistan's missiles are road mobile, with shorter range missiles based close to the border with India and the medium range ones located further inland.[92] The Strategic Plans Division, the secretariat of the NCA, administers the country's nuclear program. The three services maintain their own strategic forces commands, although operation control of the nuclear weapons delivery systems ostensibly rests with the NCA.[93] As part of this arrangement, the Army controls the nuclear-capable surface-to-surface missiles and the PAF controls nuclear-capable aircraft, while providing air defense safeguards for nuclear facilities that are specified as no-fly zones.[94]

Also included in the SFC are one or two squadrons of F-16 A/B and Mirage III and Mirage V aircraft, which are the primary aircraft meant for a nuclear delivery role, with F-16s capable of a 1,600-km range, and likely equipped with one nuclear bomb.[95] The Ra'ad air-launched cruise missile has been tested from the Mirage III and Mirage V planes. It is notable that these aircraft might soon have aerial refueling capability, which could expand the range of any nuclear delivery operation.[96] There have also been reports that the JF-17 aircraft, possibly equipped with the Ra'ad, might be prepared for nuclear missions, at least in part due to problems getting spares for the F-16s, and the diminishing possibility of getting F-16 aircraft in the future.[97]

---

90 Hans M. Kristensen, Robert S. Norris, and Julia Diamond, "Pakistani Nuclear Forces, 2018," *Bulletin of the Atomic Scientists*, 2018, Vol. 74, No. 5, 348.
91 "Chapter Six: Asia," *The Military Balance*, 2019, 119:1, 297.
92 Kristensen, Norris, and Diamond, "Pakistan Nuclear Forces, 2018," 353.
93 Aguilar et al., "An Introduction to Pakistan's Military," 35.
94 Ibid.
95 "Chapter Six: Asia," *The Military Balance*, 2019, 119:1, 297; Kristensen, Norris, and Diamond, "Pakistan Nuclear Forces, 2018," 352.
96 Kristensen, Norris, and Diamond, "Pakistan Nuclear Forces, 2018," 352.
97 Ibid., 353.

Pakistan has placed great importance on cruise missiles for a nuclear delivery role, much more so than India. Its three main cruise missile projects are:

- the Babur ground-launched, subsonic missile, which has a range of 350 kilometers, with an advanced version, Babur-2, that has a range of 700 kilometers
- the Ra'ad, an air-launched missile that has been tested from a Mirage III aircraft, while the under-development advanced Ra'ad 2 variant will have a range of 550 kilometers
- the submarine-launched Babur-3, which has been under testing since 2017, with a range of 450 kilometers and which will be based on the diesel-electric Agosta 90B submarines.[98]

Pakistan views sea-launch nuclear capability as integral to its quest for a second-strike capability.[99] But the submarine-based Babur-3 comes with problems of survivability (from Indian forces) and further instability, as well as questions about command and control.[100] It might be noted that in 2012, Islamabad established the Naval Strategic Forces Command, emphasizing the increased focus on a naval nuclear deterrent.

Pakistan's nuclear posture is officially described as a credible minimum deterrent, focusing primarily on India.[101] Islamabad maintains a first-use posture, and although it has not officially declared its nuclear doctrine, elements of that doctrine can be gleaned from public statements by military and government officials, and from the direction of the country's nuclear weapons program.[102] For Pakistan, nuclear weapons are primarily

---

98 Kristensen, Norris, and Diamond, "Pakistan Nuclear Forces, 2018," 355; Ankit Panda and Vipin Narang, "Pakistan Tests New Sub-Launched Nuclear-Capable Cruise Missile. What Now?" *Diplomat*, Jan. 10, 2017.

99 "Pakistan Says Second Strike Capability Attained," *Dawn*, Mar. 30, 2018.

100 Panda and Narang, "Pakistan Tests New Sub-Launched Nuclear-Capable Cruise Missile. What Now?"

101 Naeem Salik, "Pakistan's Nuclear Force Structure in 2025," Carnegie Endowment for International Peace, June 30, 2016, <https://carnegieendowment.org/2016/06/30/pakistan-s-nuclear-force-structure-in-2025-pub-63912>; Syed, "Pakistan to Retain Full Spectrum Deterrence Policy."

102 Sadia Tasleem, "Pakistan's Nuclear Use Doctrine," Carnegie Endowment for International Peace, Regional Insight, June 30, 2016, <https://carnegieendowment.org/2016/06/30/pakistan-s-nuclear-use-doctrine-pub-63913>.

meant to deter a conventional or nuclear attack by India.[103] They also alle-viate Pakistan's perceived lack of strategic depth. Its "red lines" reportedly include a wide range of scenarios. According to Lt. General Khalid Kid-wai (retired), long-serving former director general of the Strategic Plans Division (SPD) and now an adviser to its secretariat, the NCA, nuclear deterrence failure would lead to nuclear weapons use if India attacks Paki-stan and captures a large part of its territory, destroys a major part of its military, economically blockades Pakistan, or foments internal destabili-zation of Pakistan.[104] Although these assertions were made in 2002, there is little reason to assume the thinking has necessarily changed, even if it is unclear if Islamabad would use nuclear weapons in the broader scenarios mentioned by Kidwai, such as perceptions of internal destabilization and economic pressures.

In recent years, Pakistan has proclaimed a "full spectrum deter-rence" strategy. As mentioned earlier, the NCA, the apex nuclear weapons decision-making organization in Pakistan, states that the country faces three primary threats: increasing conventional arms imbalance with India, India's quest for BMD, and nuclearization of the Indian Ocean region.[105] Full spectrum deterrence, which would be part of the credible minimum deterrence strategy, aims to incorporate weapons of sufficient yield across three categories—strategic, operational, and tactical—against Indian counter-value, counter-force, and battlefield targets either on the mainland or outside territories.[106]

Pakistan's contention is that the gap with India's conventional mil-itary capabilities increases Islamabad's dependence on nuclear weapons (at various levels of confrontation). At least publicly, an important mes-sage from Pakistani defense planners is that there is little gap between conventional military operations and escalation to a nuclear level. In part this is meant to infuse sufficient uncertainty among Indian decision-makers about Islamabad's "red lines." Such close proximity between conventional and nuclear delivery operations has implications for Pakistani air power,

---

103 Ibid.
104 "Nuclear Safety, Nuclear Stability and Nuclear Strategy in Pakistan," A Concise Report of a Visit by Landau Network – Centro Volta, Jan. 14, 2002, <https://pugwash.org/2002/01/14/report-on-nuclear-safety-nuclear-stability-and-nuclear-strategy-in-pakistan/>.
105 Syed, "Pakistan to Retain Full Spectrum Deterrence Policy."
106 Ibid.

especially nuclear delivery systems, which would have to be kept at a higher state of readiness, if only to reinforce resolve and deterrence against India and, possibly, the United States.

One of the most concerning developments surrounding Pakistan's nuclear weapons program has been the lowering of its nuclear threshold by inducting battlefield (or tactical) nuclear weapons, such as subkiloton warheads. For Pakistan, battlefield nuclear weapons are ostensibly meant to counter India's Cold Start strategy, which plans quick military thrusts across the border in response to provocations such as terrorist attacks on Indian territory backed by Pakistan. Pakistani attack scenarios include a PAF preemptive strike with nuclear-tipped cruise missiles on IAF bases in Rajasthan, across the border.[107] This expansion of Islamabad's nuclear posture also means a greater focus on short-range missiles in a delivery role. The 350-km-range Ra'ad cruise missile and the 70-km-range Nasr ballistic missile, which was inducted in 2017, are a part of Pakistan's full spectrum deterrent and seek to deter limited attacks by India, and, according to the Pakistani military, can defeat any Indian BMD.[108] Pakistani analysts regard battlefield nuclear weapons as Islamabad's "contribution to peace" through tactical deterrence.[109] Yet that argument disregards the inherently destabilizing potential of reducing the nuclear use threshold.

Pakistan has also voiced discontent over India's BMD project due to its implications for Islamabad's credible minimum deterrent. According to Lt. Gen Khalid Kidwai, Islamabad has "cost-effective solutions" in the form of multiple independent reentry vehicle (MIRV) capability for its ballistic missiles, different categories of cruise missiles against future Indian BMD systems (including the proposed Russian S-400 system), and a full spectrum deterrence strategy that would counter India's nuclear powered submarines.[110] Thus, according to the NCA, the under-development

107 Sannia Abdullah, "Nuclear Ethics? Why Pakistan Has Not Used Nuclear Weapons … Yet," *Washington Quarterly*, Winter 2019, 159.
108 Rajagopalan, "Pakistan's Nasr Missile: 'Cold Water' over India's 'Cold Start'?"; Khan, "Cold Start Doctrine: The Conventional Challenge to South Asian Stability," 585.
109 "Pakistan Has Cost-Effective Solution to India's Latest Ballistic Missile Defence System: Report."
110 Ibid.

2,200-km-range Ababeel missile system with MIRV technology would counter India's BMD capability.[111]

For Pakistan, air power also performs a signaling role. In the subcontinent's nuclear era, there have been periodic crises sparked off mainly by terrorist attacks in India by Pakistan-backed groups, including the Kargil conflict in 1999; Operation Parakram, India's exercise in coercive diplomacy following the attack on the Indian parliament by Pakistan-backed JeM terrorists in 2001; the 2008 Mumbai attacks by the Pakistan-backed Lashkar-e-Taiba; and the 2019 Pulwama attack by JeM. Through movements of delivery systems such as its nuclear-capable fighter aircraft, Pakistan has sought to create the belief among Indian (and American) defense leaders that it would consider nuclear use if India responded to terrorist attacks with its own conventional military offensives. Most notably, since the Kargil war, there has been a debate over whether Pakistan took measures to ready nuclear warheads and delivery systems as a threat to escalate to a nuclear level, with at least one senior U.S. intelligence official confirming that this was indeed what happened, with a firm denial by a Pakistani nuclear expert.[112] There have been unconfirmed reports based on interviews with senior Pakistani military (including Air Force) officers that during the Kargil war, Islamabad had moved its strategic forces, without a full-scale alert, involving sorties by nuclear-armed F-16 aircraft, following pressure on the Pakistan army after the Indian army's artillery barrages disrupted Pakistani supply lines across the LOC.[113]

Finally, apart from these threat perceptions and utilities, Pakistan views air power as instruments of status. Missiles and fighter aircraft are among major defense weaponry that can capture the public imagination. It should be noted that Pakistan seeks parity with India with respect to weapons and strategic technology benefits accorded to India by the international community and major powers. The Indo-U.S. civilian nuclear cooperation agreement, concluded in 2009, and the waiver granted by the Nuclear Suppliers Group (NSG) in 2008 provoked consistent demands from Pakistan to be accorded the same treatment. A consistent pillar of Pakistani

---

111 Syed, "Pakistan to Retain Full Spectrum Deterrence Policy."
112 Gill, "Military Operations in the Kargil Conflict," 111–112.
113 Abdullah, "Nuclear Ethics? Why Pakistan Has Not Used Nuclear Weapons … Yet," 161–62.

foreign policy since its creation has been hyphenation with India, including especially on the nuclear level. But the NSG waiver for India and the Indo-U.S. nuclear deal have effectively de-hyphenated the Indian and Pakistani nuclear programs for the international community. This status anxiety on the part of Islamabad can also be viewed in other sectors, including missiles and aircraft. Therefore, major Indian projects to strengthen its nuclear deterrence with respect to China, such as a nuclear-powered submarine and BMD, then pushes Islamabad to consider countermoves.[114] These countermoves have especially been visible in the domain of air power.

Nuclear weapons and air power in Pakistan, especially ballistic and cruise missiles, have also acquired a great deal of symbolic and cultural importance in Pakistani society, especially since the 1998 nuclear tests. Visual representation of the Ghauri and Shaheen missiles (and the Chagai Hills nuclear test site) included their replicas and monuments in the public space as well as souvenir items depicting their representations.[115] Thus, these instruments of air power are not just meant to fulfill a security need, but also to influence popular culture and the public sphere for domestic political reasons. It is also worth mentioning that the names of Pakistani ballistic and cruise missiles, Ghauri, Ghaznavi, Babur, and Abdali, appear to be attempts at projecting a sense of dominance vis-à-vis India. Muhammad Ghori, Mahmud of Ghaznavi, and Babur, the first Mughal emperor, invaded and prevailed in present-day India (to differing consequences) and therefore naming important nuclear-capable missiles after them seeks to send a certain message of power and potency.[116]

PAF's External Alliances and Coalitions

As with numerous instances of newly independent countries of the post–World War II period, the development of Pakistan's military and its constituent air power have been influenced by alliances with major powers such as the United States and China. Indeed, Washington and Beijing are the

114 Indian ballistic missile defense thinking has also been driven by the threat of terrorist actors armed with nuclear-tipped missiles.
115 Iftikhar Dadi, "Nuclearization and Pakistani Popular Culture since 1998," in Itty Abraham (ed.), *South Asian Cultures of the Bomb* (Bloomington, IN: Indiana University Press, 2009).
116 See also "Pakistan Will Not Rename Missiles," BBC News, Feb. 23, 2006, <https://bbc.in/2BVb2PX>.

two most important defense suppliers for Pakistan over the last 50 years, with China and Pakistan regarded as "all-weather" allies. Between 1950 and 2014, China provided 40 percent of Pakistan's military equipment, with the United States coming in second at 23 percent.[117]

The PAF emerged from the British Indian air force when Pakistan was created as an independent state in August 1947. From the 1950s, with increased military cooperation between Pakistan and the United States, including through political alliances such as the Central Treaty Organization and South East Asian Treaty Organization, the Pakistan military became more and more influenced by its U.S. counterparts in doctrine and practice.[118] Pakistan and the United States have a long history of defense collaboration, beginning in 1954 when the two concluded the Mutual Defense Agreement, and followed up with the transfer of 128 F-86 Sabre fighter planes, 26 Canberra bombers, and 12 F-104 aircraft in the subsequent years. The U.S. Air Force (USAF) and the Central Intelligence Agency jointly ran the Peshawar air station from 1959 to 1970.[119] During the anti-Soviet campaign in Afghanistan, with Pakistan as a key participant and ally, Washington allowed the sale of 40 F-16 Block A fighters to Islamabad.[120] Pakistani F-16s fought and downed Soviet and Afghan air force aircraft even as at least one F-16 was shot down by a Soviet MiG23.[121] The air force component was thus at the forefront of closer U.S.-Pakistan ties since the 1950s and continued to be so during subsequent periods of defense cooperation and assistance in the 1980s and in the post-9/11 period.

For Pakistan, one of the most sensitive aspects of its overall relationship with the United States is the sense of anger and grievance in Islamabad caused by the denial of 28 remaining F-16 aircraft in 1990 by the George H.W. Bush administration, for which Pakistan had already paid a significant amount and continued to pay the installments, along with the cancellation of sale of spare parts and International Military Education Training

---

117 Blank et al., *Vector Check: Prospects for U.S. and Pakistan Air Power Engagement*, 14, 37.
118 Stephen P. Cohen, *The Idea of Pakistan* (Washington DC: Brookings Institution Press, 2004), 97–106.
119 Blank et al., *Vector Check: Prospects for U.S. and Pakistan Air Power Engagement*, 14–15.
120 Ibid., 17.
121 Richard M. Weintraub, "Afghanistan Says It Downed F16 Fighter from Pakistan," *Washington Post*, May 1, 1987.

(IMET) funding for Pakistan.[122] Under the Pressler Amendment, Washington denied these aircraft to Pakistan due to Islamabad's pursuit of nuclear weapons. Pakistan's complaint has been that periodically, including during the 1990s, the PAF suffered through lack of upgrades and acquisitions due to U.S. sanctions.[123]

After the 9/11 attacks, as the United States commenced military operations against the Taliban and al-Qaeda in Afghanistan, the Musharraf regime reluctantly agreed to military cooperation between Washington and Islamabad. This cooperation included Pakistan as a major supply route for the U.S. military, as well as the use of Pakistani air bases. The cooperation also resumed major military assistance to Pakistan, which had been suspended in 1990 under the provisions of the Pressler Amendment and under the sanctions imposed after the October 1999 coup that brought General Musharraf to power. In 2004, Pakistan was designated a major non-NATO ally, and received 14 F-16 Block As in 2005 and 18 F-16 Block Cs in 2007.[124] During those years, Washington supported Islamabad's quest for more advanced F-16s for counterterrorism operations, and with U.S. foreign military financing.[125] Aerial weapons and platforms such as F-16s, P-3C Orion anti-submarine planes, and Harpoon anti-ship missiles were provided to Pakistan ostensibly as part of broader defense cooperation for counterterrorism, but they can easily be used against India.[126] In fact, it is not clear how anti-submarine aircraft, and anti-ship missiles can be used for counterinsurgency and counterterrorism operations against groups such as the Pakistan Taliban.

In 2018, the Trump administration suspended military aid to Pakistan, which had been an important and controversial aspect of its counterterrorism and counterinsurgency campaign against the Taliban and al-Qaeda in Afghanistan. This estrangement between Washington and Islamabad came about due to two main reasons. First, the United States and Pakistan view

---

122 Aguilar et al., "An Introduction to Pakistan's Military," 26.
123 Shamsi, "Indian Air Force Modernization Plan of 2020: Challenges for Regional Air Forces," 18.
124 Blank et al., *Vector Check: Prospects for U.S. and Pakistan Air Power Engagement*, 21.
125 Donald Camp, "Defeating al-Qaeda's Air Force: Pakistan's F-16 Program in the Fight Against Terrorism," Statement before the U.S. House of Representatives Foreign Affairs Committee on South Asia, Washington, DC, Sept. 16, 2008.
126 Aguilar et al., "An Introduction to Pakistan's Military," 18.

India very differently—for Washington, New Delhi is a long-term strategic partner, while for Islamabad, New Delhi is an implacable rival.[127] This disagreement is also visible in Afghanistan, where Pakistan seeks dominance to counter Indian influence in the country. Second, the continued alliance between Pakistan and jihadist terrorist groups, including the Afghan Taliban, Haqqani Network, and Lashkar-e-Taiba, is an even more crucial reason for the break between Washington and Islamabad. This does not mean that military ties between them will completely cease, but that they would become more transactional, given fundamental disagreements on issues such as Pakistan's support for jihadist terrorist groups.

For Pakistan, India's increasingly close defense partnership with the United States has been a source of anger, because it threatens Pakistan's attempts to create operational parity with India, and because it signals a long-term strategic partnership between New Delhi and Washington. For instance, Pakistan opposed the impending sale of armed Sea Guardian drones by the United States to India in 2018 because, according to the Pakistani official sources, it would "lower the threshold for conflict" and negatively affect strategic stability in the subcontinent.[128] But the problem with this argument is that the flashpoints that lead to crises and conflict between India and Pakistan largely emerge from revisionist policies from Islamabad involving its terrorist proxies. In fact, a much more serious lowering of conflict threshold has emerged from Pakistan's incorporation of battlefield nuclear weapons into its security strategy. Pakistan also regards the United States as attempting to constrain Islamabad's strategic air power capabilities, especially its nuclear-capable missiles.[129]

The downgrading of U.S.-Pakistan ties is reflected in defense assistance. In 2016, the U.S. Congress ended financial support through foreign military financing for acquisition of eight proposed new F-16 planes, which increased their unit price from $270 million to $700 million. This in turn made Pakistani decision-makers conclude that Islamabad could no longer depend on Washington as its main supplier of major weapons systems, which has pushed Pakistan toward relying even further on Chinese

127 Pamela Constable, "Despite Trump's Punitive Military Aid Cut, Pakistan and U.S. Are Still Intertwined," *Washington Post*, Jan. 25, 2018.
128 "Pakistan Opposes Supply of US Armed Drones to India," *Economic Times*, Jul. 13, 2018.
129 Munir Akram, "A Strategic Challenge," *Dawn*, Apr. 29, 2018.

weaponry, such as the JF-17 joint production.[130] Unlike with the United States, Pakistan does not have any fundamental disagreements with China, with which it has had a long strategic military and nuclear alliance, and which is now the biggest arms supplier to Pakistan. Crucially, China's disputes with India make it an even more reliable partner for Islamabad.[131] Moreover, in 2018, Washington also cut the IMET for Pakistani military officers to study at U.S. military education institutions (about 66 officers for 2018–2019, presumably including from the PAF).[132]

Thus, it is unlikely that Pakistan will acquire any further F-16 fighter aircraft (or any other fighters) from the United States in the foreseeable future because of political reasons and budgetary constraints. As a stop-gap measure, Islamabad did buy an F-16 A/B squadron from Jordan in 2014 after the aircraft's mid-life upgrade, but under the current circumstances the planes do not represent the PAF's future.[133] According to some analysts and senior Pakistani officers, it is possible that the JF-17 would replace the F-16, although military officials and reports in 2018 reiterated that Islamabad would continue to seek more F-16s.[134]

The JF-17 and F-16 aircraft are important symbols of the direction of Pakistani foreign and defense policy especially regarding strategic alliances with major powers. The shift from F-16 to JF-17 is emblematic of the close military partnership between Pakistan and China, which does not involve any fundamental disagreements. China's military alliance with Pakistan is part of Beijing's objective of restricting India's strategic ambitions to South Asia by building up Islamabad as a counterweight that would then tie down New Delhi.[135] Defense cooperation assistance such as the JF-17 project and nuclear-capable ballistic and cruise missiles further this aim. For

---

130 Stacey, "Pakistan Shuns US for Chinese High-Tech Weapons"; Mateen Haider, "Pakistan Rejects US Conditions Attached to Sale of F-16s," *Dawn*, May 7, 2016.
131 Stacey, "Pakistan Shuns US for Chinese High-Tech Weapons"; Aparna Pande, "Friends of Last Resort: Pakistan's Relations with China and Saudi Arabia," in C. Christine Fair and Sarah J. Watson (eds.), *Pakistan's Enduring Challenges* (Philadelphia, PA: University of Pennsylvania Press, 2015), 258.
132 Anwar Iqbal, "US Cuts Military Training Programme for Pakistan," *Dawn*, Aug. 11, 2018.
133 "Pakistan Inducts 13 F-16 Fighter Jets into Air Force," LiveMint.com, May 21, 2014.
134 Blank et al., *Vector Check: Prospects for U.S. and Pakistan Air Power Engagement*, 37; "Pakistan Trying to Get More F-16 Jets from US: Report," *Economic Times*, Jul. 12, 2018.
135 Lisa Curtis, "China's Military and Security Relationship with Pakistan," Testimony before the U.S.-China Economic and Security Review Commission on May 20, 2009, The Heritage Foundation.

Pakistan, the JF-17s, along with the technological cooperation, represent an opportunity for even closer ties with China.[136]

This is not to say that the Sino-Pakistani defense partnership is particularly new. The two allies have a long history of nuclear and missile cooperation. Pakistan's missile fleet has substantially depended on direct assistance from China, including the supply of the 300-km-range M-11 missiles in the early 1990s, which became the Ghaznavi (Hatf 3).[137] More recently, China provided Pakistan with optical tracking and measurement systems in 2018, which would be essential for its Ababeel MIRV missile.[138] But it is notable that this alliance has deepened even further in areas of air power, including fighter aircraft and drones. Pakistan and China have also worked on the joint manufacture of four ZDK-03 AWACS aircraft for Islamabad's needs.[139] Apart from the JF-17, China and Pakistan have also jointly produced the K-8 Karakorum jet trainer, and there have been reports of Islamabad's interest in the J-10 fighter. The two air forces conduct annual joint exercises, which is a rarity for China at least and increases interoperability between the two sides.[140] In recent years, the PAF has participated in exercises with air forces of countries with which it has not traditionally had close ties, including Israel (with which it does not have any diplomatic relations), alongside which it participated in the USAF's Red Flag exercise in Nevada in August 2016 (and which also included the United Arab Emirates).[141]

Sino-Pakistani air power cooperation has also been visible in sharing U.S. technology. The Babur cruise missile is reportedly based on the Chinese DH-10, which itself was a reverse-engineered Tomahawk missile that was among several launched by the United States against al-Qaeda in Afghanistan—at least one of which crashed in Baluchistan and was subsequently

---

136 Blank et al., *Vector Check: Prospects for U.S. and Pakistan Air Power Engagement*, 3.
137 Shannon N. Kile, Phillip Schell, and Hans M. Kristensen, "Pakistani Nuclear Forces," *SIPRI Yearbook 2012* (Oxford, UK: Oxford University Press, 2012), 338.
138 Stephen Chen, "China Provides Tracking System for Pakistan's Missile Programme," *South China Morning Post*, Mar. 22, 2018.
139 "China Rolls Out AWACS Aircrafts for Pakistan Air Force," *Deccan Herald*, Nov. 13, 2010.
140 Blank et al., *Vector Check: Prospects for U.S. and Pakistan Air Power Engagement*, 38.
141 Judah Ari Gross, "Israel to Fly Alongside Pakistan, UAE in US Air Force Drill," *Times of Israel*, Aug. 3, 2016.

transferred to China by Pakistan.[142] Another instance of such cooperation was when Islamabad provided Beijing access to a U.S. stealth Blackhawk helicopter that had crashed during the raid to kill al-Qaeda leader Osama bin Laden in May 2011, despite Washington's request to desist from such a step.[143] Similarly, the supposedly indigenous Burraq and Shahpur drones are reported to have been based on technology from China's Rainbow CH-3, which was reverse-engineered from U.S. drone technology.[144]

The intensification of the JF-17-focused partnership is also reflected in China's Belt and Road Initiative (BRI) and its China-Pakistan Economic Corridor (CPEC), under which the two countries reportedly concluded a military agreement in 2018 that would, among other things, expand Pakistan's production of JF-17 fighter jets.[145] The BRI's military component also seeks to integrate Pakistan into the military element of China's BeiDou satellite navigation system, which would permit precision guidance for missiles, aircraft, and ships, in order to wrest away China's allies from the American global positioning system network.[146] Under this military agreement, China and Pakistan will use a CPEC special economic zone to expand the manufacture of JF-17 fighter jets and also jointly produce navigational systems, radar, and aircraft weapons for the first time in Pakistan.[147] The agreement is also meant to be a mechanism for increased Chinese sales to countries of the Muslim world.[148]

But such military agreements as part of CPEC further increase Pakistan's indebtedness to China, already estimated to be around $90 billion over a 30-year period after 2018, and could also lead to permanent Chinese bases in Pakistan.[149] Regardless, CPEC is now part of PAF's planning. A

---

142 Dennis M. Gormley, *Missile Contagion: Cruise Missile Proliferation and the Threat to International Security* (Annapolis, MD: Naval Institute Press, 2010), 73; Blank et al., *Vector Check: Prospects for U.S. and Pakistan Air Power Engagement*, 39; Kamran Khan, "Pakistan Says It Is Studying Errant U.S. Missile," *Washington Post*, Aug. 28, 1998.

143 "Pakistan Let China See Crashed U.S. 'Stealth' Copter," Reuters, Aug. 15, 2011.

144 Blank et al., "Vector Check: Prospects for U.S. and Pakistan Air Power Engagement," 39.

145 Maria Abi-Habib, "China's 'Belt and Road' Plan in Pakistan Takes a Military Turn," *New York Times*, Dec. 19, 2018.

146 Abi-Habib, "China's 'Belt and Road' Plan in Pakistan Takes a Military Turn."

147 Ibid.

148 Ibid.

149 Salman Siddiqui, "Pakistan Will Be Paying China $90b against CPEC-Related Projects," *Express Tribune*, Mar. 12, 2017; Patrick Wintour, "All-Weather Friendship: But Is Pakistan Relying Too Heavily on China?" *Guardian*, Aug. 3, 2018.

new operational airbase, PAF Base Bholari, near Karachi, inaugurated in December 2017, is meant to play an important role in safeguarding CPEC entities.[150]

At the same time, some analysts have assessed that Chinese weapons systems are inferior to those from the West, and that the systems are unoriginal and cannot be used as a force multiplier.[151] Chinese-supplied weapons and platforms are regarded as much less useful in potentially combating India, and also have dangerous weaknesses, including in fighter jets.[152] But Pakistani analysts also say that any technological deficiencies of Chinese aircraft are outweighed by the need for greater numbers of aircraft (that Beijing will presumably supply) and "reliability of sustainable development."[153]

Historically, Pakistan has close military ties with countries in the Middle East, especially Arab states. For instance, PAF pilots flew in the Syrian Air Force during the 1973 Arab-Israel war.[154] Another important partner of Pakistan's defense services, including the Air Force, is Saudi Arabia. Riyadh has historically enjoyed a special relationship with Pakistan's rulers, which has involved major financial assistance, involvement in Pakistani domestic matters, security cooperation including on Islamabad's nuclear program, and the anti-Soviet mujahideen campaign in Afghanistan in the 1980s. The Pakistan Army has periodically been stationed in Saudi Arabia to protect the kingdom, and more recently Riyadh put pressure on Islamabad to assist the Saudi military campaign against the Houthis in Yemen, which Pakistan resisted.[155] At least 1,600 Pakistani troops are generally stationed in Saudi Arabia, ostensibly for training purposes, and 1,000 more were allocated in early 2018, after Riyadh helped prevent Islamabad from being placed in an international terrorism financing watch list by the

---

150 "PAF Inaugurates New Operational Air Base at Bholari near Karachi," Geo TV, Dec. 25, 2017, <https://bit.ly/2BDnMuy>.
151 Aguilar et al., "An Introduction to Pakistan's Military," 13.
152 Ibid.
153 Shamsi, "Indian Air Force Modernization Plan of 2020: Challenges for Regional Air Forces," 20.
154 Abdus Sattar Alvi, "50 Years On: Memories of the 1973 Arab-Israeli Conflict," *Express Tribune*, Mar. 19, 2015.
155 Blank et al., *Vector Check: Prospects for U.S. and Pakistan Air Power Engagement*, 55.

Financial Action Task Force.[156] The PAF and the Royal Saudi Air Force are long-standing partners with the PAF having played an important role in the evolution of the Saudi Air Force, and PAF pilots also flew Saudi BAC Lightning planes during the kingdom's brief conflict with South Yemen in 1969.[157] Saudi Arabia also reportedly provided Islamabad with $500 million toward its purchase of F-16s in the 1980s.[158] The two air forces have substantial interoperability, not least because both fly F-16s, although that factor might be less relevant in the coming years if the PAF does not add any more F-16 planes.[159] Pakistan is also building defense alliances with Turkey, including through an agreement for purchase of attack helicopters from Ankara. A proposal for PAF pilots to train Turkish F-16 pilots fell through in 2017 due to U.S. opposition.[160]

Pakistan has reached out to Russia in recent years. The two countries signed a military cooperation agreement in 2014, after which Islamabad purchased four Mi-35 M combat helicopters from Russia, and in 2018 the two sides signed an agreement for training Pakistani military officers in Russian military education institutions.[161] Although this is fairly low-level military cooperation, it can be assumed to be sending a signal to the United States and India, as defense ties between those two countries have expanded substantially in recent years, and as U.S.-Pakistan defense ties have been downgraded under the Trump administration. Nevertheless, the Russian government has stated that the relationship is unlikely to go much further, and that Moscow would not sell more than a few transport

---

156 Saeed Shah and Ian Talley, "Saudi Arabia Stymies U.S. over Pakistan Terror List," *Wall Street Journal*, Feb. 21, 2018.

157 Blank et al., *Vector Check: Prospects for U.S. and Pakistan Air Power Engagement*, 55. In addition, Jordan transferred ten F-104 Starfighter aircraft to Pakistan during the 1971 war; Tanham and Agmon, *The Indian Air Force: Trends and Prospects*, 39.

158 Ayesha Siddiqa, "Pakistan-Saudi Arabia Military-Ideological Dependency: Can Pakistan Afford to Loose the Middle East?" GESC Chair in International Relations Lecture, Feb. 2016, 12, <https://bit.ly/2Gvdql8>.

159 Blank et al., *Vector Check: Prospects for U.S. and Pakistan Air Power Engagement*, 55.

160 Cloughley, "Pakistan's Armed Forces."

161 Ayaz Gul, "Pakistan, Russia Sign Rare Military Cooperation Pact," Voice of America, Aug. 8, 2018; Drazen Jorgic, "With Gas and Diplomacy, Russia Embraces Cold War for Pakistan," Reuters, Mar. 5, 2018.

helicopters and fighters to Pakistan, and would certainly not sell planes such as the Sukhoi 35, to avoid threatening its defense relationship with India.[162]

Pakistan also seeks to become an exporter of major defense systems, from being primarily a recipient—an aspiration that also speaks to its quest for higher global status. Exporting the JF-17 aircraft to friendly countries such as Egypt, Turkey, and Nigeria would be an important element of this strategy, although Islamabad is not in a position to offer long-term credits and financing to countries such as Myanmar, which concluded a deal for JF-17s with China in 2015, which did offer it long-term credits.[163] It should also be noted that the military agreement concluded between China and Pakistan under the CPEC in late 2018 involves expanding high-tech military production facilities in Pakistan.

This brief summary shows that although Islamabad has various military interactions with major powers, it is with Beijing that the defense, and especially the Air Force element, is becoming even more enmeshed and interoperable. Similarly, in light of the long history of collaboration with the Saudi Air Force, Pakistan also seeks to leverage Air Force–related cooperative interactions with middle-ranking powers to expand its own profile. These external interactions are also highlighted in the next section, which discusses Pakistani defense budgetary issues.

The PAF and Defense Budgetary Issues

Pakistan's internal economic situation and levels of external financial support have, as expected, an important bearing on its defense allocations. Examining Islamabad's defense spending is challenging as available data are unreliable and obfuscated, and comes with little parliamentary accountability, and with the belief that its defense budget is greater than publicly

---

162 Dipanjan Roy Chaudhury, "Russia-Pakistan Ties to Remain 'Quite Limited'," *Times of India*, Jul. 12, 2018; Usman Ansari, "Pakistan Cosies up to Russia, but Moscow Doesn't Want to Take Sides," *Defense News*, May 2, 2018.
163 Stacey, "Pakistan Shuns US for Chinese High-Tech Weapons."

disclosed.[164] Given the dominance of the military, and specifically the army, in the country's politics and security policy making, it is no surprise that the military has a decisive role to play in the defense budgetary allocation process, rather than the finance ministry. Moreover, it is the chief of army staff who ultimately takes all policy decisions related to the three services, including the PAF, for which the chief of air staff provides input.[165] At the same time, the SPD, which is dominated by the army, manages the budget for the nuclear program, and its own allocation comes from the inter-services organizations' share of 21 percent of the defense budget (along with the Inter-Services Intelligence).[166]

An enduring characteristic is that given Pakistan's rivalry with India, Islamabad frequently views its overall economic indicators relative to India. Pakistani analysts have complained that even though Islamabad has a higher proportion of gross domestic product (GDP) allocated to defense than New Delhi, due to India's larger economy, the conventional military gap between the two countries would continue to grow.[167] In this context, a senior retired Pakistani officer said in 2016 that "the gap between Pakistani and Indian military capabilities will grow, because the economic gap between our two nations will grow," and, according to one recent Pakistani air chief, "we are a resource-constrained country. We cannot match our enemy in numbers so we focus on training quality."[168] In 2017, India's GDP was $2.6 trillion, according to the World Bank, while that of Pakistan was $305 billion. India's economic progress and higher defense allocations

---

164 Muhammad Afzar Anwar and Zain Rafique, "Defense Spending and National Security of Pakistan: A Policy Perspective," *Democracy and Security*, 2012, 8, 385; Babar Ayaz, "Cloaked in Secrecy," *Newsline*, Jul. 2011; Shane Mason, "Military Budgets in India and Pakistan: Trajectories, Priorities, and Risks," Stimson Center, 2016, 22, <https://www.stimson.org/sites/default/files/file-attachments/Military-Budgets-India-Pakistan-Trajectories-Priorities-Risks-Oct2016.pdf.pdf>.
165 Blank et al., *Vector Check: Prospects for U.S. and Pakistan Air Power Engagement*, 6.
166 Mason, "Military Budgets in India and Pakistan: Trajectories, Priorities, and Risks," 27, 35.
167 Pervaiz Iqbal Cheema, "Strategic Stability in South Asia: The Role of USA," *Journal of Contemporary Studies*, Summer 2012, Vol. 1, No. 1, 6.
168 Blank et al., *Vector Check: Prospects for U.S. and Pakistan Air Power Engagement*, 34.

are used as a justification by Pakistan for increasing its own conventional defense purchases and for greater reliance on nuclear weapons.[169]

From the total defense budget, the PAF receives 20 percent, with at least 47 percent going to the army.[170] About 10 percent of the defense budget is allocated to the nuclear program and, by illustration, that translated to $747 million in 2016.[171] For 2018–2019, the proposed defense budget was Rs. 1.1 trillion ($9.6 billion), which was an increase of 19.6 percent but that does not include other defense allocations, such as pensions.[172] This was the highest increase in the defense budget in over a decade and reflected 21 percent of the total budget and 3.2 percent of GDP, which is only a marginal decline from the 3.9 percent of GDP in 2003.[173] Other assessments have slightly different figures, with military spending at 4.1 percent of the GDP in 2003, and 3.5 percent in 2017.[174]

But these proportional increases do not highlight the extent of Pakistan's persistent economic problems, which have a bearing on its defense budget. For instance, in the past, budgetary shortfalls have posed technical problems for aircraft, including modernization of parts, leading to over a dozen crashes in 2011–12, especially involving older Mirage fighters.[175] Even though since then Pakistan is inducting more JF-17 aircraft as replacements, the issue of replacement spares is a perennial issue. By 2013, the PAF had suspended the AFDP due to lack of funding, because of which several projects were halted.[176] There is little to suggest that the situation has changed.

According to one estimate in 2018, Islamabad's annual defense budget of approximately $13.8 billion is imperiled by economic problems, which

---

169 Pervaiz Iqbal Cheema, "Security Threats Confronting Pakistan," in National Institute of Defense Studies, Joint Research Series No. 9, "Security Outlook of the Asia Pacific Countries and Its Implications for the Defense Sector," 2013, 135, <http://www.nids.mod.go.jp/english/publication/joint_research/series9/pdf/10.pdf>.
170 Kamran Yousaf, "Defense Budget Up by Around 20%," *Express Tribune*, Apr. 28, 2018.
171 Mason, "Military Budgets in India and Pakistan: Trajectories, Priorities, and Risks," 6.
172 Yousaf, "Defense Budget Up by Around 20%."
173 Baqir Sajjad Syed, "Budget 2018–19: Rs 1.1 Trillion Proposed for Defence," *Dawn*, Apr. 28, 2018; Cohen, *The Idea of Pakistan*, 2004, 255.
174 "Military Expenditure by Country as Percentage of Gross Domestic Product," Stockholm International Peace Research Institute, 2018, <https://bit.ly/2DaFdXi>.
175 "Crashes Raise Concern about Pakistan Air Force," *Dawn*, Dec. 11, 2012.
176 Mateen Haider, "PAF Suspends 'Air Force Development Plan 2025,' Says Report," *Dawn*, May 30, 2013.

jeopardize any major defense purchases by Pakistan.[177] Pakistan has faced an economic downturn in recent years, and in 2018 it had to seek loans from China, Saudi Arabia, Qatar, the United Arab Emirates, and the International Monetary Fund (IMF) even as the Bajwa doctrine, propounded by army chief General Qamar Javed Bajwa connected Pakistan's economic progress to regional security.[178] Islamabad also sought bailouts from the IMF in 2013 and in 2008, which it received despite mistrust from the agency, and only after U.S. involvement.[179] These problems are rooted in systemic economic conditions, which, over the decades have been exacerbated by mismanagement, corruption, inadequate taxation revenue, export deficiencies, and a misprioritization of revisionist security policies at the expense of development.[180] Although Pakistan's economy has generally grown at around 6 percent since 1950, it seems to regularly suffer from indebtedness and balance of payments problems, which further increases its dependence on foreign aid suppliers, such as the United States, China, and Saudi Arabia. Moreover, high defense spending and the dominant role of the military in the country's economy have had a detrimental impact on the country's overall economic situation.[181] Pakistan continues to depend on foreign aid, which is also an important determinant of economic growth rates.[182] So, by February 2019, Pakistan had just $8 billion in foreign exchange reserves, and was hoping to receive $30 billion in investments and loans from Saudi Arabia and the United Arab Emirates.[183]

The military also likely faced further budgetary pressures when the U.S. military aid to Pakistan in the form of Coalition Support Funds (CSF) to the tune of $800 million was cut in 2018 due to Islamabad's continued

---

177 F.M. Shakil, "Huge Defense Outlays Hit Funding for Pakistani Provinces," *Asia Times*, Sep. 5, 2018.

178 Maria Abi-Habib, "Pakistan's Military Has Quietly Reached Out to India for Talks," *New York Times*, Sep. 4, 2018; Salman Masood and Maria Abi-Habib, "Saudi Crown Prince Promises $20 Billion in Investments for Pakistan," *New York Times*, Feb. 18, 2019; Saeed Shah, "Pakistan Turns to Gulf Countries to Keep Economy Afloat," *Wall Street Journal*, Jan. 22, 2019.

179 Shahbaz Rana, "IMF Considers Pakistan Economic Managers Deceitful," *Express Tribune*, Apr. 26, 2011.

180 See, for instance, Atif Mian, "Why Pakistan Is Back in Trouble with Balance of Payment," *Herald*, Aug. 9, 2018.

181 Cohen, *The Idea of Pakistan*, 249, 255.

182 Ibid., 249–50.

183 Abid Hussain, "Cash-Strapped Pakistan Rolls Out Red Carpet for Saudis," BBC News, Feb. 15, 2019.

support for terrorist groups in Afghanistan. It must be noted that from 2002 to 2018 the United States gave upwards of $33 billion to Pakistan, including $14 billion in CSF.[184] This military assistance is unlikely to restart, at least not to reach 2002–2018 levels, because of which Pakistan is likely to get further enmeshed within the Chinese financial orbit.

But it is worth noting that Pakistan's trade imbalance and declining foreign exchange reserves by 2018–2019 have been caused at least in part by Islamabad's dependence on Beijing and the $62 billion CPEC, part of China's BRI, which pushed Islamabad to approach the IMF for a bailout in late 2018.[185] China's "debt-trap diplomacy" also pushed Islamabad to seek loans and investment commitments from Saudi Arabia amounting to $26 billion between October 2018 and February 2019.[186] Although these are recent economic developments, they reflect long-term economic characteristics in Pakistan, including defense spending and its role in the country's economy.

Defense procurement from external sources, especially to do with air power, depends on financing concessions offered by the supplier, and so withdrawal of foreign military financing by the United States in 2016 for the purchase of F-16 aircraft led Pakistani officials to consider other sources for fighter aircraft.[187] Unlike India, Pakistan is not in a position to solicit open applications for purchase of major air weaponry and platforms, especially fighter aircraft. Its major acquisitions of fighter aircraft in substantial numbers have been the F-16s and JF-17s, which have come about as a result of political and military alliances with the United States and China, which provided financial assistance and favorable conditions. But this also means that Pakistan is unable to negotiate with private aircraft manufacturing companies or even issue formal requests for proposals from them.

Ostensibly, the PAF's procurement strategy has been to acquire technologically advanced aircraft, rather than try to match the IAF in numbers.[188]

---

184 Phil Stewart and Idrees Ali, "Exclusive: Pentagon Cancels Aid to Pakistan Over Record on Militants," Reuters, Sep. 1, 2018.

185 Saeed Shah, "Under Pressure, Pakistan Turns to IMF," *Wall Street Journal*, Oct. 9, 2018.

186 Joseph Hincks, "Saudi Arabia Is Investing $20 Billion in Pakistan. Here's What It's Getting in Return," Time.com, Feb. 19, 2019, <http://time.com/5531724/saudi-arabia-pakistan-mbs-imran-khan>; Shah, "Under Pressure, Pakistan Turns to IMF."

187 Khuldune Shahid, "Pay for Your F-16s, US Tells Pakistan," *Friday Times*, May 6, 2016.

188 Shamsi, "Indian Air Force Modernization Plan of 2020: Challenges for Regional Air Forces," 18.

But persistent structural economic problems make it less likely that Islamabad would be able to acquire adequate high-technology items in future, including for the PAF, especially with the end of foreign military financing and concessionary rates from the United States, and less likely to be able to keep up with India's acquisitions, leading to even greater reliance on nuclear weapons to deal with India's conventional military advantage.[189]

Pakistan's Air Power: Security Concerns

A concern in recent years has been the security of Pakistan's weapons systems and military assets from terrorist groups and proliferation networks. Jihadist groups have attacked numerous military facilities in Pakistan, and air assets have been a particularly high-profile set of targets. Attacks on bases with a nuclear component are a particularly serious concern, even if the jihadist terrorists attacking them are unaware of the nuclear connection. Over the last 10 years, elements of the Pakistani Taliban have carried out attacks on several military facilities, which have included important airbases, some of which are regarded as closely connected to Pakistan's nuclear weapons program. These targets include the Minhas airbase in Kamra, which was attacked four times between 2007 and 2012 and which is a suspected location of Pakistan's nuclear warheads, F-16 aircraft, and a JF-17 production factory.[190] In another instance, in summer 2011, Pakistani Taliban militants attacked the Pakistan Naval Station Mehran, near Karachi, destroying two P-3C Orion submarine reconnaissance aircraft that were based there. Then, in September 2015, 23 Air Force personnel were killed in a Pakistani Taliban assault on the Badaber PAF base near Peshawar, in Khyber Pakhtunkhwa province.[191]

As stated earlier, Pakistan is increasingly focused on inducting battlefield nuclear weapons into its arsenal, which involves short-range missiles. Such a development is regarded as destabilizing because it delegates authority for nuclear launch to lower level battlefield commanders, rather than the highest decision-makers (i.e., army chief, prime minister)

---

189 Mason, "Military Budgets in India and Pakistan: Trajectories, Priorities, and Risks."
190 Declan Walsh, "Militants Attack Pakistani Air Force Base," *New York Times*, Aug. 16, 2012.
191 "23 PAF Men among 42 Killed in Taliban Attack on Pakistan Air Base," *Economic Times*, Sep. 18, 2015.

in whom the authority for strategic nuclear launch is vested. In crisis situations, this creates a greater risk of nuclear weapon use. The possibility (however remote) of rogue commanders with control of warheads and delivery systems also increases in such circumstances. Relatedly, battlefield nuclear weapons also increase the vulnerability of nuclear weapons from terrorist groups.

Conclusion

This review of Pakistan's air power strategy highlights some broad patterns linked to Islamabad's overall foreign and defense relationships with major powers, as well as its own resource constraints. As part of the origin story of the PAF, Pakistani views have constructed a narrative that a plucky PAF has always operated against the odds and has been denied aircraft and spares by major powers.[192] But this assertion disregards crucial defense assistance from China, the United States, and other suppliers at various points.[193] Pakistan was able to leverage its usefulness to extra-regional powers into extracting major concessions and alliances.

An important takeaway is that Pakistan is undergoing shifts in international military partnerships from the United States to China, which is especially visible in high-profile defense projects such as fighter aircraft. This is part of the broader deterioration of bilateral relations between the United States and Pakistan, primarily due to Islamabad's continued support for jihadist terrorist groups in the region, such as the Afghan Taliban and Lashkar-e-Taiba.[194] The historically close defense ties between Pakistan and China have become even tighter on account of the decline of U.S.-Pakistan ties, but it also shows the extent of Pakistan's dependence and indebtedness to Beijing. Persistent and deep-rooted economic problems and their impact on the defense budget will especially be felt in the

---

192 Jamal Hussain, "Pakistani Air Power," in John Andreas Olsen (ed.), *Routledge Handbook of Air Power* (New York, NY; Routledge, 2018).

193 Saurabh Joshi, "How an Ex-Nazi Arms Dealer Sold Fighters to India and Pakistan during an Arms Embargo," StratPost.com, Aug. 31, 2016, <https://www.stratpost.com/ex-nazi-arms-dealer-sold-fighters-india-pakistan-arms-embargo>.

194 According to one opinion, in future, defense ties between Islamabad and Washington will once again be on the upswing, due to some or the other external issues—see Blank et al., *Vector Check: Prospects for U.S. and Pakistan Air Power Engagement*, 13.

PAF, which traditionally has been the recipient of major defense equipment, especially fighter aircraft.

In the nuclear weapons realm, Pakistan continues to expand its ballistic and cruise missile arsenal for launch from different platforms, including submarines. But the expansion of Pakistan's nuclear delivery systems has implications for civil-military relations, but also especially for inter-services relations. For instance, if Islamabad does deploy nuclear-capable submarine-launched ballistic or cruise missiles that would mean an expanded role for the Navy in nuclear decision-making, in which that service has historically had lesser influence than the other two services.

Additionally, the PAF (as well as the broader Pakistan military) has to consider whether it is focusing adequately on the threat from the Pakistani Taliban and other terrorist groups. Its military doctrine remains focused on India, but Islamabad's long-standing revisionist policies have historically provoked crises in the subcontinent. And the PAF plays an important role in providing Islamabad the space in which it can persist with supporting terrorist groups as part of its defense strategy.

References

"23 PAF Men among 42 Killed in Taliban Attack on Pakistan Air Base," *Economic Times*, Sep. 18, 2015.

Abdullah, S. "Nuclear Ethics? Why Pakistan Has Not Used Nuclear Weapons … Yet," *Washington Quarterly*, Winter 2019.

Abi-Habib, M. "China's 'Belt and Road' Plan in Pakistan Takes a Military Turn," *New York Times*, Dec. 19, 2018.

Abi-Habib, M. "Pakistan's Military Has Quietly Reached Out to India for Talks," *New York Times*, Sep. 4, 2018.

Aguilar, F., R. Bell, N. Black, S. Falk, S. Rogers, and A. Peritz. "An Introduction to Pakistan's Military" Harvard Kennedy Center, Belfer Center for Science and International Affairs, Jul. 2011. <https://bit.ly/2AtiW3e>.

Ahmed, A. "Towards a Proactive Military Strategy: 'Cold Start and Stop'," *Strategic Analysis*, May 2011, Vol. 35, No. 3.

Akram, M. "India's Pakistan Strategy," *Dawn*, Apr. 3, 2016.

Akram, M. "A Strategic Challenge," *Dawn*, Apr. 29, 2018.

Alvi, A.S. "50 Years On: Memories of the 1973 Arab-Israeli Conflict," *Express Tribune*, Mar. 19, 2015.

Ansari, U. "Pakistan Cosies Up to Russia, but Moscow Doesn't Want to Take Sides," *Defense News*, May 2, 2018.

Anwar, M.A., and Z. Rafique, "Defense Spending and National Security of Pakistan: A Policy Perspective," *Democracy and Security*, 2012, 8.

Ayaz, B. "Cloaked in Secrecy," *Newsline*, Jul. 2011.

Baig, M.A. and H. Iqbal, "A Comparative Study of Blitzkrieg and Cold Start Doctrine: Lessons and Countermeasures for Pakistan," *IPRI Journal*, Winter 2018.

Basrur, R. "India and Pakistan: Persistent Rivalry." *Routledge Handbook of Asian Security Studies*, edited by S. Ganguly, A. Scobell and J.C. Liow. New York, NY: Routledge, 2018.

Behravesh, M. "What Does Iran Want in Afghanistan?" Al Jazeera.com, Feb. 4, 2019. <https://www.aljazeera.com/indepth/opinion/190204092658549.html>.

Blank, J., R.S. Girven, A. Tarapore, J.A. Thompson, and A. Chan. *Vector Check: Prospects for U.S. and Pakistan Air Power Engagement*. Santa Monica, CA: RAND, 2018.

Camp, D. "Defeating al-Qaeda's Air Force: Pakistan's F-16 Program in the Fight Against Terrorism." Statement before the U.S. House of Representatives Foreign Affairs Committee on South Asia, Washington, DC, Sept. 16, 2008.

"Chapter Six: Asia," *The Military Balance*. London: International Institute for Strategic Studies, 2019.

Chaudhury, D.R. "Russia-Pakistan Ties to Remain 'Quite Limited'," *Times of India*, Jul. 12, 2018.

Cheema, P.I. "Security Threats Confronting Pakistan," in National Institute of Defense Studies, Joint Research Series No. 9, "Security Outlook of the Asia Pacific Countries and Its Implications for the Defense Sector," 2013. <http://www.nids.mod.go.jp/english/publication/joint_research/series9/pdf/10.pdf>.

Cheema, P.I. "Strategic Stability in South Asia: The Role of USA," *Journal of Contemporary Studies*, Summer 2012, Vol. 1, No. 1.

Chen, S. "China Provides Tracking System for Pakistan's Missile Programme," *South China Morning Post*, Mar. 22, 2018.

"China Rolls Out AWACS Aircrafts for Pakistan Air Force," *Deccan Herald*, Nov. 13, 2010.

"China to Sell 48 High-End Military Drones to Pakistan," *Economic Times*, Oct. 10, 2018.

Cloughley, B. "Pakistan's Armed Forces," *Asian Military Review*, May 7, 2018. <https://asianmilitaryreview.com/2018/05/pakistans-armed-forces>.

Cohen, S.P. *The Idea of Pakistan*. Washington, DC: Brookings Institution Press, 2004.

Constable, P. "Despite Trump's Punitive Military Aid Cut, Pakistan and U.S. Are Still Intertwined," *Washington Post*, Jan. 25, 2018.

Constable, P. "Pakistan Military Prepares for a Possible Indian Attack," *Washington Post*, Sept. 22, 2016.

"Crashes Raise Concern about Pakistan Air Force," *Dawn*, Dec. 11, 2012.

Curtis, L. "China's Military and Security Relationship with Pakistan," Testimony before the U.S.-China Economic and Security Review Commission, The Heritage Foundation, May 20, 2009.

Curtis, L. "The Reorientation of Pakistan's Foreign Policy Toward Its Region," *Contemporary South Asia*, June 2012, Vol. 20, No. 2.

Dadi, I. "Nuclearization and Pakistani Popular Culture Since 1998." *South Asian Cultures of the Bomb*, edited by Itty Abraham. Bloomington, IN: Indiana University Press, 2009.

Fair, C.C. *Fighting to the End: The Pakistan Army's Way of War*. New York, NY: Oxford University Press, 2014.

Fair, C.C. "Studying Drones: The Low Quality Information Environment of Pakistan's Tribal Areas." *Preventive Force: Drones, Targeted Killing, and the Transformation of Contemporary Warfare*, edited by K. Fisk and J.M. Ramos. New York, NY: New York University Press, 2016.

Faruqui, A. "Failure in Command: Lessons from Pakistan's Indian Wars, 1947–1999," *Defense Analysis*, 2001, Vol. 17, No. 1.

Faruqui, A. "Imran's Foreign and Defense Policy: An Initial Assessment," *Daily Times*, Aug. 1, 2018.

Fedorchak, V. "Air Warfare." *Routledge Handbook of Defence Studies*, edited by D.J. Galbreath and J.R. Deni. New York, NY: Routledge, 2018.

Gady, F.S. "Confirmed: Pakistan Air Force Now Operates 70 JF-17 Fighter Jets," *Diplomat*, Dec. 13, 2016.

Gady, F.S. "Pakistan Stands Up New Fighter Squadron," *Diplomat*, Mar. 8, 2018.

Gady, F.S. "Pakistan to Arm Super Mushak Aircraft for Counterinsurgency Operations," *Diplomat*, May 24, 2017.

Gill, J.H. "Military Operations in the Kargil Conflict." *The Causes and Consequences of the Kargil Conflict*, edited by P.R. Lavoy. Cambridge, UK: Cambridge University Press, 2009.

Gormley, D.M. *Missile Contagion: Cruise Missile Proliferation and the Threat to International Security*. Annapolis, MD: Naval Institute Press, 2010.

Goulter C. and H.V. Pant. "Realignment and Indian Airpower Doctrine: Challengers in an Evolving Strategic Context," *Journal of Indo-Pacific Affairs*, Fall 2018.

Gross, J.A. "Israel to Fly Alongside Pakistan, UAE in US Air Force Drill," *Times of Israel*, Aug. 3, 2016.

Gul, A. "Pakistan, Russia Sign Rare Military Cooperation Pact," Voice of America, Aug. 8, 2018.

Gupta, A. *Building an Arsenal: The Evolution of Regional Power Force Structures*. Westport, CT: Praeger, 1997.

Gupta, A. "The February Flashpoint: Lessons Learned," *Geopolitics*, Mar. 2019.

Haider, M. "PAF Suspends 'Air Force Development Plan 2025,' Says Report," *Dawn*, May 30, 2013.

Haider, M. "Pakistan Rejects US Conditions Attached to Sale of F-16s," *Dawn*, May 7, 2016.

Haider, E. "S-400: Pakistan Faces a Major Asymmetric Threat," *Newsweek Pakistan*, Oct. 11, 2018. <https://newsweekpakistan.com/s-400-pakistan-faces-major-asymmetric-threat>.

Hincks, J. "Saudi Arabia Is Investing $20 Billion in Pakistan. Here's What It's Getting in Return," Time.com, Feb. 19, 2019. <http://time.com/5531724/saudi-arabia-pakistan-mbs-imran-khan>.

Hussain, A. "Cash-Strapped Pakistan Rolls Out Red Carpet for Saudis," BBC News, Feb. 15, 2019.

Hussain, J. "Pakistani Air Power." *Routledge Handbook of Air Power*, edited by John Andreas Olsen. New York, NY: Routledge, 2018.

Hussain M.Z. and R.Q. Ahmed, "Space Programs of India and Pakistan: Military and Strategic Installations in Outer Space and Precarious Regional Strategic Stability," *Space Policy*, Vol. 47, Feb. 2019.

Irish, J. "Pakistan Wants Air Force Upgrade for Prolonged Militant Fight," Reuters, Apr. 7, 2016.

Iqbal, A. "US Cuts Military Training Programme for Pakistan," *Dawn*, Aug. 11, 2018.

Jorgic, D. "With Gas and Diplomacy, Russia Embraces Cold War for Pakistan," Reuters, Mar. 5, 2018.

Joshi, S. "How an Ex-Nazi Arms Dealer Sold Fighters to India and Pakistan during an Arms Embargo," StratPost.com, Aug. 31, 2016. <https://www.stratpost.com/ex-nazi-arms-dealer-sold-fighters-india-pakistan-arms-embargo>.

Joshi, S. "IAF Admits Unprepared for 2-Front Defense," StratPost.com, Mar. 12, 2016, https://www.stratpost.com/iaf-admits-unprepared-for-2-front-defense/.

Joshi, S. "LMT Tata to Build F-16 Wings in India," StratPost.com, Sept. 4, 2018. <https://bit.ly/2BEKqn1>.

Joshi, S. "Lockheed Martin to Announce New F-16 Block 50 Variant," StratPost.com, Feb. 15, 2012. <https://bit.ly/2IgTEfd>.

Joshi, S. "PAF Road Test: No Big Deal," StratPost.com, Apr. 2, 2010. <https://www.stratpost.com/paf-road-test-no-big-deal/>.

Joshi, S. "The Story of a Glacial Trust Deficit," StratPost.com, Dec. 23, 2008. <https://www.stratpost.com/the-story-of-a-glacial-trust-deficit>.

Joshi, S. "Video: PAF JF-17 at #PAS 15," StratPost.com, Jul. 5, 2015. <https://www.stratpost.com/video-paf-jf-17-at-pas15/>.

"Kayani Spells Out Threat Posed by Indian Doctrine," *Dawn*, Feb. 4, 2010.

Khan, K. "Pakistan Says It Is Studying Errant U.S. Missile," *Washington Post*, Aug. 28, 1998.

Khan, M.U.H. "Pakistan-China Defence Cooperation," *Defence Journal*, Mar. 31, 2014.

Khan, O.F. "Pakistan Air Force Chief: Shoot Down US Drones," *Times of India*, Dec. 7, 2017.

Khan, Z. "Cold Start Doctrine: The Conventional Challenge to South Asian Stability," *Contemporary Security Policy*, Dec. 2012, Vol. 33, No. 3.

Kile, S.N., P. Schell, and H.M. Kristensen. "Pakistani Nuclear Forces," *SIPRI Yearbook 2012*. Oxford, UK: Oxford University Press, 2012.

Kristensen, H.M., R.S. Norris, and J. Diamond. "Pakistani Nuclear Forces, 2018," *Bulletin of the Atomic Scientists*, 2018, Vol. 74, No. 5.

Lambeth, B.S. "Airpower in India's 1999 Kargil War," *Journal of Strategic Studies*, June 2012, Vol. 35, No. 3.

Mansoor, R. "Pakistan is Losing the Space Race," *Diplomat*, Feb. 1, 2018.

Mashal M., and S. Masood. "Cutting Off Pakistan, U.S. Takes Gamble in Complex Afghan War," *New York Times*, Jan. 5, 2018.

Mason, S. "Military Budgets in India and Pakistan: Trajectories, Priorities, and Risks," Stimson Center, 2016. <https://www.stimson.org/sites/default/files/file-attachments/Military-Budgets-India-Pakistan-Trajectories-Priorities-Risks-Oct2016.pdf.pdf>.

Masood S., and M. Abi-Habib, "Saudi Crown Prince Promises $20 Billion in Investments for Pakistan," *New York Times*, Feb. 18, 2019.

Mian, A. "Why Pakistan Is Back in Trouble with Balance of Payment," *Herald*, Aug. 9, 2018.

"Military Expenditure by Country as Percentage of Gross Domestic Product," Stockholm International Peace Research Institute, 2018. <https://bit.ly/2DaFdXi>.

Moore M., and K. Khan. "Pakistan Moves Nuclear Weapons," *Washington Post*, Nov. 11, 2001. <https://wapo.st/2todE57>.

"Navy Test-Fires Anti-Ship Missile," *Nation*, Mar. 5, 2018.

Negi, M.S. "Pakistan Builds Airbase near Gujarat Border, Deploys Chinese JF-17 fighters," *India Today*, Jul. 9, 2018. <https://bit.ly/2HQrt6A>.

"Nuclear Safety, Nuclear Stability and Nuclear Strategy in Pakistan," A Concise Report of a Visit by Landau Network – Centro Volta, Jan. 14, 2002. <https://pugwash.org/2002/01/14/report-on-nuclear-safety-nuclear-stability-and-nuclear-strategy-in-pakistan>.

"PAF Inaugurates New Operational Air Base at Bholari near Karachi," Geo TV, Dec. 25, 2017. <https://bit.ly/2BDnMuy>.

"Pakistan Air Force Is Seeking a Lead-in Fighter Trainer," Quwa.org, Dec. 12, 2018. <https://quwa.org/2018/12/12/ideas-2018-pakistan-air-force-is-seeking-a-lead-in-fighter-trainer-2>.

"Pakistan Air Force Needs to Replace 190 Planes by 2020," *Dawn*, Mar. 15, 2016.

"Pakistan Could Use F-16 Jets against India: US Lawmakers to Obama," *Times of India*, Apr. 28, 2016.

"Pakistan Eyes Export Variant of China's First Stealth Aircraft," *Times of India*, Nov. 1, 2016.

"Pakistan Has Cost-Effective Solution to India's Latest Ballistic Missile Defence System: Report," *Economic Times*, Nov. 7, 2018."Pakistan Inducts 13 F-16 Fighter Jets into Air Force," LiveMint.com, May 21, 2014.

"Pakistan Let China See Crashed U.S. 'Stealth' Copter," Reuters, Aug. 15, 2011.

"Pakistan Opposes Supply of US Armed Drones to India," *Economic Times*, Jul. 13, 2018.

"Pakistan Says Second Strike Capability Attained," *Dawn*, Mar. 30, 2018.

"Pakistan Successfully Test-Fires 'Beyond Visual Range' Missile from JF-17 Thunder," *Express Tribune*, Feb. 2, 2018.

"Pakistan Trying to Get More F-16 Jets from US: Report," *Economic Times*, Jul. 12, 2018.

"Pakistan Will Not Rename Missiles," BBC News, Feb. 23, 2006. <https://bbc.in/2BVb2PX>.

Panda A., and V. Narang, "Pakistan Tests New Sub-Launched Nuclear-Capable Cruise Missile. What Now?" *Diplomat*, Jan. 10, 2017.

Pande, A. "Friends of Last Resort: Pakistan's Relations with China and Saudi Arabia." *Pakistan's Enduring Challenges*, edited by C.C. Fair and S.J. Watson. Philadelphia, PA: University of Pennsylvania Press, 2015.

Rajagopalan, R.P. "Pakistan's Nasr Missile: 'Cold Water' over India's 'Cold Start'?" *Diplomat*, Jan. 31, 2019.

Rana, S. "IMF Considers Pakistan Economic Managers Deceitful," *Express Tribune*, Apr. 26, 2011.

Roy, D. "Russia-Pakistan Ties to Remain 'Quite Limited'," *Times of India*, Jul. 12, 2018.

Salik, N. "Pakistan's Nuclear Force Structure in 2025," Carnegie Endowment for International Peace, June 30, 2016. <https://carnegieendowment.org/2016/06/30/pakistan-s-nuclear-force-structure-in-2025-pub-63912>.

Schmitt, E. "Pakistan Injects Precision Into Air War on Taliban," *New York Times*, Jul. 29, 2009.

Shah, S., and I. Talley, "Saudi Arabia Stymies U.S. over Pakistan Terror List," *Wall Street Journal*, Feb. 21, 2018.

Shah, S. "Under Pressure, Pakistan Turns to IMF," *Wall Street Journal*, Oct. 9, 2018.

Shah, S. "Pakistan Turns to Gulf Countries to Keep Economy Afloat," *Wall Street Journal*, Jan. 22, 2019.

Shahid, K. "Pay for Your F-16s, US Tells Pakistan," *Friday Times*, May 6, 2016.

Shakil, F.M. "Huge Defense Outlays Hit Funding for Pakistani Provinces," *Asia Times*, Sep. 5, 2018.

Shamsi, Z.U.H. "Indian Air Force Modernization Plan of 2020: Challenges for Regional Air Forces." Australian Defence College, Oct. 2012. <http://www.defence.gov.au/ADC/Publications/Commanders/2012/04_IAF%202020%20AWC.pdf>.

Shukla, M. "Pakistan Eyes 62 JF-17 Jets from China as Rafale Dogfight Continues in India," *Zee News*, Jan. 7, 2019.

Siddiqa, A. "Pakistan-Saudi Arabia Military-Ideological Dependency: Can Pakistan Afford to Lose the Middle East?" GESC Chair in International Relations Lecture, Feb. 2016. <https://bit.ly/2Gvdql8>.

Siddiqui, N. "JF-17B Fighter Jet Takes Maiden Test Flight," *Dawn*, Apr. 28, 2017.

Siddiqui, N. "Iranian Drone Shot Down by PAF, Confirms FO," *Dawn*, June 21, 2017.

Siddiqui, S. "Pakistan Will Be Paying China $90b against CPEC-Related Projects," *Express Tribune*, Mar. 12, 2017.

Singh, S. "China Builds Air Assets in Tibet, Indian Missile Units Head East," *Indian Express*, Dec. 20, 2018.

Singh, J. "IAF's Ground Reality," *Sunday Tribune*, May 6, 2007.

Singh, S. "IAF Worry: In Next Two Years, Only 26 Fighter Squadrons, Short by 16," *Indian Express*, Jan. 24, 2019.

Sirohi, S. "India-Pakistan Crisis: F-16s and the Missile Mystery," TheWire.in, Mar. 11, 2019. <https://thewire.in/security/india-pakistan-crisis-f-16s-and-the-missile-mystery>.

Stacey, K. "Pakistan Shuns US for Chinese High-Tech Weapons," *Financial Times*, Apr. 17, 2018.

Stewart, P., and I. Ali, "Exclusive: Pentagon Cancels Aid to Pakistan over Record on Militants," Reuters, Sep. 1, 2018.

Stone, M. "Exclusive: U.S. Offers India Armed Version of Guardian Drone – Sources," Reuters, Jul. 18, 2018. <https://reut.rs/2Ax74gy>.

Subramaniam, A. "Doctrinal Evolution in the Indian Air Force." *Handbook of Indian Defence Policy: Themes, Structures, and Doctrines*, edited by H.V. Pant. New York, NY: Routledge, 2016.

Syed, B.S. "Pakistan to Retain Full Spectrum Deterrence Policy," *Dawn*, Dec. 22, 2017.

Syed, B.S. "Budget 2018-19: Rs 1.1 Trillion Proposed for Defence," *Dawn*, Apr. 28, 2018.

Tanham, G.K. and M. Agmon. *The Indian Air Force: Trends and Prospects*. Santa Monica, CA: RAND, 1995.

Tasleem, S. "Pakistan's Nuclear Use Doctrine," Regional Insight, Carnegie Endowment for International Peace, June 30, 2016. <https://carnegieendowment.org/2016/06/30/pakistan-s-nuclear-use-doctrine-pub-63913>.

Tellis, A.J. "India's ASAT Test: An Incomplete Success," Carnegie Endowment for International Peace, Apr. 15, 2019. <https://carnegieendowment.org/2019/04/15/india-s-asat-test-incomplete-success-pub-78884>.

"Thrifty at 50: How the Pakistan Air Force Keeps ageing Mirages Flying," *Dawn*, Apr. 29, 2018.

Trimble, S. "F-16s Powered up Pakistan's Counter-Insurgency Strikes," *FlightGlobal*, Nov. 13, 2011. <https://bit.ly/2GGCEhr>.

Walsh, D. "Militants Attack Pakistani Air Force Base," *New York Times*, Aug. 16, 2012.

Warnes, A. "Pakistan Air Force Builds for the Future," *Asian Military Review*, Jan. 10, 2019. <https://asianmilitaryreview.com/2019/01/pakistan-air-force-builds-for-the-future>.

Weinbaum, M. "Insurgency and Violent Extremism in Pakistan," *Small Wars and Insurgencies*, 2017, Vol. 28, No. 1.

Weintraub, R.M. "Afghanistan Says It Downed F16 Fighter from Pakistan," *Washington Post*, May 1, 1987.

Wintour, P. "All-Weather Friendship: But Is Pakistan Relying Too Heavily on China?" *Guardian*, Aug. 3, 2018.

Yousaf, K. "Defense Budget Up by Around 20%," *Express Tribune*, Apr. 28, 2018.

Zhen, L. "Did Pakistan Use Its Chinese JF-17 Jets to Shoot Down Indian Planes?" *South China Morning Post*, Mar. 1, 2019.

11 INDONESIA
Small Air Force, Big Nation
Adhi Priamarizki and Tiola Javadi

Introduction

In February 2018, in attempt to replace its aging fleets, the Indonesian Air Force (Tentara Nasional Indonesia Angkatan Udara (TNI AU)) signed a deal with the Russian government to procure 11 Su-35s, scheduled to be delivered in three phases between August 2019 and mid-2020. The Indonesian government planned for the Su-35s to replace the already obsolete TNI AU's F-5 Tiger aircrafts, which have been used since 1980. Besides procuring the Sukhoi, TNI AU also seeks to enhance various facilities to support its platform rejuvenation, among others, by adding and improving a number of fighter shelters across the nation. This revitalization has been gradually implemented, particularly in the context of the relatively stable economic growth that Indonesia has been enjoying in the past decade.

Despite the rejuvenation plan, it is widely known that the TNI AU has always been perceived as "the stepson" in the Indonesian military (Tentara Nasional Indonesia (TNI)). This perception is rooted in an unfair allegation that the service was involved in a coup attempt in 1965 or the so-called September 30, 1965, affair, which has subsequently become a great disgrace for the service.[1] Prior to this allegation, during the administration of Indonesia's first president, Sukarno (1945 to 1967), the TNI AU had received large amounts of Soviet cutting-edge technology armaments, such as the MiG-15,

---

1 The September 30, 1965, affair refers to a failed coup attempt allegedly carried out by the Indonesian Communist Party (PKI). The real perpetrator nonetheless remains hazy.

MiG-17, MiG-19, and MiG-21, which made the service one of the most advanced in the South East Asian region. However, the fall of Sukarno and the rise of President Suharto (1967 to 1988) marked the end of TNI AU's golden era. The Army (Tentara Nasional Indonesia Angkatan Darat (TNI AD)) emerged as the only dominant service within the Indonesian military and served as a political tool of Suharto's regime. The Army, through its territorial command structure, has been the spearhead of Indonesia's defense system.[2] Since then, TNI has further galvanized itself into an army-centric military, with the Navy and Air Force playing subordinate roles.

The fall of Suharto in 1998 forced the military to launch a reform, particularly to withdraw itself from politics and to focus solely on defense matters. However, at the same time, the detrimental impacts of the 1997 Asian financial crisis robbed the country's financial capability, which consequently prevented the TNI's modernization plan. Nevertheless, Indonesia's steady economic growth in the past decade altered the arduous circumstance and supplied adequate cash capital. The situation also resulted in the readiness of the Indonesian government to embark on its delayed military modernization plan, which was followed by plenty of studies dissecting TNI's modernization plan. Scholarly literature focusing exclusively on TNI AU, however, is still very limited. This chapter, therefore, aims to augment the discussion on the Indonesian Air Force. To serve this purpose, this study discusses TNI AU's current force structure and doctrine. Additionally, it attempts to examine Indonesia's strategic environment and its implications for the Air Force. Potential doctrinal change,

---

2 Territorial command is part of Indonesian military structure that expresses the policy of total people's defense and supports the military domestic security role with a structure stretched from provincial to village level. It provides intelligence data for the military and part of the armed forces' defense structure. Furthermore, territorial command exists across the archipelago with a territorial command covering one or a few provinces. This territorial command is also often involved in shady business activities with local politicians and businesspeople. Under the authoritarian New Order regime, the system helped the military to oppress the regime's political opponents. See Robert Lowry, *The Armed Forces of Indonesia.* (New South Wales: Allen & Unwin, 1996); Marcus Mietzner, "Business as Usual? The Indonesian Armed Forces and Local Politics in the Post-Soeharto Era," in Edward Aspinall and Greg Fealy (eds.), *Local Power and Politics in Indonesia: Decentralisation & Democratisation* (Singapore: ISEAS, 2003); Jun Honna, "The Politics of Securing Khaki Capitalism in Democratizing Indonesia," in Paul Chambers and Napisa Waitoolkiat (eds.), *Khaki Capital: The Political Economy of the Military in Southeast Asia* (Copenhagen: Nias Press, 2017).

menacing challenges, and future development of TNI AU will also be discussed in this chapter.

Like other air forces, the Indonesian Air Force has been engaging in various joint training missions with other countries' air forces. Discussing TNI AU involvement in those training efforts will be another focus of the chapter. In order to examine the abovementioned elements, this chapter will be structured as follows. It starts with an examination of the genesis of the Air Force as well as the service's doctrine and force structure. The chapter then continues by elaborating Indonesia's strategic environment and how TNI AU contributes to secure the nation. The next section focuses on the expansion and procurement plan of the TNI AU to support its operations. Finally, the last part of the chapter analyses prospects and potential challenges of the service, particularly within the context of Indonesia's ambition to be a maritime power, put forward by President Joko Widodo.

Small Air Force, Big Nation

TNI AU Genesis and Predicament

The genesis of the TNI can be traced back to the formation of the Indonesian People's Security Body (Badan Keamanan Rakyat (BKR)). BKR's air service, BKR Udara, in the early days of the republic, operated former Japanese aircrafts such as a fighter (Nakajima Ki-43 Hayabusa), a bomber (Mitsubishi Ki-51 Guntai), reconnaissance aircraft (Nakajima series), trainers (Ki-79 B Nishikoren and Tachikawa Ki-36 or Cukiu), and many more.[3] The Indonesian military's inferior numbers and backward technology forced the TNI to conduct guerrilla warfare against the Netherlands during the Revolutionary War (1945–1949). BKR Udara itself only launched one air operation in the Revolutionary War, which was on July 29, 1947, attacking three cities in Central Java (Semarang, Salatiga, and Ambarawa). The Indonesian Air Force also joined the guerrilla operations against the Dutch military, particularly through the Garuda Mulya troops that focused their actions around Surakarta and Karanganyar in Central Java.

---

3 TNI AU, *50 Tahun Emas Pengabdian TNI Angkatan Udara (TNI AU's 50 Years of Service)* (Jakarta: TNI AU Information Desk, 1996), 25.

Indonesia's military emphasis on guerrilla tactics made the Air Force play a marginal role within the military. It was only during the Sukarno presidency that TNI AU gained greater significance, especially when Indonesia was enjoying a close tie with the Soviet Union. By 1962, the Indonesia Air Force received 41 Mi-4s, 9 Mi-6s, 8 SM-1s, 6 An-12Bs, 30 MiG-15s, 49 MiG-17s, 10 MiG-19s, 20 MiG-21s, 12 Tu-2s, 14 Tu-16s, and 22 IL-28s from the Soviet Union.[4] The massive expansion of aircraft forced the Air Force to introduce new squadrons and wings to exercise command and control over a number of squadrons with the same or similar roles.[5] However, the rise of Suharto, who preferred forging close relations with the United States and its allies, and Indonesia's economic crisis in the mid-1990s made the Air Force scrap many of its Soviet-made platforms, mostly due to obsolesce and shortage of technical expertise.[6]

Suharto's tenure, labeled as the New Order, utilized the military, specifically the army, to serve his political interests. Indonesian military's *dwifungsi* or dual function paradigm became a strong justification for the armed forces for engaging in socio-politico realms.[7] The Army's territorial command played a tantamount role for suppressing political opposition to the regime. TNI AU itself was forced to take the back seat, not only because of the Army domination, but also due to the allegation of Air Force top officers' involvement in the September 30, 1965, incident. This situation had galvanized the Indonesian military as an army-centric force with the Air Force and Navy merely playing supporting duties. Below the examination continues with a discussion on TNI AU doctrine.

---

4 TNI AU, *Bakti TNI Angkatan Udara 1946–2004* (*The Contribution of TNI AU 1946–2004*) (Jakarta: TNI AU, 2004), 58.

5 Ibid., 103.

6 Lowry, *The Armed Forces of Indonesia*, 106.

7 It was the late Indonesian army general A.H. Nasution who introduced the concept of *dwifungsi* in order to give the military an opportunity to be involved on the country's political stage, insisting that military duties should not be confined only to defense matters because soldiers were freedom fighters during the Revolutionary War (1945–1949) who should take responsibility for the post-independence nation-building. Under Suharto's authoritarian regime, the concept became a justification for wider involvement of the military in various aspects, including social and economic activities.

TNI AU Doctrine

The Indonesian armed forces stratify their doctrine into the basic doctrine, the master plan doctrine, and the implementation doctrine:

- The *State Defense Doctrine* is the basic doctrine that is the foundation for all defense-related doctrine and valid at the national level.
- The *Master Plan Doctrine* is divided into military defense doctrine and non-military defense doctrine. TNI holds the responsibility for the first one, while the Department of Defense is responsible for the non-military defense-related doctrine.
- The *Implementation Doctrine* is also segmented into military defense-related doctrine and non-military defense-related doctrine.[8]

Likewise, the Department of Defense is responsible for the non-military defense-related implementation doctrine. Meanwhile, the military defense-related implementation doctrine consists of a doctrine for each service within the TNI (the Army, the Navy, and the Air Force).[9]

TNI AU positioned its doctrine *Swa Bhuwana Paksa* as a foundation for strategic concepts and as a source for handbooks within the Air Force structure.[10] The purpose of the doctrine is to provide an understanding about the role, function, and duty of the TNI AU. Further, it unites the thinking, behavior, and action regarding the development and the use of TNI AU capabilities, in order to ensure the success of the TNI AU, both in military operation and military operation other than war (MOOTW).[11] The service's doctrine realizes the origin of the Air Force as *tentara rakyat* (people's army), but also asserts its professionalism as soldiers who focus only on defense-related activities and avoid political practices.[12] TNI AU positions itself as the frontline and the primary component of national air power in protecting the nation from the threats in and through the air.[13] The

---

8 Department of Defense, *Doktrin Pertahanan Negara (State Defense Doctrine)* (Jakarta: Department of Defense, 2007).

9 Ibid.

10 *Swa Bhuwana Paksa* literally means the "wings of motherland."

11 Air Force Chief of Staff Decree No. 3, Year 2007, in *Doktrin TNI Angkatan Udara Swa Bhuwana Paksa (Indonesian Air Force Doctrine Swa Bhuwana Paksa)*, article 2.

12 *Indonesian Air Force Doctrine Swa Bhuwana Paksa,* article 8.

13 Ibid., article 13.

Indonesian Air Force also recognizes the significance of technology and the high price of it as some of the weaknesses of the TNI AU.[14]

TNI AU adopts a centralized command at the policy level and a decentralized command at the implementation level. The policy level regulates operational responsibilities, operational targets and purposes, control over air power assets, and the use of air power resources. The command in the Air Force is stratified into a strategic command, an operational command, and a tactical command:

- *Strategic command* refers to the application of military power to achieve national purposes in implementing political, economic, social, defense, and law policies.
- *Operational command* holds the duty to plan, execute, and control the military campaign.
- *Tactical command* handles the implementation of power concentration in order to fulfill the operational purposes and maximize the use of armament.[15]

Based on the current doctrine, the TNI AU is mandated with development functions (*fungsi pembinaan*) and force utilization functions (*fungsi penggunaan*). Development functions consist of force development that focuses on organization, personnel, facility and service material, system and method, and budget to support the duties of the service in order to create a relevant posture for the TNI AU. Capability development centers on building the air combat, air control, and air support abilities. Meanwhile, force utilization functions are divided into a deterrent, an attacker, and a recoverer.[16] The service determines the use of its force based on three division of theaters (*palagan*) with the sphere beyond Indonesia's exclusive economic zone (EEZ) as the outermost *palagan*. TNI AU prevents adversaries from attacking the homeland by the use of strategic fighters or jet fighters (*pesawat tempur strategis*) to destroy the enemies' abilities in their base and intercept them. The second theater is located outside the territorial seas to the end of EEZ. Every threat that enters this territory will be destroyed by jet fighters and any aircraft or air instruments will be attacked by interceptor aircraft

14 Ibid., article 16.
15 Ibid., article 18.
16 Ibid., article 21.

(*pesawat sergap*) and medium-range missiles. The next theater is the land and sea inside the territorial sea line in which the service employs short-range missiles, jet fighters, and a short-range air defense system (*baterai penangkis serangan udara*) to counter adversaries.[17]

Since May 2018, TNI AU has three Air Force Operational Commands (Komando Operasi Angkatan Udara ((KOOPSAU)); previously the service only had two KOOPSAUs. Each of those commands is led by an air vice-marshal. KOOPSAU I is located in Jakarta, covering the western part of Indonesia, KOOPSAU II is in Makassar, Sulawesi island, safeguarding the central area of the country, and KOOPSAU III is in Biak Papua, bulwarking the east side. The following section will discuss TNI AU's force structure, including the role of KOOPSAU in sustaining the service's day-to-day activities.

Force Structure

The operational duties of the Indonesian Air Force are mainly supported by the existence of its three KOOPSAUs. Each of these operational commands administers eight to 19 Air Force bases:[18]

1. KOOPSAU I operates 19 Air Force bases in the western part of Indonesia, such as Halim Perdanakusuma, Atang Sendjaja, Roesmin Nurjadin, Husein Sastranegara, and Supadio Air Force bases.
2. KOOPSAU II manages 11 Air Force bases, including Iswahyudi, Abdul Rachman Saleh, Sultan Hasanuddin, Iskandar, and Tarakan. The Iswahyudi air base (Magetan, East Java) will be home for the upcoming 11 Su-35s, and is planned to be under Squadron 14. Meanwhile, Sultan Hasanuddin air base (Makassar, South Sulawesi) is the current home for TNI AU's Su-27s and Su-30s of Squadron 11.
3. KOOPSAU III handles Jayapura, Manuhua, Merauke, Timika, Pattimura Ambon, Leo Wattimena Morotai, Eltari Kupang, and Dumatubun air bases.

---

17 Ibid., article 38.
18 By October 2018, the newly formed KOOPSAU III did not yet have any fighters and airlift squadrons, though it had been operating officially.

Based on its doctrine, TNI AU's military operations are divided into Strategic Air Attack Operation (OSUS), Air Offensive Fight Operation (OLUO), Air Defense Operation (Opshanud), Air Support Operation (Opsdukud), and Information Operation (Opsinfo). Meanwhile, in the joint-operation missions, TNI AU implements OSUS, OLUO, and Opsdukud. Additionally, for MOOTWs, the service assigns Opshanud, Opsdukud, and Opsinfo.[19] Furthermore, TNI AU through its National Air Defense Command (Komando Pertahanan Udara Nasional (KOHANUDNAS)) and KOOPSAUs provides radar coverage and air supremacy to support the overall military operations. As previously discussed, the KOOPSAUs manage several air force bases that are posts for TNI AU's squadrons, with each of them covering different types of aircraft, namely airlift, fighters, and training.

In January 2005, Indonesia's defense minister, Juwono Sudarsono (2004–2009), proposed the minimum essential force (MEF) program, aiming to revitalize and transform the country's military structure as well as its defense industry, to President Susilo Bambang Yudhoyono (2004–2014).[20] The scheme, which officially started in 2007, aims to achieve the objective by 2024. The rejuvenation of TNI AU platforms became a necessary step as more than a third of the service's aircraft have been operating for more than 30 years and only about 60 percent are ready for service.[21] The Air Force has been procuring aircraft from various sources, including its jet fighters. From the 1980s to the end of the 1990s, Indonesia was known as an avid user of U.S. and Western Europe products, in particular the U.K.'s Hawk Mk 209 and the U.S.'s F-5 and F-16.[22]

---

19 See Air Force Chief of Staff Decree No. Skep/475/XII/2006 about *Buku Petunjuk Induk Operasi TNI AU (Guidance on TNI AU Operations)* and Air Force Chief of Staff Decree No. Skep/167/VIII/2004 about *Bujuklak TNI AU tentang Operasi Udara Pada Operasi Gabungan TNI (Guidance on TNI AU's Air Operations during TNI's Joint-Operations).*

20 S. Rajaratnam School of International Studies (RSIS), *Transforming the Indonesian Armed Forces: Prospects and Challenges* (Singapore: RSIS, 2011), <https://www.rsis.edu.sg/wp-content/uploads/2014/07/ER111125_Transforming_Indon_Armed_Forces.pdf>.

21 RSIS, *Rethinking TNI AU's Arms Procurement: A Long-run Projection* (Singapore: RSIS, 2014), <https://www.rsis.edu.sg/wp-content/uploads/2014/09/PR_140709_Rethinking-TNI-AU.pdf>.

22 In May 2016, the Indonesian Air Force officially stopped the utilization all of its F-5 fighters. Indonesia received 32 Hawk Mk 209s in 1996, but nowadays only 21 of them operate and eight F-16As and four F-16Bs in 1989, with only seven of F-16A and three of F-16B still in service. See RSIS, "Rethinking TNI AU's Arms Procurement."

The bad memory of embargoes nevertheless forced Indonesia to diversify its arsenal. Jakarta procured its first batch of Russian Sukhoi (two Su-27 SKs and two Su-30 MKs) in 2003. In 2008 and 2010, TNI AU procured two Su-30 MK2s and three Su-27 SKMs respectively. The service finally started to operate a full Sukhoi squadron with the arrival of six Su-30 MK2s in 2013, in total 15 Sukhoi fighters. Indonesia also added 24 F-16 C/Ds that underwent upgrade and regeneration through the Excess Defense Articles program.[23] Besides the 24 upgraded F-16 C/Ds, TNI AU also received four F-16 Block 25s and two F-16 Block 15s for use as spare parts. In 2015, Indonesia procured 16 Korean made KAI T-50s to replace the already obsolete Hawk fighters.[24]

The Indonesian government runs several airlift and transportation aircraft in addition to the jet fighters. TNI AU obtained nine second-hand C-130H from the Australian Air Force, which were delivered at the end of 2017.[25] Besides that, the Indonesian Air Force has another nine Hercules aircraft, the first batch of which was received by TNI AU in 1960.[26] For surveillance purposes, TNI AU procured CN-235 MPA in 2016 and CN-235 in 2018 from PT Dirgantara Indonesia (Indonesian Aerospace (PT DI)). Indonesia also acquired 16 EMB-314 Super Tucano, which can be used as additional muscle for surveillance duty and trainer aircraft. Since the early 2010s, TNI AU has shown an interest in building an unmanned aerial vehicle (UAV) squadron. The service currently owns several Israel-made Aerostar tactical UAVs that are operated by Squadron 51 in Pontianak, West Kalimantan province. Indonesia added four Chinese Wing Loong I UAVs in February 2018 to strengthen Squadron 51.[27]

---

23 One of the F-16s, however, experienced engine failure and combusted in 2015, making only 23 available in service. See Reska K. Nistanto, "Profil Pesawat F-16 yang Terbakar di Halim Perdanakusuma" ("Profile of the F-16 that Burned at Halim Perdanakusuma"), *Kompas*, Apr. 14, 2015. <https://tekno.kompas.com/read/2015/04/16/10443747/Profil.Pesawat.F-16.yang.Terbakar.di.Halim.Perdanakusuma>.

24 Since December 2015, there have been only 15 T-50s ready in service as the other one experienced total loss due to a training incident.

25 However, one of the aircraft experienced an incident in December 2016 in Papua, in the eastern part of Indonesia.

26 Nistanto, "Profil Pesawat F-16 yang Terbakar di Halim Perdanakusuma."

27 Ridzwan Rahmat, "Indonesia Acquires Four Wing Loong I UAVs from China," *IHS Jane's Defence Weekly*, Feb. 25, 2018. <https://defence.pk/pdf/threads/indonesia-acquires-four-wing-loong-i-uavs-from-china.546238/>

TNI AU and Indonesia's Strategic Environment

Having discussed the brief history, doctrine, and current posture of TNI AU, this chapter now examines Indonesia's strategic environment and its implications to the TNI AU. Geostrategic location, strategic vulnerabilities, geopolitical aspiration, and constitutional mandate are the four building blocks that shape Indonesia's core interests.[28] First, Indonesia is an archipelagic country that is located between two continents (Asia and Australia) and two oceans (Indian Ocean and Pacific Ocean). Second, having vast land and maritime domains makes the country vulnerable to various threats, such as military incursion and natural disaster. Third, Jakarta's free and active foreign policy approach is sacrosanct in how the country navigates its position in global politics. Fourth, the country's constitution requires Indonesia to contribute to world peace and order. These four blocks therefore require Indonesia to actively engage in maintaining regional peace and security, which necessitates adequate military force to support this objective.

Indonesia also aspires to become a regional maritime power in the East Asian region by pledging an active defense principle to achieve this objective.[29] Although there is an objective to become a maritime power, Indonesia's defense ministry perceived the existence of hybrid threats, such as terrorism and cyber warfare, as the most prominent challenges for the country.[30] Indonesia's 2015 Defense White Paper acknowledged eight immediate threats for the country in the form of terrorism, separatism, natural disaster, border intrusion, piracy and natural encroachment, disease, cyber attack and espionage, and drugs.[31] Moreover, the White Paper elucidated Indonesia's objective to establish a global partnership that is

---

28 Iis Gindarsah and Adhi Priamarizki, *Indonesia's Maritime Doctrine and Security Concerns* (Singapore: RSIS, 2015), 133–6. <http://www.rsis.edu.sg/rsis-publication/idss/indonesias-maritime-doctrine-and-security-concerns>.

29 The active defense principle requires Indonesia to avoid building an aggressive and expansive defense posture, but without neglecting its national interests. See Presidential Regulation No. 97 Year 2015 on *Kebijakan Umum Pertahanan Negara Tahun 2015–2019 (General Policies on State Defense Year 2015–2019)*.

30 See Defense Minister Regulation No. 19 Year 2015 on *Kebijakan Penyelenggaraan Pertahanan Negara Tahun 2015–2019 (State Defense Policy Guidance)*.

31 Department of Defense, *Buku Putih Pertahanan (Defense White Paper)*, Jakarta: Department of Defense, 2015, 22–23.

perceived by the country as no dominant power in the region.[32] Based on this view, Jakarta perceives China's involvement in the ongoing disputes in the South China Sea as a potential threat to the security of the nation as well as regional stability. Iis Gindarsah identified at least four key dimensions of Indonesia's interests toward the disputes that involve China and a number of Southeast Asian states.[33] First, China's claim over the South China Sea could overlap with Indonesia's entitlement over the waters adjacent to the Natuna Islands. Second, Indonesia perceives the competition over natural resources in the South China Sea may spread to the Natuna area where considerable natural gas reserves are allegedly located. Third, the South China Sea disputes can potentially destabilize Indonesia's border and maritime security. Fourth, prolonged tension, or potential armed friction, in the South China Sea could jeopardize Indonesia's maritime border and security.

Besides securing its country from potential external threats, TNI AU also played a paramount part in MOOTWs, particularly disaster relief. Indonesia's 2015 Defense White Paper acknowledged the country's territory as a disaster-prone area in which the military must be ready to tackle inherent risks of the situation. TNI, including the TNI AU, itself has long been developing its disaster relief capability under the umbrella of MOOTW. For instance, TNI AU actively engaged in disaster relief operations following the September 2018 earthquake in Palu City, Central Sulawesi. The Air Force utilized at least 10 C-130s, three C-295s, and a Boeing B-737 for airlift operations in Palu City. Prior to the 2018 Palu earthquake, the Indonesian Air Force was involved in many disaster relief duties, such as the July 2018 Lombok earthquake and 2004 Indian Ocean earthquake and tsunami that affected the Aceh province. Moreover, TNI AU's surveillance capability came in handy in finding the lost Air Asia QZ8501 aircraft in December 2014. The service, however, aspires to improve its capability in order to sustain daily operations. The section below elaborates the expansion and procurement plan of TNI AU for serving such purpose.

---

32 Ibid., 2.
33 Iis Gindarsah, "Indonesia's Security Review: Complex but Stable," in *Security Outlook of the Asia Pacific Countries and Its Implications for the Defense Sector* (Tokyo: National Institute for Defense Studies, 2016).

Expansion and Procurement

The current Indonesian military commander-in-chief, Air Chief Marshal (ACM) Hadi Tjahjanto, in January 2018 pledged continuation of the MEF program and arms procurement based on an effect-based concept and interoperability as part of the armed forces' top priorities.[34] This idea could be interpreted into a plan to establish an acquisition plan that emphasizes more on the tactical level needs and eagerness to improve joint-force capability of TNI. In addition, ACM Hadi Tjahjanto stated that TNI AU's arm acquisitions will focus mainly on airlift aircraft and helicopters, drone technologies, and radars, not to mention educating adequate numbers of pilots, for the near future. In addition, enhancing humanitarian assistance capability will be another target for Indonesian Air Force.[35] The MEF plan has determined TNI AU's procurement plan until 2024, which mainly focuses on acquiring multirole jet fighters to enhance its defense capability (see Table 11.1).

Unit	Platforms	Quantity
Fixed Wings	Jet Fighter	128
	Transport	40
	Surveillance	16
	Maritime – Patrol	3
	Trainer – Basic	46
	Trainer – Advanced	16
UAV	n/a	28
Rotary Wings	Helicopter – Logistics	68
	Helicopter – Trainer	11
Radar	n/a	32

Table 11.1   Rethinking TNI AU 2024
Source: RSIS (2014).

---

34 Kristian Erdianto, "Panglima Hadi Tjahjanto Tetapkan 11 Program Prioritas Pembangunan TNI" ("Chief Hadi Tjahjanto Pledged 11 TNI Development Priorities"), *Kompas*, Jan. 1, 2018, <https://nasional.kompas.com/read/2018/01/25/16441141/panglima-hadi-tjahjanto-tetapkan-11-program-prioritas-pembangunan-TNI>.
35 Iwan Santosa, "Pemersatu Bangsa, Penggentar Lawan" ("The Nation Unifier, Enemies Deterrer"), *Kompas*, Jul. 2018. <https://interaktif.kompas.id/hari_bakti_TNI>.

Although the TNI AU has been showing a significant progress to reach the MEF goal, the service needs to expedite the pace as there is still a gap toward the objective. For example, TNI in October 2018 had only 85 jet fighters or less than 70 percent of the MEF target. In order to fill the gap, the Indonesian government announced the procurement of 11 Russian Su-35 in 2018 with two of them arriving in August 2019 and the rest within two years.[36] Those Su-35s will be included in Squadron 14 in Iswahyudi air base in Magetan, East Java, which was previously home for the already retired F-5 Tigers. In addition, the Indonesian government is currently negotiating with the Korean government to jointly produce 4.5-generation jet fighters, Korean Fighter Xperiment/Indonesia Fighter Xperiment (KFX/IFX), which will replace TNI AU's Hawk fighters. The deal for the project was actually announced in December 2015 with the Korean side funding 80 percent of the budget while Indonesia the remaining. The program aimed to produce the first batch of KFX/IFX fighters by 2026.[37] However, the deal faced some difficult challenges, such as a transfer of technology agreement between the two governments and license issues. The procurement of Su-35s has complicated the license problems as many components of the KFX/IFX will use American products, namely its electronically scanned array radar, infrared search and track, electronic optics targeting pod, and radio frequency jammer.[38]

With the completion of the MEF program approaching, Indonesia's defense minister and TNI are now discussing the next phase of the rejuvenation of military platforms. It is deemed that the MEF program will be upgraded into a standard essential force scheme in which the development will be focused not only on platform procurement, but also on combat capability of each of the military services, although power projection beyond

---

36 Ramadhan Rizki, "Menhan Pastikan Dua Jet Sukhoi SU 35 Atraksi di HUT TNI 2019" ("Defense Minister Guarantees an Attraction of Sukhoi Su-35 on TNI Anniversary"), *CNN Indonesia*, Sep. 11, 2018, <https://www.cnnindonesia.com/nasional/20180910203213-20-329272/menhan-pastikan-dua-jet-sukhoi-su-35-atraksi-di-hut-TNI-2019>.

37 Syafiul Hadi, "Produksi Pesawat Tempur KFX/IFX Ditunda, Menhan Akan Temui Jokowi" ("KFX/IFX Fighters Production Delayed, Defense Minister Will Meet Jokowi"), *Tempo*, Jul. 26, 2018, <https://nasional.tempo.co/read/1110752/produksi-pesawat-tempur-kfxifx-ditunda-menhan-akan-temui-jokowi>.

38 Merdeka, "Menko Polhukam Bentuk Tim Kecil Tuntaskan Pengadaan Sukhoi dan KFX/IFX" ("Coordinating Minister for Politics, Law, and Security Forms a Team to Handle Sukhoi and KFX/IFX Procurement"), *Merdeka*, Aug. 3, 2018, <https://www.merdeka.com/peristiwa/menko-polhukam-bentuk-tim-kecil-tuntaskan-pengadaan-sukhoi-dan-kfxifx.html>.

the country's border is still off the table. The scheme also intends to realize Indonesia's aspiration to play a greater role, together with other nations, in the Indo-Pacific region. In addition, improving humanitarian assistance and disaster relief capability has become another focus of TNI AU's near future procurement plan, especially in acquiring amphibious aircraft.

Interoperability and Joint Exercises

TNI AU, like the Navy and Army, has been dealing with interoperability issues ever since Indonesia decided to diversify its arms suppliers, following the trauma of an arms embargo imposed by western countries. In the past, the embargo caused logistical scarcity for the TNI AU, particularly for its F-16s acquired in 1989 and C-130 Hercules transport aircraft. Since then, as discussed, the TNI AU has procured a mixed balance of East- and West-made fighters, leaving questions about their interoperability. Borrowing the definition from the RAND Corporation, interoperability—in its broadest sense—means the ability of components within an organization to provide and accept services from each other, to enable them to operate effectively together.[39] Within this context, the interoperability of the TNI AU is assessed in three main dimensions: interoperability between platforms within the TNI AU; interoperability between the TNI AU and its Navy and Army fellows; and interoperability between the TNI AU and its regional and global partners.

Within the TNI AU, the interoperability between all of the platforms has yet to be fully achieved. Its components of jet fighters, for instance, are relatively diverse, which hampers their ability to effectively work together. In late 2005, Indonesia retired all 16 of its F-5E/F fighter jets, putting them in reserve for future use, and budgeted $1.5 billion to purchase 16 new aircraft to replace each unit. After a long deliberation on the potential replacement for the units—which included the JAS39 Gripen from Sweden and the F-16 Block 60 Viper from the United States—the government decided to sign the procurement of Su-35 fighters from Russia. However, separately, between 2011 and 2017, under an acquisition project code-named Peace Bima

---

39 Myron Hura et al., "Interoperability: A Continuing Challenge in Coalition Air Operations" (Santa Monica: CA: RAND Corporation, 2000) <https://www.rand.org/pubs/monograph_reports/MR1235.html>.

Sena 11, Indonesia received grants comprising 24 units of F-16 jet fighters from the United States, with the last two units delivered in January 2018.[40] These aircraft are currently based in Magetan, East Java, and Pekanbaru, Riau. As discussed, Indonesia is also renegotiating the procurement of the KFX/IFX fighters, and it was initially agreed that a few prototypes and around 120 aircraft would be produced by 2026 and 2032, respectively.

Although the rationale behind the diversified acquisitions is under-standable—which is not to be dependent on a single country—it inevitably reduces the ability of TNI AU's platforms to operate together effectively. As noted by Benjamin Schreer, the TNI AU possesses fighter squadrons with largely incompatible aircraft, requiring the institution to run different trainings, logistics, and maintenance—and complicates networking systems across different types of aircraft.[41] Additionally, Indonesia has yet to possess an airborne early warning and control (AEWC) system—a system to help air combat operations to extend the range and loiter capability of its tactical aircraft. However, the TNI AU has been able to leverage some of its platforms to operate together during its two major, regular internal joint exercises: the Jalak Sakti—focusing on combat abilities, organized by KOOPSAU I—and Angkasa Yudha—focusing on battle survival, organized by KOOPSAU II. During the 2016 Jalak Sakti exercise, for instance, TNI AU used its F-16s, Hawk 100/200, C-130 Hercules, Boeing B-737, as well as SA-330 and NAS-332 helicopters to engage in an integrated attack and sur-veillance simulation. In the 2016 Angkasa Yudha exercise in Natuna, TNI AU utilized 73 of its planes, including Su-27, Su-30, F-16, Hawk 109, T-50i Golden Eagle, and EMB-314 Super Tucano types.

Similarly, the TNI AU is not fully interoperable with Army and Navy assets; the other two services also source their armaments from different countries. However, for the past decade, TNI commanders, including today's commander, Hadi Tjahjanto, have emphasized interoperability as an essential element in arms procurement. Moreover, the TNI AU routinely participates in joint training with the Army and the Navy. In June 2017,

---

40 Prashanth Parameswaran, "What Does Indonesia's New Air Force Chief Pick Mean?", *Diplomat*, Jan. 19, 2017, <https://thediplomat.com/2017/01/what-does-indonesias-new-air-force-chief-pick-mean>.

41 Benjamin Schreer, "Moving Beyond Ambitions? Indonesia's Military Modernisation" (Barton: Australian Strategic Policy Institute, Nov. 2013), <https://s3-ap-southeast-2.amazonaws.com/ad-aspi/2017-07/Strategy_Moving_beyond_ambitions.pdf>.

the three TNI services were engaged in one of their biggest joint exercises, held in Situbondo, East Java. During the drill, which aimed to test the preemptive attack doctrine and the interoperability within the three services, the TNI displayed a wide range of its platforms, such as TNI AU's 40 jet fighters—including 8 Su-27/30, 6 F-16s, 10 Hawk 100/200s, 2 F-5 Tigers, 12 T-50 Golden Eagles, and 2 EMB-314 Super Tucanos—as well as 32 cargo aircraft—including 16 C-130 Hercules and 11 Super Puma helicopters.

At the same time, the Navy brought 32 ships, along with tactical vehicles and rocket launchers, and the Army was present with its Scorpion, Stormer APC, and Stormer Commando tanks, as well as its armored vehicles. The extent of interoperability among these platforms, however, remains unclear. Similarly, during the 72nd anniversary of the TNI in October 2016, the three services displayed a wide range of platforms, including the Army's Leopard 2RI tanks and Apache helicopters, the Navy's 17 FIX Wing planes, and TNI AU's F-16s and Sukhoi, although there was again very limited simulation on their interoperability.

Meanwhile, overseas, the TNI AU has attempted to achieve interoperability with its regional and global partners, although it largely relies on the U.S.-made platforms in doing so. Since the early 1970s, the TNI AU has been involved in various joint military exercises. Notably, TNI AU and the Republic of Singapore Air Force have been conducting a joint exercise since 1980 code-named Elang Indopura. In the latest exercise in December 2016, Indonesia deployed five of its F-16 jet fighters based in Pekanbaru for a joint combat exercise with Singapore's F-16s. TNI AU has also been engaged in joint exercises with air forces from other Association of Southeast Asian Nations (ASEAN) countries, including with the Royal Malaysian Air Force since 1975, code-named the Elang Malindo exercise, and with the Royal Thai Air Force since 1981, code-named Elang Thainesia. In the latest exercise in July 2017, both the TNI AU and the Royal Thai Air Force used their F-16s.

Besides its ASEAN partners, the TNI AU is also involved in regular joint training efforts with other air forces, notably those in Australia and the United States. TNI AU has been working closely with Australia, having sent its F-16s to Australia for training in 2009 and Su-27/30s in 2012.[42] More recently, in November 2017, both countries engaged in their routine bilateral peacekeeping exercise, Garuda Kookaburra. In October 2017, the TNI AU sent six F-16s based in the Iswahyudi air base to participate in Exercise Elang AUSINDO 17, held at a Royal Australian Air Force base in Darwin, where the Australians used up to eight of their F/A-18A Hornets. Additionally, the TNI AU cooperates intensely with the United States. In November 2016, the TNI AU engaged in Cope West, a joint exercise with the U.S. Marine Corps held in Manado, North Sulawesi, where it was involved in a 10-day exercise, including a fighter-to-fighter exercise—the first of such exercises in 19 years of U.S.-Indonesia bilateral military relations. During the exercise—which aimed to enhance interoperability between the TNI AU and the United States, and to exchange techniques related to combined air operations specific to the United States and Indonesian aircraft—Indonesia used six of its F-16s, while the U.S. Marine Corps brought six of its F-18s. In March 2018, the TNI AU and the U.S. Air Force also had a two-week joint training in Manado which involved parachuting, infiltration, combat search and rescue, and evacuation drills. In the exercise, Indonesia again used its F-16 fighters to engage in dog fight simulation with American F-16s.

Prospects and Challenges

In December 2017, the TNI AU, through its official website, announced its plan to have "no area left without air cover." The plan focuses both on the acquisition of new equipment as well as redeployment of existing resources. According to the plan, the TNI AU will have an additional three squadrons of combat aircraft to supplement the existing eight squadrons; additional heavy, medium, and tactical transport squadrons; the standing up of two squadrons of UAVs; and a new helicopter squadron. TNI AU will also acquire four AEWC systems; counterterrorist helicopters; four aerial refueling tankers, 12 ground-based radar systems, and an undisclosed number of amphibious aircrafts. In addition, Hadi Tjahjanto, who has since

---

42 Ade Marboen, "Flight Sukhoi Su-27/30 TNI AU Latihan ke Australia," *Antara News*, Jul. 26, 2012, TNI, <https://www.antaranews.com/berita/820848/wing-penerbang-kehormatan-untuk-ksad-dan-ksal>.

ascended to TNI Commander, stated that a number of bases that did not have any combat aircraft will be equipped with them, in order to provide total air defense coverage in line with Indonesia's plans to be a maritime power. That said, the TNI still faces a number of challenges to achieve this goal, namely the uncertainty over its role in Indonesia's maritime ambition, budgetary concerns, Indonesia's focus on internal security, and questions over the sustainability of its diverse procurement pattern.

Since Joko Widodo's ascent to the presidency in 2014, Indonesia has reiterated its ambition to be a maritime power. In many of his official addresses, President Widodo has put forward the concept of Indonesia as *Poros Maritim Dunia* (Global Maritime Fulcrum (GMF)) as the country's strategic positioning, emphasizing Indonesia's identity as a maritime nation and an archipelagic state. In his presidential inaugural speech, he highlighted the importance of the waters surrounding the Indonesian archipelago as the country's potential economic and national strength resources rather than a natural disadvantage.[43] The long-term vision of the concept pictured the country as a maritime power that is respected by extra-regional powers, friendly to all regional partners, and feared by any potential regional and international entity that intends to disrupt Southeast Asia's security. President Widodo urged the nation to look at itself as "a power between two oceans" which stressed geographical, geopolitical, and geo-economic reality of Indonesia. The concept projected Indonesia as an "Indo-Pacific power" and foresaw the increasingly interconnected Pacific and Indian Oceans as the main arena of Indonesian foreign policy engagement.[44]

However, the president has not yet given a clear direction for the TNI, let alone the TNI AU, to fit into this GMF concept. Despite the hazy set of directions, the Indonesian military under the Widodo administration has started to add a Primary Command for Combat (*Kotama Tempur*) for the

---

43 Kompas, "Ini Pidato Perdana Jokowi Sebagai Presiden ke-7 RI" ("Jokowi's First Speech as the President"), *Kompas*, Oct. 20, 2014, <http://nasional.kompas.com/read/2014/10/20/1318031/artikel-detail-komentar-mobile.html>.

44 Iis Gindarsah & Adhi Priamarizki, "Politics, Security and Defense in Indonesia: The Pursuit of Strategic Autonomy," in *Indonesia's Ascent: Power, Leadership, and the Regional Order*, eds. Christopher B. Roberts, Ahmad. D. Habir, & Leonard C. Sebastian (Hampshire: Palgrave Macmillan, 2015).

eastern part of Indonesia.[45] This is in line with the president's development aspiration to build the region. The establishment of TNI AU's KOOPSAU III is part of this *Kotama Tempur* expansion. The Army meanwhile created the third division of Strategic Command Reserve (*Kostrad*), in Gowa, South Sulawesi province, to cover the eastern side of the country. While the other services only enjoyed one additional *Kotama Tempur*, the Navy received the opportunity to add two, namely the third Fleet Command (Koarmada III) and third Marine Troop (Pasmar III), both located in Sorong, West Papua province and handling provinces in Indonesia's eastern area. These *Kotama Tempur* developments may hint at an initial effort to embed the GMF concept in the Indonesian military, although the attempt was rather an acceleration of execution tardiness than GMF-inspired advancement.[46]

Second, budget remains a major issue across all TNI services, particularly the TNI AU. Although the Indonesian government increased its defense budget by almost 400 percent between 2000 and 2010, it has yet to meet the country's overall military expenses. Between 2000 and 2004, the Indonesia authority only managed to fund less than 75 percent of the defense needs, where almost three-quarters of the budget was used for daily needs and 35 percent of it was allocated for development.[47] In addition, the MEF program mandated Indonesia allocate at least 1.5 percent of its gross domestic product (GDP) for defense, in order to sustain the development of the military. However, until 2018, Indonesia only allocated about one percent of its GDP as defense budget. Defense itself turned out to be a tertiary priority for Jakarta compared to other sectors, particularly education which consumed about 20 percent of the country's total budget.

The Indonesian government normally divided the defense budget equally among the three services. This split, however, failed to accommodate

---

45 *Kotama Tempur* refers to the TNI's primary combat force in taming potential threats in which each service has different units. KOOPSAU is considered as the TNI AU's *Kotama Tempur*.

46 The idea of establishing various *Kotama Tempur* in Indonesia's eastern provinces appeared as early as 2006, around the same time with the promulgation of MEF program. See Andi Widjajanto and Makmur Keliat, *Reformasi Ekonomi Pertahanan Indonesia (Indonesia's Defense Economic Reform)* (Jakarta: Pacivis UI, 2006); Alexandra Wulan, "Military Transformation," in Shafiah F. Muhibat (ed.), *For Indonesia 2014–2019: Social-Politics Security Agenda* (Jakarta: Center for Strategic and International Studies, 2014).

47 Indonesia's defense budget mainly covers: 1) organic activities, such as personnel needs; 2) functional activities, such as research and development, as well as exercises; and 3) daily operational routines.

the needs of the Indonesian Air Force. In order to accommodate its budgetary requirement, the TNI AU is compelled to focus most of its budget
on maintenance and the procurement of spares. For instance, cutting the
spending for TNI AU officers' welfare (particularly housing and business
trip support) is now a common practice for the service. While it is true that
the military in general has been spending the defense budget mostly for
paying manpower, the TNI AU in reality has to readjust the welfare components to support the day-to-day operations of the service.

Third, the TNI AU has relatively limited support from the national
defense industry. One of Indonesia's landmark defense industry players is
PT DI, which handles aircraft production. Nonetheless, the company has
not yet met the expectation of the TNI AU for fulfilling the service's needs.
For example, the Air Force preferred to procure Augusta Westland (AW-
101) rather than PT DI–made helicopters, such as its Superpuma Family.[48]
ACM (ret.) Agus Supriatna (2015–2017), pointed out that PT DI's lack of
ability to fulfill the TNI AU's requirements and its tardiness in delivering
its products to the service as reasons for not procuring from the company.[49]

Fourth, the TNI is still largely fixated on an internal security mindset, which put the Army as the first priority. The Army has been using a
territorial command structure as a tool to respond to potential threats both
from internal and external realms. The New Order administration utilized
territorial command due to the system's vast intelligence network, which
was extremely useful to undermine dissenting voices against the authoritarian regime.[50] In addition, the Indonesian military has been preserving its
centrality in an internal security role with the Army as the spearhead to do
the task. Such a situation did not only result in giving the service an inferior

---

48 The procurement later caused a stir as there is an indication of budgetary
markup. Several high-ranking officers have been put into custody due to this case.
See Dylan Aprialdo Rachman, "Kasus Heli AW101, KPK Akan Dukung TNI AU
Hadapi Gugatan Perdata" ("AW 101 Case: KPK Supports TNI AU in the Court"),
Kompas, Jul. 8, 2018, <https://nasional.kompas.com/read/2018/05/08/06310051/kasus-heli-
aw101-kpk-akan-dukung-TNI-au-hadapi-gugatan-perdata>.
49 Abraham Utama, "KSAU: Saya Komisaris PTDI, Tahu Kenapa Beli Helikopter
AW101" ("Air Force Chief: I Am PT DI Commissioner, and Know the Reason for Buying
AW 101 Helicopter"), *CNN Indonesia*, Nov. 25, 2018, <https://www.cnnindonesia.com/
nasional/20151125101959-20-93911/ksau-saya-komisaris-ptdi-tahu-kenapa-beli-
helikopter-aw101>.
50 Makmur Supriyatno, *Tentang Ilmu Pertahanan (On Defense Studies)* (Jakarta: Pustaka Obor,
2014).

position, but also in making technological development a secondary option for the military.

Last, the procurement pattern of the TNI AU, which involves diverse countries to supply its armaments, has not only impeded its interoperability, as discussed, but also may not be sustainable in the long run, within the context of great power rivalries. For instance, Indonesia's program to procure the Su-35 may face challenges due to possible consequences of U.S. legislation that aims to penalize states that procure Russian military systems. Through the Countering America's Adversaries Through Sanctions Act, the United States proposes sanctions on Russia's defense customers in response to Russia's annexation of Crimea and its alleged interference in the 2016 U.S. presidential election. According to the chief of information of the TNI AU, Air Commodore Novyan Samyoga, although contracts have been signed, Indonesia will have no option other than to terminate the deal. He further added that if the contract cancellation materializes, Indonesia may be forced to procure "Western fighter aircraft(s)."[51] Besides Russia, growing rivalries between the United States and China may also have a impact on Indonesia's options to procure certain armaments from the latter. On the other hand, Indonesia may see this growing tension as an opportunity: if the major powers perceive Indonesia as a significant partner in the region, it will push them to offer aid and joint military exercises, resulting in the TNI AU's arms modernization and enhanced combat capability.

Conclusion

Since its inception, the TNI AU has been operating primarily as a supporting component within the TNI due to its limited capability and the Indonesian military's emphasis on ground-based warfare. The changing strategic environment, particularly with China's emergence as a great power and the ongoing dispute in the South China Sea, has provided an opportunity for the Air Force to increase its prominence. President Widodo's aspiration to enhance the country's maritime capability may also enhance this prospect. For maritime defense development, however, such

---

51 Jon Grevatt, "Indo Defence 2018: Indonesia's Su-35 Procurement Faces CAATSA Hurdle," *Jane's Defence Weekly*, Nov. 9, 2018, <https://www.janes.com/article/84427/indo-defence-2018-indonesia-s-su-35-procurement-faces-caatsa-hurdle>.

a plan has not yet been significantly implemented, other than executing the delayed arrangement of the previous administration in establishing an additional fleet command and one extra marine troop, making fortuity not as favorable as it seems.

For a big nation, the TNI AU acknowledges that it needs to boost its capability to leave no area without air cover and has been undergoing phases of modernization. However, it still faces a number of challenges, mainly related to its role in Indonesia's ambition to be a maritime power and focus on internal security—which place the Navy and the Army, respectively, in more advantaged positions. The Air Force has realized this reality and positioned itself mainly as a supporting force. The looming platforms' obsolescence has forced the TNI AU to embrace rejuvenation and modernization of its posture. The plan to upgrade the platforms, however, has been experiencing some difficulties, particularly financial challenges. Although Indonesia's steady economic growth has helped to increase the country's defense budget, the figure has not yet met the equilibrium to properly support the day-to-day operations. For the Air Force, this challenging condition has forced the service to convert its welfare spending into maintenance funding.

The procurement process, meanwhile, is not without predicaments. On the one hand, within the context of growing rivalries between the United States and Russia, as well as between the United States and China, the TNI AU may gain opportunities for military grants and joint exercises to enhance its capability. On the other hand, the situation has produced a complicated situation due to bureaucratic hindrances and potential penalties, especially from the U.S. side. Jakarta may face some penalties from Washington due to the TNI AU's usage of Sukhoi, because the United States has a strict regulation against the utilization of Russian military products. Despite these complex circumstances, Indonesia seems to maintain its stance to diversify its platforms with some Russian-made fighters. The Southeast Asian country's bitter experience with arms embargos by western countries became the primary drive for the country's decision to vary its armaments, including the TNI AU's jet fighters.

The decision to diversify its platforms produces an inherent risk of logistical nightmare as well as questions over interoperability. The discussion on interoperability and joint exercises in this chapter noted the Air

Force's several joint exercises, both with other services within the TNI as well as overseas counterparts. Most of those exercises nevertheless employed products made by western countries, notably the U.S.-made F-16s, with limited participation of the TNI AU's Sukhoi squadrons, which actually projected as its spearhead platforms. This intriguing ballgame consequently generates some enquiries over the readiness of the Indonesian Air Force in operating those Sukhoi on a more regular basis. Although there is no compulsory requirement, participating in various joint exercises serves as a way to enhance knowledge and capability through learning from other air services.

References

Air Force Chief of Staff Decree No. 3, Year 2007, in *Doktrin TNI Angkatan Udara Swa Bhuwana Paksa (Indonesian Air Force Doctrine Swa Bhuwana Paksa)*.

Department of Defense. *Doktrin Pertahanan Negara*. Jakarta: Department of Defense, 2007.

Department of Defense. *Buku Putih Pertahanan*. Jakarta: Department of Defense, 2015.

Erdianto, K. "Panglima Hadi Tjahjanto Tetapkan 11 Program Prioritas Pembangunan TNI" ("Chief Hadi Tjahjanto Pledged 11 TNI Development Priorities"). *Kompas*, Jan. 1, 2018. <https://nasional.kompas.com/read/2018/01/25/16441141/panglima-hadi-tjahjanto-tetapkan-11-program-prioritas-pembangunan-TNI>.

Gindarsah, I. "Indonesia's Security Review: Complex but Stable." *Security Outlook of the Asia Pacific Countries and Its Implications for the Defense Sector*. Tokyo: National Institute for Defense Studies, 2016.

Gindarsah, I., and A. Priamarizki. "Indonesia's Maritime Doctrine and Security Concerns." Singapore: RSIS, 2015. <http://www.rsis.edu.sg/rsis-publication/idss/indonesias-maritime-doctrine-and-security-concerns>.

Grevatt, J. "Indo Defence 2018: Indonesia's Su-35 Procurement Faces CAATSA Hurdle." *Jane's Defence Weekly*, Nov. 9, 2018. <https://www.janes.com/article/84427/indo-defence-2018-indonesia-s-su-35-procurement-faces-caatsa-hurdle>.

Hadi, S. "Produksi Pesawat Tempur KFX/IFX Ditunda, Menhan Akan Temui Jokowi" ("KFX/IFX Fighters Production Delayed, Defense Minister Will Meet Jokowi"). *Tempo*, Jul. 26, 2018. <https://nasional.tempo.co/read/1110752/produksi-pesawat-tempur-kfxifx-ditunda-menhan-akan-temui-jokowi>.

Honna, J. "The Politics of Securing Khaki Capitalism in Democratizing Indonesia." *Khaki Capital: The Political Economy of the Military in Southeast Asia*, edited by P. Chambers and N. Waitoolkiat. Copenhagen: Nias Press, 2017.

Hura, Myron et al. "Interoperability: A Continuing Challenge in Coalition Air Operations," Santa Monica, CA: RAND Corporation, 2000. <https://www.rand.org/pubs/monograph_reports/MR1235.html>.

Kompas. "Ni Pidato Perdana Jokowi Sebagai Presiden ke-7 RI" ("Jokowi's First Speech as the President"). *Kompas*, Oct. 20, 2014. <http://nasional.kompas.com/read/2014/10/20/1318031/artikel-detail-komentar-mobile.html>.

Lowry, R. *The Armed Forces of Indonesia*. New South Wales: Allen & Unwin Pty Ltd., 1996.

Marboen, A., "Flight Sukhoi Su-27/30 TNI AU Latihan ke Australia." *Antara News*, Jul. 26, 2012. <https://www.antaranews.com/berita/820848/wing-penerbang-kehormatan-untuk-ksad-dan-ksal>.

Merdeka. "Menko Polhukam Bentuk Tim Kecil Tuntaskan Pengadaan Sukhoi dan KFX/IFX" ("Coordinating Minister for Politics, Law, and Security Forms a Team to Handle Sukhoi and KFX/IFX Procurement"). *Merdeka*, Aug. 25, 2018. <https://www.merdeka.com/peristiwa/menko-polhukam-bentuk-tim-kecil-tuntaskan-pengadaan-sukhoi-dan-kfxifx.html>.

Mietzner, M. "Business as Usual? The Indonesian Armed Forces and Local Politics in the Post-Soeharto Era." *Local Power and Politics in Indonesia: Decentralisation & Democratisation*, edited by E. Aspinall and G. Fealy. Singapore: ISEAS, 2003.

Nistanto, R.K. "Profil Pesawat F-16 yang Terbakar di Halim Perdanakusuma" ("Profile of the F-16 that Burned at Halim Perdanakusuma"). *Kompas*, Apr. 16, 2015. <https://tekno.kompas.com/read/2015/04/16/10443747/Profil.Pesawat.F-16.yang.Terbakar.di.Halim.Perdanakusuma>.

Parameswaran, P. "What Does Indonesia's New Air Force Chief Pick Mean?" *Diplomat*, Jan. 19, 2017. <https://thediplomat.com/2017/01/what-does-indonesias-new-air-force-chief-pick-mean>.

Presidential Regulation No. 97 Year 2015 on *Kebijakan Umum Pertahanan Negara Tahun 2015–2019 (General Policies on State Defense Year 2015–2019)*.

Rachman, D.A. "Kasus Heli AW101, KPK Akan Dukung TNI AU Hadapi Gugatan Perdata" ("AW 101 Case: KPK Supports TNI AU in the Court"). *Kompas*, Jul. 8, 2018. <https://nasional.kompas.com/read/2018/05/08/06310051/kasus-heli-aw101-kpk-akan-dukung-TNI-au-hadapi-gugatan-perdata>.

Rahmat, R. "Indonesia Acquires Four Wing Loong I UAVs from China." *IHS Jane's Defence Weekly*, Feb. 25, 2018. <https://defence.pk/pdf/threads/indonesia-acquires-four-wing-loong-i-uavs-from-china.546238/>.

Rizki, R. "Menhan Pastikan Dua Jet Sukhoi SU 35 Atraksi di HUT TNI 2019" ("Defense Minister Guarantees an Attraction of Sukhoi Su-35 on TNI Anniversary"). *CNN Indonesia*, Sep. 11, 2018. <https://www.cnnindonesia.com/nasional/20180910203213-20-329272/menhan-pastikan-dua-jet-sukhoi-su-35-atraksi-di-hut-TNI-2019>.

S. Rajaratnam School of International Studies. "Transforming the Indonesian Armed Forces: Prospects and Challenges." Singapore: RSIS, 2011. <https://

www.rsis.edu.sg/wp-content/uploads/2014/07/ER111125_Transforming_
Indon_Armed_Forces.pdf>.

S. Rajaratnam School of International Studies. "Rethinking TNI AU's Arms Procurement: A Long-run Projection." Singapore: RSIS, 2014. <https://www.rsis.
edu.sg/wp-content/uploads/2014/09/PR_140709_Rethinking-TNI-AU.pdf>.

Santosa, I. "Pemersatu Bangsa, Penggentar Lawan" ("The Nation Unifier, Enemies
Deterrer"). *Kompas*, Jul. 2018. <https://interaktif.kompas.id/hari_bakti_TNI>

Schreer, B. *Moving Beyond Ambitions? Indonesia's Military Modernisation.* Barton:
Australian Strategic Policy Institute, Nov. 2013. <https://s3-ap-southeast-2.
amazonaws.com/ad-aspi/2017-07/Strategy_Moving_beyond_ambitions.pdf>.

Supriyatno, M. *Tentang Ilmu Pertahanan (On Defense Studies).* Jakarta: Pustaka Obor,
2014.

Tentara Nasional Indonesia Angkatan Udara. *50 Tahun Emas Pengabdian TNI Angkatan Udara.* Jakarta: TNI AU Information Desk, 1996.

Tentara Nasional Indonesia Angkatan Udara. *Bakti TNI Angkatan Udara 1946–2004.*
Jakarta: TNI AU, 2004.

Utama, A. "KSAU: Saya Komisaris PTDI, Tahu Kenapa Beli Helikopter AW101" ("Air
Force Chief: I Am PT DI Commissioner, and Know the Reason for Buying AW 101
Helicopter"). *CNN Indonesia*, Nov. 25, 2015. <https://www.cnnindonesia.com/
nasional/20151125101959-20-93911/ksau-saya-komisaris-ptdi-tahu-kenapa-
beli-helikopter-aw101>.

Widjajanto, A., and M. Keliat. *Reformasi Ekonomi Pertahanan Indonesia (Indonesia's
Defence Economic Reform).* Jakarta: Pacivis UI, 2006.

Wulan, A. "Transformasi Militer" ("Military Transformation"). *Untuk Indonesia
2014–2019: Agenda Sosial-Politik Keamanan (For Indonesia 2014–2019: Social-Politics
Security Agenda)*, edited by Shafiah F. Muhibat. Jakarta: Center for Strategic and
International Studies, 2014.

12 TAIWAN
The Strait Challenge

Wu Shang-su

Introduction

Since 1949, air power has been a critical factor in the military balance across the Taiwan Strait, with Taipei always having made its air force a top priority in defense investment. After seven decades of concentrated effort, Taiwan has built a remarkable air defense network composed of fighters, surface-to-air missiles (SAMs), ground-based and airborne radars, and other facilities. Nevertheless, the other side of the Strait has built up much larger forces that would overwhelm Taiwan's air defense network using conventional and non-conventional means. Taiwan's air power has thus been much constrained, making certain changes necessary at structural, strategic, and other levels, despite all efforts. In order to form a comprehensive view on Taiwan's air power, this chapter will be divided into four parts, historical review, dynamic challenges, current plans and difficulties, and conclusion.

Historical Review

Since 1945, the Air Force of the Republic of China (ROC) regime, also known as the Republic of China Air Force (ROCAF), has taken up the role of air power in Taiwan. Before that, the air services of the Imperial Japanese Army and Navy were the backbone of the air power on the island, but were overwhelmed by their Allied counterparts in the latter phase of the Pacific

War.[1] Air power became critical after the ROC regime fled to Taiwan as a result of its defeat by the Chinese communists in 1949. The ROC subsequently became known as the People's Republic of China (PRC). Politically, aerial operations, such as dropping propaganda leaflets and harassment attacks, were designed to support the ROC's major agenda of "retaking the mainland," a posture meant to challenge the PRC's legitimacy.[2] Militarily, the PRC's emerging air power was gradually substantialized from 1949 onwards. Although the communist People's Liberation Army (PLA) did not significantly apply air power during the civil war, it rapidly established the PLA Air Force (PLAAF) using Japanese and American aircraft left respectively by Manchukuo and the ROC, then expanded with Soviet aircraft provided by Moscow.[3] In the early 1950s, the PLAAF and then the PLA Navy Air Force (PLANAF) achieved technological superiority by introducing MiG-15 jet fighters, which were much more advanced than the ROCAF's World War II legacies, such as P-51D and P-47N piston engine fighters.[4] When the ROCAF finally introduced F-84G and F-86F jet fighters from the United States in the mid-1950s, its PRC counterpart obtained MiG-17s.[5] It is therefore evident that air power has been the key factor in cross-Strait relations from the initial phase.

The PRC's challenge across the Strait was somewhat moderated during the Quemoy Crisis of 1958 as a result of military technologies and the following international situation. During the crisis, the United States supplied the ROCAF with the AIM-9B air-to-air missiles (AAM) and created the first combat record on September 24, 1958. That year, Taiwan also

---

1   Thomas E. Griess, Thomas B. Buell, John H. Bradley, and Jack W. Dice, *The Second World War: Asia and the Pacific* (New York: Square One, 2002), 189, 192; "Aerial Battle of Taiwan–Okinawa," Weapons and Warfare, Aug. 22, 2015, <https://weaponsandwarfare.com/2015/08/22/aerial-battle-of-taiwan-okinawa/>.
2   Herbert A. Friedman, "Nationalist and Communist Chinese Propaganda Leaflets," Psywarrior, <http://www.psywarrior.com/NationalistChinesePropaganda.html>; Xiaoming Zhang, *Red Wings over the Yalu: China, the Soviet Union, and the Air War in Korea* (College Station, TX: Texas A&M University Press, 2002), 82.
3   Ralph N. Clough, *Island China* (London: Harvard University Press, 1978), 100–01.
4   Xiaoming Zhang, "Air Combat for the People's Republic: The People's Liberation Army Air Force in Action, 1949–1969," in Mark A. Ryan, David M. Finkelstein, and Michael A. McDevitt (eds.), *Chinese Warfighting: The PLA Experience since 1949* (London: M.E. Sharpe, 2003), 281.
5   Stockholm International Peace Research Institute (SIPRI), "Arms Transfers Database," Database, accessed Aug. 24, 2018, <http://armstrade.sipri.org/armstrade/page/trade_register.php>; "MiG-17 FRESCO," Federation of American Scientists, June 17, 2000, <https://fas.org/nuke/guide/russia/airdef/mig-17.htm>.

received the first SAMs, MIM-14, so that air thus evolved into a network combined with airborne and land-based assets. Apart from missiles, the ROCAF also demonstrated its capability of retaining air superiority during the crisis, as evidenced in its landslide victories in aerial combat. After the crisis, both the Soviet Union and the United States continuously supplied further fighters to take both sides of the Strait into the supersonic era. The ROCAF received supersonic F-100F, F-104, and F-5A/B fighters from the late 1950s to the early 1960s, while its PLA counterparts acquired supersonic MiG-19s and MiG-21s.[6] However, the Sino-Soviet split in 1964 paused the modernization of China's air power, and the subsequent Cultural Revolution further disrupted training of most PLA units and the defense industry that supplied most aircraft.[7] In other words, Taiwan faced a less challenging environment for its air power, as the final aerial combat in the Strait occurred in 1967 with a victory for the ROCAF.[8]

Apart from the need to respond to the threats, Taiwan's air power was also shaped by its main arms supplier, the United States. Most aircraft received by the ROCAF were defensive, intended to prevent Chiang Kai-shek's strategic goal of retaking the mainland. For example, the ROCAF's request for F-4 fighters was rejected because of the range and capacity.[9] The ROCAF's F-100Fs might have been seen as offensive, but their capacity and range are inferior to the F-4.[10] Although Washington provided some PB4Y2 bombers in the early 1950s, such slow propeller aircraft were unlikely to achieve offense vis-à-vis the PLAAF and PLANAF jet fighters.[11] In contrast, Washington provided a range of defensive assets for Taipei. MIM-23 SAMs

---

6   Clough, *Island China*, 108; Zhang, X., "Air Combat for the People's Republic: The People's Liberation Army Air Force in Action, 1949–1969", 288; SIPRI, "Arms Transfers Database."

7   Baichun Zhang, Jiuchun Zhang, and Fang Yao, "Technology Transfer from the Soviet Union to the People's Republic of China: 1949–1966," *Comparative Technology Transfer and Society*, Vol. 4, No. 2 (2006), 109, 115; *The Military Balance 1965* (London: International Institute for Strategic Studies (IISS), 1965), 8.

8   The ROCAF's 4 F-104Gs engaged with the PLAAF's 12 Mig-19s on Jan. 13, 1967 and the outcome was 2:1. Martin W. Bowman, *Lockheed F-104 Starfighter* (Wiltshire, UK: Crowood Press, 2000), 165.

9   Nancy Bernkopf Tucker, *Strait Talk: United States–Taiwan Relations and the Crisis with China* (Cambridge, MA: Harvard University Press, 2009), 94–95.

10 Francis Crosby, *The World Encyclopedia of Bombers* (London: Anness, 2004), 218–9, 228–9.

11 Yi-ming Chen and Wen-shao Liu, *Military Aircraft in the Chinese Civil War* (Taipei: The Wing of China Press, 1992), 68, 90; Stewart Wilson, *Aircraft of WWII* (Fyshwick, ACT, Australia: Australian Aviation, 1998), 43.

and M-42 self-propelled anti-aircraft guns were introduced to strengthen the air defense network in the mid-1960s.[12] When the United States assisted Taiwan to develop its aviation industry, under the influence of the Nixon Doctrine, the chosen type of fighter was defense-oriented F-5E/F.[13]

The fluctuating international environment in the 1970s and 1980s reshaped Taiwan's air power, in particular, the termination of the official ties between Taipei and Washington in 1979, followed by the communiqué of August 17, 1982, that would gradually reduce the latter's arms sales to the former.[14] This left the ROCAF facing the serious challenge of securing further supplies of fighters, but with U.S. assistance Taiwan managed to work out several solutions with assistance from the United States. Taiwan extended its project of manufacturing F-5E/Fs locally and obtained retired F-104s from Germany, Japan, and the United States to meet immediate needs. For the next generation of fighters, Taipei requested F-16s or F-5Gs, later F-20s, but none of these requirements were permitted by Washington.[15] Taiwan consequently boosted its aviation industry, with American technological assistance, to develop the Indigenous Defense Fighter (IDF) project in the 1980s.[16] Simultaneously, the PRC was only able to upgrade the J-7, a copy of the MiG-21, and developed its larger version, the J-8, with some western technological assistance. The upgrade project for the J-8 was based on American technology, thus endowing Washington with some advantages in maintaining the balance of air power across the Strait.[17] In addition to the indigenous fighter, from 1981 Taipei also developed

---

12 SIPRI, "Arms Transfers Database."

13 "F5E/F Supersonic Fighter," Aerospace Industrial Development Corporation <http:// www.aidc.com.tw/en/military/f5e>; Tucker, *Strait Talk: United States–Taiwan Relations and the Crisis with China*, 61, 64, 94.

14 "Joint Communiqué of the People's Republic of China and the United States of America," Embassy of the People's Republic of China in the United States of America, Aug. 17, 1982, <http://www.china-embassy.org/eng/zmgx/doc/ctc/t946664.htm>.

15 Tucker, *Strait Talk: United States–Taiwan Relations and the Crisis with China*, 134–43, 150–1; SIPRI, "Arms Transfers Database."

16 T.S. Jan and C.G. Jan, "Development of Weapon Systems in Developing Countries: A Case Study of Long Range Strategy in Taiwan," *Journal of Operational Research Society*, Vol. 51, No. 9, Sep. 2000, 1043; Peter R. March, *Directory of Military Aircraft of the World* (London: Cassell & Co., 2001), 11.

17 Bates Gill and Taeho Kim, *China's Arms Acquisitions from Abroad: A Quest for "Superb and Secret Weapons,"* SIPRI Research Report No. 11 (Oxford: Oxford University Press, 1995), 42, 74, 92; Tucker, *Strait Talk: United States–Taiwan Relations and the Crisis with China*, 160.

indigenous SAMs, the Tien-Kung (TK) series.[18] Aside from weapon systems, some runways were built on superhighways and highways in the 1970s as alternative airstrips.[19] In short, Taiwan was able to retain certain niches in its air power vis-à-vis the PLA during the Cold War, despite its international isolation from the 1970s.

Aside from fighters, the ROCAF also developed other elements of its air power, including airlift, ground attack, reconnaissance, and anti-submarine, in addition to some aviation units in the ROC Army (ROCA) and Navy (ROCN). The airlift capacity was important for Taipei due to its offensive military structure and supply for the offshore islands. Under the strategic guidance of "retaking the mainland," the airlift was indispensable for the ROCA's airborne brigades. As the sea lines of communication to the islands under ROC control, such as Quemoy (Kinmen) and Matsu, were vulnerable to the PLA's firepower and the sea denial means of the People's Liberation Army Navy (PLAN), airlift was the main alternative available for supply during a crisis, such as that of 1958. There were three waves of modernization: C-119s introduced in the 1960s to replace the World War II legacy, such as the C-46s and C-47s, and then the C-123s and C-130s acquired in the 1970s and 1980s respectively.[20]

The ROCAF's reconnaissance units carried out the most frequent missions across the Strait, with several periods of modernization corresponding to the fighters, from F-6 (P-51 reconnaissance version) during the Korean War, the RT-33A, RF-86F, RF-84F, RF-100A, and RF-101A in the 1950s, and then the RF-104G in the 1960s. Furthermore, based on the cooperation with the U.S. Central Intelligence Agency, the ROCAF also operated electronic surveillance platforms, such as the B-17, P-2V, C-54, P-3A, and optical reconnaissance aircraft, the famous U-2.[21] Since the ROCAF received

---

18 "Tien-Kung (Sky Bow)," Missile Defense Advocacy Alliance, Jul. 18, 2018, <http://missiledefenseadvocacy.org/air-defense/air-defense-of-u-s-partners/allied-air-defense-systems/tien-kung-sky-bow/>.

19 "Military Stages Highway Landing, Takeoff Drill in Tainan," *Taiwan Today*, Apr. 12, 2011, <https://taiwantoday.tw/news.php?unit=2,23,45&post=1752>.

20 Clough, *Island China*, 104; SIPRI, "Arms Transfers Database"; "Bombardment of Quemoy," *Taiwan Today*, Aug. 1, 1988, <https://taiwantoday.tw/news.php?unit=4&post=4657>.

21 SIPRI, "Arms Transfers Database"; Jay Taylor, *The Generalissimo's Son* (London: Harvard University Press, 2000), 206–09, 239, 286; Xiaoming Zhang, "The PLAAF's Evolving Influence within the PLA and upon National Policy," in Richard P. Hallion, Roger Cliff, and Philip C. Saunders (eds.), *The Chinese Air Force: Evolving Concepts, Roles and Capabilities* (Washington DC: National Defense University Press, 2013), 74.

S-2A/E anti-submarine aircraft, anti-submarine warfare was added into the ROCAF. In parallel, the ROCN established its aviation units in the late 1970s with the Hugh-500 anti-submarine warfare (ASW) helicopters, and the ROCA obtained some light fixed-wing aircraft, such as the U-6A and O-1, for liaison and observation purposes in the 1950s and 1960s.[22] It was expanded with numerous locally assembled UH-1 utility helicopters from 1969.[23]

The Evolving Strategic Circumstances

In the post–Cold War era, Taiwan's air power was reshaped as a result of a range of changing external and internal factors. First, dramatic changes to China's international circumstances reshaped its aerial threats on Taiwan. The Tiananmen massacre of 1989 severed China's access to American and European military aviation industries, but its rapprochement with the Soviet Union in the same year provided an alternative of external aviation technology for the PLAAF and PLANAF.[24] The collapse of the Soviet Union and the subsequent economic difficulties made Russia sell Flanker series fighters, including a license for the production of the Su-27S. Furthermore, parts and technologies, as well as reverse-engineering from Russia, Ukraine, and other former Soviet countries helped China to progress its aviation industry with several indigenous models of combat aircraft, such as the J-10, J-11, J-15, and J-16 fighters, with their strong Soviet/Russian influence.[25] Beijing's improvement of its aviation industry has led it to

---

22 Chuanren Chen, "Taiwan May Acquire Seahawk Helos from U.S.," *AIN Online*, June 6, 2018, <https://www.ainonline.com/aviation-news/defense/2018-06-06/taiwan-may-acquire-seahawk-helos-us>; SIPRI, "Arms Transfers Database."

23 Pierre Delrieu, "Taiwan Orders Additional UH-60s," *Asian Military Review*, Jul. 24, 2017, <https://asianmilitaryreview.com/2017/07/taiwan-orders-additional-uh-60s/>; "UH-1H Helicopter," Aerospace Industrial Development Corporation, <http://www.aidc.com.tw/en/military/uh1h>.

24 Philip C. Saunders and Joshua K. Wiseman, "China's Quest for Advanced Aviation Technologies," in Richard P. Hallion, Roger Cliff, and Philip C. Saunders (eds.), *The Chinese Air Force: Evolving Concepts, Roles and Capabilities* (Washington DC: National Defense University Press, 2013), 298.

25 Robert Farley, "Russia's Deadly Su-27 Fighter: Everything You Need Know," *National Interest*, June 20, 2017, <https://nationalinterest.org/blog/the-buzz/russias-deadly-su-27-fighter-everything-you-need-know-21239>; Saunders and Wiseman, "China's Quest for Advanced Aviation Technologies," 302–09.

being able to supply aerial refueling aircraft and airborne warning and command system (AWACS), based on former Soviet models of platforms, such as the An-12 (Y-8) and Tu-16 (H-6), to support fighter operations.[26] This dramatic shift has provided China with more freedom to develop its air power and other military means against Taiwan, whereas the United States and other western suppliers, in the pre-Tiananmen era, were more cautious than Russia and other sources in transferring their aviation technologies.[27] Taiwan subsequently faced heavier pressure in contrast to the last phase of the Cold War.

In the first decade after the Cold War, Taiwan launched a wave of modernization for the ROCAF and even achieved some superiority over its Chinese counterparts. Taipei successfully procured three models of new fighters, 130 Indigenous Defensive Fighters (IDFs), 60 French Mirage 2000-5s, and 150 American F-16 A/B MLUs with beyond-visual-range capability of respective mid-range AAMs, the indigenous Tien-Chien (Sky Arrow) II, MICA-EM, AIM-7M, and then AIM-120C. These fighters are supported by six E-2T AWACSs obtained in the 1990s and 2000s. With such assets, Taipei temporarily created certain qualitative and quantitative superiority over its counterparts across the Strait. The PLAAF and the PLANAF only had Flanker fighters with beyond-visual-range capability in the 1990s, until the indigenous PL-12 AAM went into service in 2004 and they did not have AWACS until the mid-2000s.[28]

However, in the mid-2000s, the political deadlock in Taiwan impeded further modernization, such as the failed procurement of 66 F-16 C/Ds.[29] In contrast, Beijing caught up with a massive production of its J-10s, J-11s, and J-16s, in addition to purchasing Russian Su-30MKs and Su-35s. By 2019, the

---

26 David Shlapak, "Equipping the PLAAF: The Long March to Modernity," in Richard P. Hallion, Roger Cliff, and Philip C. Saunders (eds.), *The Chinese Air Force: Evolving Concepts, Roles and Capabilities* (Washington DC: National Defense University Press), 197–8, 206, 208.

27 Saunders and Wiseman, "China's Quest for Advanced Aviation Technologies," 295–8.

28 Shlapak, "Equipping the PLAAF: The Long March to Modernity," 200; Shirley A. Kan, *Taiwan: Major U.S. Arms Sales Since 1990* (Washington DC: Congressional Research Service, 2014), 20; *Military Balance 2003* (London: IISS, 2003), 155; *The Military Balance 2004* (London: IISS, 2004), 172.

29 Michael S. Chase, "Taiwan's Arms Procurement Debate and the Demise of the Special Budget Proposal: Domestic Politics in Command," *Asian Survey*, Vol. 48, No. 4, 2008, 703–04, 710–21; Jim Wolf, "Taiwan Overdue for F-16 Jets, Ex U.S. Official Say," *Reuters*, Jul. 7, 2010, <https://www.reuters.com/article/us-taiwan-usa-fighters-interview/taiwan-overdue-for-f-16-jets-ex-u-s-official-say-idUSTRE6655BL20100706>.

total complement of fourth-generation fighters reached 988. The PLANAF has further obtained an aircraft carrier that can launch air strikes from a mobile location, and has plans to commission more carriers in the coming years.[30] Although the reported poor performance of the Chinese J-11Bs vis-à-vis the Thai JAS-39C fighters during an exercise in 2015 has thrown doubt on the quality of the Chinese fourth-generation fighters, it can be said that the PLAAF and the PLANAF have achieved overwhelming quantitative superiority over the ROCAF with support of AWACS and aerial refueling capabilities.[31]

It is, however, China's unconventional tactics, which are aimed at asymmetrically neutralizing the ROCAF's capability, that pose a more serious challenge to Taiwan's air power than the expansion of the PLAAF and PLANAF fighter fleets. Despite the fact that Beijing has gradually established superiority in terms of fighters and supportive capabilities, it would take some time to annihilate Taipei's 300+ fighters supported by an intense SAM network.[32] In contrast, disruption or even paralysis of the ROCAF air bases could facilitate immediate air superiority for the Chinese air forces. Taiwan's SAM network is supposed to take over air defense, but it too might fall prey to unconventional tactics on the part of China.[33]

Such unconventional means against Taiwan's air power might be twofold: projection of firepower and sabotage. Since the early 1990s, the PLA has deployed ballistic missiles followed by cruise missiles in the 2000s aimed at the ROCAF's bases.[34] Due to the limited width of the Strait, many Chinese tactical projectiles with 150-km range and above are capable of striking Taiwanese air bases in the western part, whereas other bases in the east are exposed to sub-launched missiles and longer ranged muni-

30 SIPRI, "Arms Transfers Database"; *The Military Balance 2019* (London: IISS, 2019), 260–61.

31 Fei-Yu Jiang, "Zao taiguo wanbai zhonggong kongjun zhan li bei jiantao" ("Defeated by the Thai Air Force, the PLAAF's Capability Is Under Review"), *China Times*, May 3, 2016, <https://www.chinatimes.com/realtimenews/20160503006607-260417>.

32 *The Military Balance 2018* (London: IISS), 304–05.

33 Michael J. Lostumbo, David R. Frelinger, James Williams, and Barry Wilson, *Air Defense Options for Taiwan: An Assessment of Relative Costs and Operational Benefits* (Santa Monica, CA: RAND Corporation, 2016), 12–13.

34 Richard C. Kagan, *Taiwan's Statesman: Lee Teng-hui and Democracy in Asia* (Annapolis: Naval Institute Press, 2007), 122; "Hong Niao Series (HN-1/-2/-3)," CSIS Missile Threat Project, Center for Strategic and International Studies, Aug. 12, 2016, <https://missilethreat.csis.org/missile/hong-niao>; *Annual Report to Congress: Military and Security Developments Involving the People's Republic of China 2018* (Washington DC: Office of the Secretary of Defense, 2018), 101.

tions.[35] If the PLA adopts the concept of Russian container missiles and launches attacks from cargo ships in offshore or even inside Taiwanese ports, the alarm time would be further shortened.[36] Using such standoff weapons with conventional warheads may not completely destroy air bases, but the disruption if not paralysis caused by initial salvos would considerably decrease if not pause the ROCAF's capacity for launching interception sorties.[37] By the same token, Beijing may send special forces for sabotage purposes. The trends in urbanization mean that most ROCAF bases are gradually being surrounded by civilian buildings, which provide good cover for observation, sniping, and other attacks, such as mortars or rockets. Furthermore, proper espionage measures could lead to successful penetration into the bases for even more efficient sabotage. The Chinese special forces could also attack the chain of command in order to create an even broader impact in terms of paralysis.[38] To sum up, it is evident that Taiwan's air power gradually began to face multiple forms of threat from China from the 1990s.

In the face of China's strengthening air power, the role of the United States has become increasingly important as a supplier and a potential protector. Despite acquisition of some non-American aircraft, such as the French Mirage 2000-5, Washington has been the essential supplier of aircraft, surveillance facilities, SAMs, and other elements of air power to Taipei. Undeniably, Taiwan has established a sizable aviation industry, but American technology is still indispensable.[39] Apart from the unsuccessful acquisition for F-16C/Ds, Washington did sell several types of aircraft, such as the P-3C ASW aircraft, and provided upgrades for transforming the F-16A/B fighters to F-16Vs, but it has not supplied any additional fighters.[40]

35 Steven Stashwick, "2 US Warships Transit Taiwan Strait," *Diplomat*, Oct. 23, 2018, <https://thediplomat.com/2018/10/2-us-warships-transit-taiwan-strait/>.

36 Michael Stott, "Deadly New Russian Weapon Hides in Shipping Container," *Reuters*, Apr. 26, 2010, <https://www.reuters.com/article/us-russia-weapon/deadly-new-russian-weapon-hides-in-shipping-container-idUSTRE63P2XB20100426>.

37 Lostumbo, Frelinger, Williams, and Wilson, *Air Defense Options for Taiwan: An Assessment of Relative Costs and Operational Benefits*, 5–6, 16–17; *Annual Report to Congress: Military and Security Developments Involving the People's Republic of China 2018*, 29.

38 Darryl J. Lavender, *China's Special Operations Forces Modernization, Professionalization and Regional Implications* (Carlisle Barracks, PA: U.S. Army War College, 2013), 14–15.

39 "IDF Fighter," Aerospace Industrial Development Corporation, last accessed June 6, 2019, https://www.aidc.com.tw/en/military/idf.

40 SIPRI, "Arms Transfers Database"; Kan, *Taiwan: Major U.S. Arms Sales Since 1990*, 59.

As a result, the ROCAF faces difficulties in acquiring new fighters, especially fifth-generation ones, without U.S. approval. Aside from supply, it is possible that the U.S. air bases adjacent to Taiwan, such as the Kadena and Futenma bases in Okinawa, and U.S. aircraft carriers could reduce the strategic pressure on the ROCAF by posing threats on the flanks of Chinese aerial operations, and thus creating an operational dilemma. If Beijing were to go for a quick and compact operation against Taipei with less damage and a strategic surprise without massive preparation, exposing the flank to the United States would make such tactics highly uncertain. Given that China would plan to secure the flank, an escalation for confrontation or even a preemptive strike on the U.S. bases would be inevitable. Despite Beijing's expanding strike capability covering those American bases, a further escalation into a major war between superpowers may in fact act as a deterrent.[41]

However, Washington's fluctuating foreign policies under various leaderships as well as the lack of bilateral alliance or defense treaty with Taipei would also make the use of U.S. air power or intervention uncertain. Moreover, the fear of a war between superpowers decision-makers may similarly deter American from such actions. Lack of joint exercises would also impede the interoperability between the U.S. air forces and the ROCAF, although the latter retains a F-16 attachment in the United States for training.[42]

Finally, Taiwan's domestic politics play another obvious role in managing its air power. Before the democratization in the 1990s, the ROC regime during martial law governance allowed the same group of decision-makers to determine all defense policies, including procurements. Since the first popular presidential election in 1996, four presidents and their respective administrations from the Chinese Nationalist Party, also known as Kuomintang (KMT), and the Democratic Progressive Party (DPP) have had distinct approaches to defense, having their respective effects on air power.

When the KMT administration was under Li Teng-hui's leadership during the 1990s, the party's political dominance backed up by Taiwan's

---

41 Eric Heginbotham et al., *The U.S.-China Military Scorecard: Forces, Geography, and the Evolving Balance of Power, 1996–2017* (Santa Monica, CA: RAND Corporation, 2015), 51–9.
42 Wendell Minnick, "F-16 Crash in US Part of Taiwan's Secretive 'Gamblers' Program," *Defense News*, Jan. 25, 2016, <https://www.defensenews.com/air/2016/01/25/f-16-crash-in-us-part-of-taiwans-secretive-gamblers-program>.

booming economy was able to conduct various major projects to strengthen air power, such as the introduction of three types of fighters and procurement of the Patriot Advanced Capacity (PAC) II SAMs.[43] Between 2000 and 2008, the Chen Shui-bian administration of the DPP attempted to strengthen deterrence by developing cruise missiles to be used against China—another form of air power. However, the Chen administration's plan to purchase 66 F-16C/Ds to replace the F-5E/Fs was postponed as a result of the political deadlock with the KMT and other opposition parties. Subsequently, the Ma Ying-jeou administration of the KMT focused on harmonious cross-Strait relations without any major procurement of fighters but merely upgrading the existing IDFs and F-16A/Bs as well as acquiring PAC III SAMs. In addition, indigenous TK-III SAMs were put into service in 2011.[44] Since 2016 the incumbent Tsai Ing-wen administration of the DPP has highlighted the importance of defense industries and launched an indigenous project of advanced trainers, based on the IDF. The trainers are expected to replace F-5E/Fs for certain aerial combat functions.[45] The United States under the Trump administration also provided more airborne munitions such as AGM-154C air-to-surface missiles (ASM) and AGM-88B anti-radar missiles for better ground attack capability, but so far no deal for the acquisition of fighters or any other major reform of the ROCAF has been put in place.[46]

Taiwan's air power has faced multiple challenges, such as China's increasing threats, both conventional and unconventional, unstable supply and intervention from the United States, and the influence of domestic politics in the last three decades. As a result, Taipei has to deal with the increasingly constrained role of its fighters due to their vulnerability and the deterioration in their numbers, while the SAMs also face conventional and unconventional attacks. Under such serious threats, Taiwan's efforts

---

43 Bernard D. Cole, *Taiwan's Security: History and Prospects* (Abingdon: Routledge, 2006), 98–99, 107–9; SIPRI, "Arms Transfers Database."

44 "Tien-Kung III (Sky Bow III) Surface-to-Air Missile System," Army Technology, <https://www.army-technology.com/projects/tien-kung-iii-sky-bow-iii-surface-to-air-missile-system>.

45 Chris Pocock, "Taiwan Confirms Indigenous Jet Trainer Development," *AIN Online*, Feb. 16, 2017, <https://www.ainonline.com/aviation-news/defense/2017-02-16/taiwan-confirms-indigenous-jet-trainer-development>.

46 "Taipei Economic and Cultural Representative Office (TECRO) in the United States," Defense Security Cooperation Agency, <http://www.dsca.mil/tags/taipei-economic-and-cultural-representative-office-tecro-united-states>.

in other aspects of air power, such as ASW and attack helicopters, would prove rather futile without air superiority. Although the incumbent administration seems to want to rejuvenate Taiwan's air power, the various difficulties have not been sufficiently addressed. Based on existing policies, Taipei's current approach to air power could be observed in three main directions: operational offense, resilience, and modernization.

Taiwan's Current Policies for Air Power

As air power can serve both defensive and offensive purposes, shifting toward offense in operational and tactical levels would be a solution to Taiwan's vulnerability to China's attack. Since the 1950s, defense has been the main guiding principle for Taipei in developing its air power, not only for its own security but also in order for Washington to be able to control the cross-Strait situation. However, based on continuous improvements in firepower from China, a purely defensive strategy from Taiwan would be less effective than undertaking some additional offensive means in order to create strategic depth.[47] If Taipei were able to project firepower to the adjacent Chinese provinces, particularly air bases and other key military facilities, Beijing would either take the risk of being preemptively attacked or set assembly locations away from Taiwan. Either outcome would allow Taiwan, and also the United States, to have a greater response time and to present China with more challenges to achieve air superiority, a crucial condition for other operations, such as landing and airborne. Beijing may adopt fortification measures to absorb attacks from Taipei, but the effect would be limited; since force projection is necessary, its extensive and exposed facilities such as runways and wharves are unlikely to be fully fortified. Taiwan's offense capability could also extend the warfare on Chinese territory, particularly to various major cities such as Shanghai, thus having economic effects on China as well as psychological impacts.

The ROCAF is presently developing several offensive capabilities. In terms of offense, the ROCAF has armed its IDF fighters with indigenous Wan Chien (Ten Thousand Swords) standoff munitions with a range of about 200 kilometers, compared to the American AGM-154C and AGM-88B

---

47 Tucker, *Strait Talk: United States-Taiwan Relations and the Crisis with China*, 227–28, 238–39.

ASMs available for the F-16Vs.[48] The ROCAF also possesses other ground attack weapons, such as AGM-65G missiles and Paveway guided bombs, but their relatively short ranges would expose platforms to China's air defense firepower.[49] Conventional wisdom would maintain that it is not wise for the ROCAF to squander precise assets such as fighters for offensive missions in the initial phase of warfare, but the vulnerability of its air bases might suggest limited or no opportunity for second sorties for many fighters. Compared to being destroyed on the ground, launching one-time attacks might be a better option. However, the ROCAF bases are also vulnerable to preemptive strikes that may disrupt or even deny such offensive missions. Furthermore, the ROCAF's preparations for offense, such as changing personnel schedule, are unlikely to remain hidden from China's close surveillance using instruments and espionage. In contrast to launching air strikes, using cruise missiles, mainly indigenous Hsiung Feng (Brave Wind) IIE cruise missiles, and even Yun Feng (Cloud Peak) surface-to-surface missiles (SSM) in the future, is probably more feasible for offensive purposes. Nevertheless, similar to China's ballistic and cruise missiles, the limited payloads of conventional warheads may not create sufficient damage to achieve strategic or operational goals. Furthermore, it is possible for the PLAAF to intercept the HF IIE missiles with subsonic speed.[50]

Based on such technological and tactical conditions, Taiwan's air power for offense, whether using aircraft or cruise missiles, would only achieve disruption or harassment rather than annihilation of the PLA deployments. Despite the limited impact, it would lower the pressure on the ROCAF fighter wings and the SAM units. Furthermore, time would not be on China's side strategically, because the longer the warfare against Taiwan, the longer the economic loss, not to mention other uncertainties, such as those of the Indo-China borders and the South China Sea. As such, Taiwan's offensive operations may have subtle but critical impacts on any subsequent campaign. It must be noted that offense is not a panacea to

---

48 "Taiwan Unveils 'Wan Chien' Air-to-Ground Stand-Off Weapon," *Defense Update*, Jan. 20, 2014, <https://defense-update.com/20140120_taiwan-unveils-wan-chien-air-ground-cruise-missile.html>.
49 SIPRI, "Arms Transfers Database"; *The Military Balance 2018*, 305.
50 "Missiles of Taiwan," CSIS Missile Threat Project, Center for Strategic and International Studies, June 15, 2018, <https://missilethreat.csis.org/country/taiwan>.

Taipei's strategic challenges; the advantages of a first strike to gain air superiority may indeed incite Beijing to take preemptive action as well. Furthermore, the PLA may or will have enough capacity to absorb such losses from Taiwan's operational offense and proceed with plans to suppress the ROCAF and use other defensive means. Moreover, taking an offensive stance, especially in a preemptive manner, is a serious political decision that would challenge Taiwan's leadership and risks loss of legitimacy, the mainstay of forces, and the most fundamental value, peace.

Taiwan's efforts regarding survival are based on surveillance, command, and SAMs. Due to its lack of strategic depth and exposed air bases, early warning is crucial for Taipei to retain strategic initiation. Aside from the six E-2 Hawkeye-2000 AWACSs, a certain level of early warning would be provided by the Pave Paws and more than a dozen TPS-77 radars. Taiwan's SAMs and other mobile radars would provide alternative and supplementary surveillance, especially when fixed ones might be lost, but there is little room for optimism regarding their sustainability in the face of the PLA's various strike means.[51]

Jointness has been a major goal for Taiwan's military organizational reform in recent years. As the ROC Ministry of National Defense (MND) established the Information, Communications, and Electronic Force Command (ICEFCOM), the ROCAF has merged all air defense units under a unified command. Ideally, such a concentrated command chain would increase the efficiency of response and coordination, an essential factor for the resilience on air defense. Taiwan's defense industry has supplied a range of mid- and short-ranged SAMs, while the United States has delivered PAC-3 and FIM-92 Stinger missiles. A more comprehensive SAM network composed of different missiles enhances the survivability of each system for covering all others.[52] Due to the vulnerability of its air bases and fighters, the ROCAF has shown an interest in purchasing vertical and/or

---

51 SIPRI, "Arms Transfers Database"; *National Defense Report 2017* (Taipei: Ministry of National Defense, 2017), 80; Lostumbo, Frelinger, Williams, and Wilson, *Air Defense Options for Taiwan: An Assessment of Relative Costs and Operational Benefits*, 12–13.
52 *National Defense Report 2017*, 75, 79, 83–84;

short takeoff and landing (V/STOL) fighters with the only available option being the F-35B, but no deal has been achieved to date.[53]

Modernizing its fighters has been a high priority of Taiwan's defense investment, but the failure to purchase F-16C/Ds leaves the ROCAF in a multi-faceted dilemma. The fact that it has three models of fighters from diverse sources means a considerable challenge in the area of logistics, especially when it comes to the relatively high costs of operating Mirage-2000s.[54] As they are coming up to two decades of servicing, a decision as to whether they are replaced or upgraded might be viewed as an opportunity for the ROCAF to simplify its fighter models. Aside from issues of availability, the vulnerability of fighters and air bases makes the ROCAF's modernization more than a question of simple upgrades or replacements, and in fact provides it with a potential turning point to reform its force structure from fighter centered to SAM centered considering issues of overall improved survivability and sustainability.

Despite the various strategic aspects, the ROCAF's modernization of fighters has so far remained conventional. Taipei has therefore launched upgrade projects for all three fighter models, and is rumored to be acquiring F/A-18s or F-15s.[55] Upgrading existing fighters is a short- to mid-term aim, complemented by the upgraded F-16Vs delivered in 2018 and the IDF upgrade project finished in the same year, and followed by the long-term goal of the procurement of F-35Bs.[56] It can be seen that Taiwan is taking a conventional approach to maintaining its air power with some potential

---

53 *The 2017 Quadrennial Defense Review* (Taipei: Ministry of National Defense, 2017), 44; Franz-Stefan Gady, "Defense Minister: Taiwan Is Seeking F-35 Stealth Fighter," *Diplomat*, May 15, 2018, <https://thediplomat.com/2018/05/defense-minister-taiwan-is-seeking-f-35-stealth-fighter>.

54 Minnie Chan, "Cost of Taiwan's Ageing Mirage Jets in Spotlight Again as Fighter Goes Missing," *South China Morning Post*, Nov. 22, 2017, <https://www.scmp.com/news/china/diplomacy-defense/article/2119000/cost-taiwans-ageing-mirage-jets-spotlight-again-fighter>.

55 Hsin-hui Lu and Kuan-lin Liu, "Air Force to Upgrade Capabilities According to Needs, Threats," *Focus Taiwan*, Feb. 13, 2018, <http://focustaiwan.tw/news/aipl/201802130008.aspx>; Tien-pin Lo and Jonathan Chin, "Deals Inked for Mirage Parts, Training," *Taipei Times*, Sep. 26, 2018, <http://www.taipeitimes.com/News/taiwan/archives/2018/09/26/2003701184>.

56 Chen Chuanren, "Taiwan's Ching-kuo Upgrade Complete, SEAD Role Next," *AIN Online*, Mar. 20, 2018, <https://www.ainonline.com/aviation-news/defense/2018-03-20/taiwans-ching-kuo-upgrade-complete-sead-role-next>; Gavin Phipps, "Taiwan Takes Delivery of First F-16V Aircraft," *Jane's Defence Weekly*, Oct. 22, 2018, <www.janes.com/article/83947/taiwan-takes-delivery-of-first-f-16v-aircraft>.

for gradual and structural changes, such as expanding its number of SAMs and introducing VTOL fighters. It must be noted that these policies may not be adequate to deal with the severe threats posed by China.

Challenges and Potential Solutions

Taiwan's present policies regarding air power are more useful to counter conventional rather than unconventional threats. Improving command, fighters, and SAMs will help overcome Taiwan's inferiority vis-à-vis China. With a better exchange rate in aerial combat, the ROCAF would last longer and impose higher costs on the PLAAF and PLANAF. However, the progress of each side of the Strait toward fifth-generation fighters may affect Taipei's efforts in preparing conventional aerial combat.

Ideally, the ROCAF's next generation of fighters should match the demands of quality and quantity, but the reality would present a difficult trade-off. Since the United States is the only supplier of fighters to Taiwan under the latter's international isolation, the F-35, in particular the B model with VTOL capability, would be an effective solution. Furthermore, following previous deals with Australia, South Korea, and Japan, selling F-35s to Taipei would be more acceptable for Washington and its regional allies. However, there are several obstacles for the ROCAF to obtain F-35s: first, the high prices—$89.2 million for the A model and $115.5 million for the B model—would discourage Taipei with its annual defense budget of less than $11 billion.[57] Despite the promise of increasing the defense budget to three percent of gross domestic product, the reality was still not close to this goal in 2018.[58] Taiwanese decision-makers could, of course, arrange a special budget for the F-35s, but opposition to this is likely in the volatile democracy, as seen in the mid-2000s, and opposition parties, such as the KMT, are likely to repeat their refusal to approve such expenditure. In addition, Washington would also have cause for concern on the frequent espionage cases in the ROC armed forces meaning a high risk of leaking

---

57 "Pentagon and Lockheed Martin Agree to Reduced F-35 Price in New Production Contract," F-35 Lightening II, Lockheed Martin, Sep. 28, 2018, <https://www.f35.com/news/detail/pentagon-and-lockheed-martin-agree-to-reduced-f-35-price-in-new-production>; *The Military Balance 2018* (London: IISS, 2018), 302.
58 Pei-ju Teng, "Defense Budget to Increase by NT$18.3 billion in 2019: Taiwan Premier," *Taiwan News*, Jul. 27, 2018, <https://www.taiwannews.com.tw/en/news/3492793>.

classified information about the F-35 to China.[59] Finally, a key determining factor in the issue would be whether Washington were to offer Taiwan F-35s and, if so, when. Lacking an alternative to purchasing F-35s would mean that the ROCAF would fall definitively behind its PLA counterparts in terms of modernization. If a deal were to be struck, a decade for delivery and commission would be likely.[60]

The ROCAF's future procurement of new fighters is a significant indicator of Taiwan's approach to air power. Currently, the situation appears favorable in both Taipei and Washington, with the Tsai administration, in contrast to its predecessor, paying more attention on military buildups, and the Trump administration having a friendly attitude toward Taiwan. Despite having settled arms sales of munitions, subsystems, and logistical supply, there has nevertheless been no agreement on big-ticket projects. This situation might be interpreted as negotiation challenges, such as Taipei's lukewarm attitude or inability to purchase, or the ROCAF's plans for structural reform, moving from fighters to SAMs. In the meantime, Beijing has gradually put its fifth-generation fighters, the J-20s, into service as well as other potential models, such as J-31s, that may exacerbate Taiwan's inferiority in air power. Although the real stealth capability of the J-20 may be doubtful after reports of its detection by an Indian Su-30MKI, Beijing's technological challenges to Taipei's air power should not be underestimated, since the latter has a very limited margin for errors.[61]

In a case of non-conventional threats from China, Taiwan's ongoing policies related to air power would be of little use. Despite Taiwan's gradual acquisition of certain missiles for offensive purposes, their limited numbers and payloads would make their use unwise for striking the PLA's numerous ballistic, cruise missiles, and other standoff weapons that are mobile and concealed. Taiwan's PAC-3 and TK-III SAMs would be outnumbered if used to intercept approaching Chinese missiles and other projectiles and

59 Peter Mattis, "Counterintelligence Remains Weakness in Taiwan's Defense," *China Brief*, Vol. 17, No. 11, Aug. 17, 2017, <https://jamestown.org/program/counterintelligence-remains-weakness-in-taiwans-defense>.

60 Lostumbo, Frelinger, Williams, and Wilson, *Air Defense Options for Taiwan: An Assessment of Relative Costs and Operational Benefits*, 86.

61 "'Not So Invisible After All': Su-30 Manages to Detect Top Chinese Stealth Jet," *Sputnik*, May 19, 2018, <https://sputniknews.com/military/201805191064607135-chinese-stealth-jet-detected>.

their locations would then be exposed to attack.[62] Since both Taipei's offensive and defensive means are unlikely to neutralize Beijing's firepower projection, the ROCAF bases would remain vulnerable and likely left in a deteriorated condition. With vulnerable air bases, the effect of modernizing existing fighters would be reduced or even invalid.

The organizational transformation may even fall prey to sabotage by China. Although the more jointed command systems would increase efficiency of response and coordination for aerial combat, the impact of paralyzing them would be broader and deeper than the original more separated arrangement. Since computers and the Internet are indispensable in such command facilities, cyber attacks would be significant. Undeniably, the military Internet is isolated from external cyber space and the ICEF-COM is also in operation for cyber threats, but espionage cases to date prove the possibility of using defectors to deliver malware or viruses to the military network, in addition to the use of other means such as updates.[63] As cyber attacks may not be precisely synchronized with the timing of strikes, kinetic operations present a more accurate alternative. Although the command venues are undoubtedly located in guarded military institutes, their functions may not be fully secured. The MND has deployed a marine bridge to guard several key locations of the command chain, but considerable risk of espionage makes defending such facilities quite uncertain.[64]

Furthermore, antitank missiles, drones, and other portable weapons would give Chinese special forces the advantage of being able to raid or attack the command chain, in addition to their ability to plan in advance when, where, what, and how to attack. If hubs prove too hard to strike, other parts of air defense, such as communication lines, can remain exposed to sabotage. Once the command chain is disrupted or denied, Taiwan's air power would be seriously damaged.

---

62 Lostumbo, Frelinger, Williams, and Wilson, *Air Defense Options for Taiwan: An Assessment of Relative Costs and Operational Benefits*, 12.

63 Sophia Yang, "200 million Cyber Attacks Hit Taiwan's Military Networks in 2017: Expert," *Taiwan News*, May 28, 2018, <https://www.taiwannews.com.tw/en/news/3441894>; *National Defense Report 2017*, 72; *The Military Balance 2018*, 305.

64 Robert Beckhusen, "How Taiwan Would Defend Itself from a Decapitation Strike (By China)," *National Interest*, May 18, 2018, <https://nationalinterest.org/blog/the-buzz/how-taiwan-would-defend-itself-decapitation-strike-by-china-25876>.

Sabotage would pose similar threats to individual elements of Taiwan's air power. Due to the vulnerability of its air bases, SAMs are seen as the most critical means of denial for the ROCAF against its PLA counterparts. The SAMs' thin armor or lack of it, however, make them easy targets for China's "fifth column," as light weapons or improvised explosive devises (IEDs) are enough to do serious damage to SAM systems. Usually SAMs are planned to be deployed at specific tactical positions for greater detection and broader coverage than on bases. After engagement, the SAMs would move to the next tactical location to prevent it being targeted by the PLA, indicating that mobility is key to survival in wartime.[65] However, tactical movement and field deployment of SAM units would be vulnerable to attack by Chinese special forces and other penetrated personnel. The F-35B with VTOL capability is thus seen the solution to strengthen Taiwan's air power as these aircraft are able to hide in non-fixed locations, but their hideouts would be targets for sabotage, and light weapon and/or IED attacks as well.

There two main reasons for the serious threats posed by the penetration of Chinese agents and soldiers: accessibility and Taiwan's very small size. In 2017, close to 2.7 million people from the PRC visited Taiwan for various purposes, such as tourism and business, and the increasing number of marriages with spouses of PRC background has grown to more than 342,000.[66] With such large numbers, infiltration would not be difficult, with some reports stating that there are already thousands of Chinese spies on the island.[67] Since half of Taiwan is mountainous, the suitable areas for mobile deployments of SAMs, VTOL aircraft, and other systems would not be overly broad, enabling China to continuously monitor them. Furthermore, satellite pictures and onsite surveys by personnel would identify potential hideouts and plan for ambush and attack. Although China's infiltrated personnel would be outnumbered by the ROC Army, other ground

---

65 Lostumbo, Frelinger, Williams, and Wilson, *Air Defense Options for Taiwan: An Assessment of Relative Costs and Operational Benefits*, 79.
66 "Tongji ziliao" ("Statistical Data"), National Immigration Agency, Oct. 30, 2018, <https://www.immigration.gov.tw/ct.asp?xItem=1354552&ctNode=29699&mp=1>.
67 Sophia Yang, "Ex-Deputy Defense Minister: 4,000 Chinese Spies in Taiwan," *Taiwan News*, Jan. 17, 2017, <https://www.taiwannews.com.tw/en/news/3077000>; Li-Hua Chung and Jonathan Chin, "5,000 Chinese Spies in Taiwan: Source," *Taipei Times*, Mar. 13, 2017, <http://www.taipeitimes.com/News/front/archives/2017/03/13/2003666661>.

troops, paramilitary units, and police forces, the former would enjoy being able to initiate active attack whereas the latter would have to defend much broader areas, including those containing potential infiltrated spies from China. Additionally downsizing of ground troops and insufficient recruitments would also reduce Taiwan's overall capacity.[68] As long as Taipei does not deal effectively with such unconventional challenges, the overall threat to its air power will endure.

To cope with these challenges, besides its current policies, Taiwan's most effective countermeasures to China's unconventional challenges would be expanded offensive capacity, additional air bases, and improved internal security. Taipei needs to enlarge its SSM arsenal to have larger capacity for striking more targets of PLA ballistic, cruise missiles, and other means of projecting firepower. It is clear that such buildups will be costly with the expensive HF IIE and other missiles and their limited payload. At this point, however, there is almost no alternative but offense for Taipei to thwart Beijing's missiles. With fewer missiles available on the other side of the Strait, Taiwan's air power, both fighters and SAMs, would face less pressure. Before any VTOL fighters are commissioned, the existing fleets are still important to Taiwan's air power. In order to optimize the high expenditure on the existing fighters, it would be necessary to lower their level of vulnerability. Most air bases in western Taiwan are exposed to various means of firepower from China, but there are only two in the eastern part, making construction of more bases in these areas ideal, using the Central Mountains or other geographic formations to narrow the angle and direction of attacks. Land acquisition in these less populated areas would also be easier in contrast to the crowded western areas.[69] In terms of sabotage and other internal threats coming from personnel infiltrated from China, it is recommended that tighter border control and tougher security measures for military institutes and other related facilities be put into immediate action. Such preparations will affect not only air power but also Taiwan's overall defensive capability.

---

68 Scott Morgan, "Taiwan Military Working towards 169,000 Active Personnel Goal," *Taiwan News,* Nov. 6, 2018, <https://www.taiwannews.com.tw/en/news/3569018>.
69 Ibid.

Conclusion

Taiwan's successful maintenance of air power during the Cold War eventually inspired China to pursue several methods for overcoming the former. To date, however, Taipei has not developed creative solutions to its desperate situation in the face of various threats from Beijing. The incumbent Tsai administration, for example, has expressed relatively high concern about defense, and yet it still takes a predominantly conventional approach to manage Taiwan's air power, with current policies that would not effectively reduce its vulnerability. Although several solutions are available, it is highly doubtful that the incumbent administration or its successors might conduct any significant reform. As such, the ROCAF would remain consistently inferior to its PLA counterparts, with related budgets for air power and air defense failing to ever reach the optimal effect. Such continued weakness in its air power would mean that Taipei would be more reliant on Washington's extended deterrence, a factor that has remained uncertain for some time now and keeps Taiwan in a very precarious position.

References

"Aerial Battle of Taiwan–Okinawa." Weapons of Warfare, Aug. 22, 2015. <https://weaponsandwarfare.com/2015/08/22/aerial-battle-of-taiwan-okinawa/>.

Annual Report to Congress: Military and Security Developments Involving the People's Republic of China 2018. Washington DC: Office of the Secretary of Defense, 2018.

Beckhusen, R. "How Taiwan Would Defend Itself from a Decapitation Strike (By China)." National Interest, May 18, 2018. <https://nationalinterest.org/blog/the-buzz/how-taiwan-would-defend-itself-decapitation-strike-by-china-25876>.

Bowman, M.W. Lockheed F-104 Starfighter. Wiltshire, UK: Crowood Press, 2000.

Chan, M. "Cost of Taiwan's Ageing Mirage Jets in Spotlight Again as Fighter Goes Missing." South China Morning Post, Nov. 22, 2017. <https://www.scmp.com/news/china/diplomacy-defense/article/2119000/cost-taiwans-ageing-mirage-jets-spotlight-again-fighter>.

Chase, M.S. "Taiwan's Arms Procurement Debate and the Demise of the Special Budget Proposal: Domestic Politics in Command." Asian Survey, Vol. 48, No. 4 (2008), 703–24.

Chen, C. "Taiwan's Ching-kuo Upgrade Complete, SEAD Role Next." AIN Online, Mar. 20, 2018. <https://www.ainonline.com/aviation-news/defense/2018-03-20/taiwans-ching-kuo-upgrade-complete-sead-role-next>.

Chen, C. "Taiwan May Acquire Seahawk Helos from U.S." *AIN Online*, June 6, 2018. <https://www.ainonline.com/aviation-news/defense/2018-06-06/taiwan-may-acquire-seahawk-helos-us>.

Chen, Y.M., and Liu, W.S. *Military Aircraft in the Chinese Civil War*. Taipei: The Wing of China Press, 1992.

Chung, L.H., and J. Chin. "5,000 Chinese Spies in Taiwan: Source." *Taipei Times*, Mar. 13, 2017. <http://www.taipeitimes.com/News/front/archives/2017/03/13/2003666661>.

Clough, R.N. *Island China*. London: Harvard University Press, 1978.

Cole, B.D. *Taiwan's Security: History and Prospects*. London: Routledge, 2006.

Crosby, Francis. *The World Encyclopedia of Bombers*. London: Anness, 2004.

Delrieu, P. "Taiwan Orders Additional UH-60s." *Asian Military Review*, Jul. 24, 2017. <https://asianmilitaryreview.com/2017/07/taiwan-orders-additional-uh-60s/>.

"F5E/F Supersonic Fighter." Aerospace Industrial Development Corporation. <http://www.aidc.com.tw/en/military/f5e>.

Farley, R. "Russia's Deadly Su-27 Fighter: Everything You Need Know." *National Interest*, June 20, 2017. <https://nationalinterest.org/blog/the-buzz/russias-deadly-su-27-fighter-everything-you-need-know-21239>.

Friedman, H.A. "Nationalist and Communist Chinese Propaganda Leaflets." Psywarrior. <http://www.psywarrior.com/NationalistChinesePropaganda.html>.

Gady, F.S. "Defense Minister: Taiwan Is Seeking F-35 Stealth Fighter." *Diplomat*, May 15, 2018. <https://thediplomat.com/2018/05/defense-minister-taiwan-is-seeking-f-35-stealth-fighter>.

Gill, B., and T. Kim, *China's Arms Acquisitions from Abroad: A Quest for "Superb and Secret Weapons."* SIPRI Research Report No. 11. Oxford: Oxford University Press, 1995.

Griess, T.E., T.B. Buell, J.H. Bradley, and J.W. Dice. *The Second World War: Asia and the Pacific*. New York: Square One, 2002.

Heginbotham, E., et al. *The U.S.-China Military Scorecard: Forces, Geography, and the Evolving Balance of Power, 1996–2017*. Santa Monica, CA: RAND Corporation, 2015.

"Hong Niao Series (HN-1/-2/-3)." CSIS Missile Threat Project, Center for Strategic and International Studies. <https://missilethreat.csis.org/missile/hong-niao/>.

"IDF Fighter." Aerospace Industrial Development Corporation. <https://www.aidc.com.tw/en/military/idf>.

Jan, T.S., and C.G. Jan. "Development of Weapon Systems in Developing Countries: A Case Study of Long Range Strategy in Taiwan," *Journal of Operational Research Society*, Vol. 51, No. 9 (2000): 1041–50.

Jiang, F.Y., "Zao taiguo wanbai zhonggong kongjun zhan li bei jiantao" ("Defeated by the Thai Air Force, the PLAAF's Capability Is under Review"). *China Times*, May 3, 2016. <https://www.chinatimes.com/realtimenews/20160503006607-260417>.

"Joint Communiqué of the People's Republic of China and the United States of America." Embassy of the People's Republic of China in the United States of America, Aug. 17, 1982. <http://www.china-embassy.org/eng/zmgx/doc/ctc/t946664.htm>.

Kagan, R.C. *Taiwan's Statesman: Lee Teng-hui and Democracy in Asia.* Annapolis: Naval Institute Press, 2007.

Kan, S.A. *Taiwan: Major U.S. Arms Sales Since 1990.* Washington DC: Congressional Research Service, 2014.

Lavender, D.J. *China's Special Operations Forces Modernization, Professionalization and Regional Implications.* Carlisle Barracks, PA: U.S. Army War College, 2013.

Lo, T.P., and J. Chin. "Deals Inked for Mirage Parts, Training." *Taipei Times,* Sep. 26, 2018. <http://www.taipeitimes.com/News/taiwan/archives/2018/09/26/2003701184>.

Lostumbo, M.J., D.R. Frelinger, J. Williams, and B. Wilson. *Air Defense Options for Taiwan: An Assessment of Relative Costs and Operational Benefits.* Santa Monica, CA: RAND Corporation, 2016.

Lu, H.H., and K.L. Liu. "Air Force to Upgrade Capabilities according to Needs, Threats." *Focus Taiwan,* Feb. 13, 2018. <http://focustaiwan.tw/news/aipl/201802130008.aspx>.

March, P.R. *Directory of Military Aircraft of the World.* London: Cassell & Co., 2001.

Mattis, P. "Counterintelligence Remains Weakness in Taiwan's Defense." *China Brief,* Vol. 17, No. 11, Aug. 17, 2017. <https://jamestown.org/program/counterintelligence-remains-weakness-in-taiwans-defense>.

"MiG-17 FRESCO." Federation of American Scientists, June 17, 2000. <https://fas.org/nuke/guide/russia/airdef/mig-17.htm>.

The Military Balance 1965. London: International Institute for Strategic Studies, 1965.

The Military Balance 2003. London: International Institute for Strategic Studies, 2003.

The Military Balance 2004. London: International Institute for Strategic Studies, 2004.

The Military Balance 2018. London: International Institute for Strategic Studies, 2018.

The Military Balance 2019. London: International Institute for Strategic Studies, 2019.

"Military Stages Highway Landing, Takeoff Drill in Tainan." *Taiwan Today,* Apr. 12, 2011. <https://taiwantoday.tw/news.php?unit=2,23,45&post=1752>.

Minnick, W. "F-16 Crash in US Part of Taiwan's Secretive 'Gamblers' Program." *Defense News,* Jan. 25, 2016. <https://www.defensenews.com/air/2016/01/25/f-16-crash-in-us-part-of-taiwans-secretive-gamblers-program>.

"Missiles of Taiwan." CSIS Missile Threat Project, Center for Strategic and International Studies, June 15, 2018. <https://missilethreat.csis.org/country/taiwan>.

Morgan, S. "Taiwan Military Working towards 169,000 Active Personnel Goal." *Taiwan News*, Nov. 6, 2018. <https://www.taiwannews.com.tw/en/news/3569018>.

National Defense Report 2017. Taipei: Ministry of National Defense, 2017.

"'Not So Invisible After All': Su-30 Manages to Detect Top Chinese Stealth Jet." *Sputnik*, May 19, 2018. <https://sputniknews.com/military/201805191064607135-chinese-stealth-jet-detected>.

Phipps, G. "Taiwan Takes Delivery of First F-16V Aircraft." *Jane's Defence Weekly*, Oct. 22, 2018. <https://www.janes.com/article/83947/taiwan-takes-delivery-of-first-f-16v-aircraft>.

"Pentagon and Lockheed Martin Agree to Reduced F-35 Price in New Production Contract," F-35 Lightening II, Lockheed Martin, Sep. 28, 2018. <https://www.f35.com/news/detail/pentagon-and-lockheed-martin-agree-to-reduced-f-35-price-in-new-production>.

Pocock, Chris. "Taiwan Confirms Indigenous Jet Trainer Development." *AIN Online*, Feb. 16, 2017. <https://www.ainonline.com/aviation-news/defense/2017-02-16/taiwan-confirms-indigenous-jet-trainer-development>.

Saunders, Philip C. and Joshua K. Wiseman. "China's Quest for Advanced Aviation Technologies." *The Chinese Air Force: Evolving Concepts, Roles and Capabilities*, edited by R.P. Hallion, R. Cliff, and P.C. Saunders. Chapter 12. Washington DC: National Defense University Press, 2013.

Shlapak, D. "Equipping the PLAAF: The Long March to Modernity." *The Chinese Air Force: Evolving Concepts, Roles and Capabilities*, edited by R.P. Hallion, R. Cliff, and P.C. Saunders. Chapter Eight. Washington DC: National Defense University Press, 2013.

Stashwick, S. "2 US Warships Transit Taiwan Strait." *Diplomat*, Oct. 23, 2018, <https://thediplomat.com/2018/10/2-us-warships-transit-taiwan-strait/>.

Stockholm International Peace Research Institute (SIPRI). "Arms Transfers Database." Database. Accessed Aug. 24, 2018. <http://armstrade.sipri.org/armstrade/page/trade_register.php>.

Stott, M. "Deadly New Russian Weapon Hides in Shipping Container," *Reuters*, Apr. 26, 2010. <https://www.reuters.com/article/us-russia-weapon/deadly-new-russian-weapon-hides-in-shipping-container-idUSTRE63P2XB20100426>.

"Taipei Economic and Cultural Representative Office (TECRO) in the United States." Defense Security Cooperation Agency. <http://www.dsca.mil/tags/taipei-economic-and-cultural-representative-office-tecro-united-states>.

"Taiwan Unveils 'Wan Chien' Air-To-Ground Stand-Off Weapon." *Defense Update*, Jan. 20, 2014. <https://defense-update.com/20140120_taiwan-unveils-wan-chien-air-ground-cruise-missile.html>.

Taylor, J. *The Generalissimo's Son*. London: Harvard University Press, 2000.

Teng, Pei-ju. "Defense Budget to Increase by NT$18.3 billion in 2019: Taiwan Premier." *Taiwan News*, Jul. 27, 2018. <https://www.taiwannews.com.tw/en/news/3492793>.

"Tien-Kung (Sky Bow)." Missile Defense Advocacy Alliance. <http://missiledefenseadvocacy.org/air-defense/air-defense-of-u-s-partners/allied-air-defense-systems/tien-kung-sky-bow/>.

"Tien-Kung III (Sky Bow III) Surface-to-Air Missile System." Army Technology. <https://www.army-technology.com/projects/tien-kung-iii-sky-bow-iii-surface-to-air-missile-system>.

"Tongji ziliao" ("Statistical Data"). National Immigration Agency. Oct. 30, 2018. <https://www.immigration.gov.tw/ct.asp?xItem=1354552&ctNode=29699&mp=1>.

Tucker, Nancy Bernkopf. *Strait Talk: United States–Taiwan Relations and the Crisis with China*. Cambridge, MA: Harvard University Press, 2009.

"UH-1H Helicopter." Aerospace Industrial Development Corporation. <http://www.aidc.com.tw/en/military/uh1h>.

Wilson, Stewart. *Aircraft of WWII*. Fyshwick, Australia: Australian Aviation, 1998.

Wolf, J. "Taiwan Overdue for F-16 Jets, Ex U.S. Official Say." *Reuters*, Jul. 7, 2010. <https://www.reuters.com/article/us-taiwan-usa-fighters-interview/taiwan-overdue-for-f-16-jets-ex-u-s-official-say-idUSTRE6655BL20100706>.

Yang, S. "200 million Cyber Attacks Hit Taiwan's Military Networks in 2017: Expert." *Taiwan News*, May 28, 2018. <https://www.taiwannews.com.tw/en/news/3441894>.

Yang, S. "Ex-Deputy Defense Minister: 4,000 Chinese Spies in Taiwan." *Taiwan News*, Jan. 17, 2017. <https://www.taiwannews.com.tw/en/news/3077000>.

Zhang, B., J. Zhang, and F. Yao. "Technology Transfer from the Soviet Union to the People's Republic of China: 1949–1966," *Comparative Technology Transfer and Society*, Vol. 4, No. 2 (2006): 105–67.

Zhang, X. "Air Combat for the People's Republic: The People's Liberation Army Air Force in Action, 1949–1969." *Chinese Warfighting: The PLA Experience since 1949*, edited by Mark A. Ryan, David M. Finkelstein, and Michael A. McDevitt. London: M.E. Sharpe, 2003.

Zhang, X. "The PLAAF's Evolving Influence within the PLA and upon National Policy." *The Chinese Air Force: Evolving Concepts, Roles and Capabilities*, edited by R.P. Hallion, R. Cliff, and P.C. Saunders. Chapter Three. Washington DC: National Defense University Press, 2013.

Zhang, X. *Red Wings over the Yalu: China, the Soviet Union, and the Air War in Korea*. College Station, TX: Texas A&M University Press, 2002.

13 KENYA
Niche from the Start

Donovan C. Chau

> *"Flying is one of the best tests of nationality which exists. It is a combination of science and skill, of organisation and enterprise."*[1]
> Winston Churchill

> *"All air forces are products of their history, their successes and failures at the operational level, and the understanding that the force as a whole has regarding its role in national security."*[2]
> Sanu Kainikara

Introduction

The story of the Kenya Air Force (KAF) is a uniquely African one: legacy western support has resulted in its positive reputation; a diversity of influences due to Kenya's geography has helped shape its composition; and a bright future lies ahead for the KAF—if professionalism can check corruption. This chapter begins with a brief overview of Kenya's strategic context and the current capabilities of the KAF. The next two sections provide the historical and political backgrounds necessary for understanding the current force structure of the KAF. Following these is a forward-looking discussion of the KAF's potential challenges and opportunities that lie ahead. The conclusion highlights salient and unique aspects of the KAF.

---

1 Winston Churchill, "Unconquerable and Incomparable," speech given at Royal Aero Club dinner, Savoy Hotel, London, Mar. 4, 1914. *Never Give In! The Best of Winston Churchill's Speeches* (New York: Hachette Books, 2015), 56.
2 Sanu Kainikara, "The Future Relevance of Smaller Air Forces," Air Power Development Centre, Royal Australian Air Force, Working Paper 29, Apr. 2009, 14.

Context and Capabilities

The KAF, the air component of the Kenya Defense Forces (KDF), is a representation of the nation it protects. Thus, before describing its current capabilities, it is imperative for one to understand Kenya's strategic landscape, from its national interests to its foreign and defense policies. The *Defence White Paper* of 2017 lays bare Kenya's national interests. "Our political independence, the integrity of our national territory, and the security and socio-economic well-being of our people comprise our basic national interests."[3] The *White Paper* clarifies further Kenya's interests as sovereignty and territorial integrity, national security, economic prosperity and well-being of Kenyans, and national prestige.[4] Kenya's national interests are not unique, from border security to economic growth. The KDF is charged with protecting these national interests.

Kenya's foreign policy objectives make additions to the country's straightforward national interests. Additional objectives include promoting subregional and regional integration and cooperation, enhancing regional and global peace and security, promoting international cooperation and multilateralism, promoting and protecting the interests of Kenyans abroad, and enhancing partnership with the Kenya diaspora and descendants.[5] Here, the strategic context of Kenya becomes decidedly more African in nature, particularly the emphasis on regional and subregional contexts as well as multilateralism and international cooperation. These points, in particular, should be taken into consideration when examining the KAF force structure.

The objectives of Kenya's defense policy are also spelled out clearly in the 2017 *White Paper*. These aims include the following: quick response to varying degrees of threat without risking escalation of the level of conflict; in peacetime, a balance of power that does not encourage an arms race and a defense capability that deters coercion or aggression; and, in times of crisis, a readiness for effective defense that is easily and immediately demonstrated in order to solve crises without recourse to the use of lethal

---

3 *Defence White Paper* (Nairobi, Kenya: Ministry of Defence, 2017), 2, <http://www.mod.go.ke/wp-content/uploads/2017/05/White-Paper.pdf>.
4 Ibid.
5 Ibid., 3.

force.[6] Kenya's defense policy highlights the need for rapid capabilities as well as a minimum use of force. These also should be taken into account as we consider the KAF's capabilities.

In terms of Kenya's threat landscape, terrorism is highlighted prominently in the 2017 *White Paper*. For example, "the international threat posed by ISIL [the Islamic State of Iraq and the Levant] with its training cells in neighbouring Somalia poses one of the key external security challenges for Kenya."[7] Additionally, "at the regional level, the Al-Shabaab operatives affiliated to ISIL in Somalia continue to propagate Somali irredentism across the Horn of Africa, while they masquerade as Islamists."[8] At the same time, the *White Paper* raises concerns about more conventional threats, notably the reorganization and modernization of regional military forces. Clearly, though, the main security threat to Kenya is described as transnational terrorism.

The *White Paper*'s discussion of the threats to Kenya closes with a point that merges the aforementioned threats to the nation: "The territorial integrity of the country is threatened because the borders that were inherited at independence are yet to be conclusively delineated and demarcated."[9] President Uhuru Kenyatta reemphasized this point in his 2018 State of Nation address: "Beyond terrorism, we remain vulnerable to other security threats; many of them, from terrorism to trafficking, across borders."[10] Here, then, we see an historical issue that has been raised as both a current national security and foreign policy challenge to Kenya. This points to a clear need for defensive capabilities.

With this understanding of Kenya's strategic landscape, we may now turn to the KAF and its force structure. According to its official website, the vision of the KAF is to be a "premier" air force that delivers air defense 24/7.[11] Its primary mission is to defend and maintain the sovereignty of

---

6 Ibid., 7.

7 Ibid., 11.

8 Ibid.

9 Ibid., 12.

10 Uhuru Kenyatta, "Remarks During the State of Nation Address," Parliament Buildings, Nairobi, Kenya, May 2, 2018, <http://www.president.go.ke/2018/05/02/remarks-by-his-excellency-hon-uhuru-kenyatta-c-g-h-president-and-commander-in-chief-of-the-defence-forces-of-the-republic-of-kenya-during-the-state-of-the-nation-address-at-parliament-buildings-n/>.

11 "Kenya Air Force," Ministry of Defense <http://kaf.mod.go.ke>.

Kenya's airspace at all times, while its secondary missions include supporting related services in pursuance of national objectives as well as assistance and cooperating with other Kenyan civil authorities in situations of emergency or disaster when called upon by the national government.[12] Furthermore, the KAF "may be called upon to restore peace in any part of Kenya affected by unrest or instability only with the approval of the National Assembly [Kenya's legislative branch]."[13] The KAF vision and mission set are not particularly unique, although its additional call to be ready to "restore peace" is distinctly African. But taking into full account Kenya's geography, which occupies 582,650 square kilometers (224,962 square miles) plus land and maritime borders with Somalia, Ethiopia, South Sudan, Uganda, and Tanzania, the mission set becomes daunting for any air force, let alone the KAF.[14]

Given its strategic landscape and responsibilities, the KAF is best described as a niche air force. Niche air forces, according to Kainikara, include these five characteristics, all of which apply to the KAF:

- Does not have the full spread of air power capabilities
- Able to carry out only a few dedicated roles and/or functions in limited capacity
- Possesses only a restricted number of systems because of the lack of resource capacity to acquire and maintain larger numbers of sophisticated airborne systems
- Depends, almost completely, on external sources for industrial developments, operational support, and infrastructure
- Is of limited importance in the highest, strategic level of national security.[15]

The following sections describe why the KAF is a niche air force and how this came to be. But, for now, let us examine the KAF's current force structure. The KAF comprises approximately 2,500 personnel, over double the size of Kenya's navy but only about 12 percent the size of the army. The KAF has an assortment of approximately 130–140 fixed and rotary aircraft,

---

12 Ibid.
13 *Defence White Paper*, Ministry of Defence, 20.
14 For more on the significance of Kenya's geography in relation to its politics, see Donovan C. Chau, *Global Security Watch – Kenya* (Santa Barbara, CA: Praeger, 2010), 1–20.
15 Kainikara, "The Future Relevance of Smaller Air Forces," 4.

with at least a dozen or more on order.[16] Notable among the KAF arsenal are approximately 21 F-5E/F combat aircraft, four of which are trainer aircraft. The KAF also possesses a mix of small and medium transport aircraft, from the approximately 11 Y-12s to four DHC-5s to three DHC-8s. In terms of rotary aircraft, the dominant light attack helicopters are the approximately 40 MD 500s, which historically have been operated by the Kenya Army's 50th Air Cavalry Division. The KAF has a mix of utility and transport rotary aircraft, from the 14 SA 330s to six Z-9s to two Mi-171Es.[17]

More recently, the KAF has been adding to its multi-mission rotary aircraft inventory in response to security challenges in and around Somalia. In December 2016, for example, the KAF received the first six of eight UH-1H-II medium-lift helicopters.[18] It also acquired at least two AH-1F attack helicopters.[19] In addition, it has acquired eight H125M light attack and armed reconnaissance helicopters, at least four of which were delivered in the latter half of 2018.[20] Given the nature of these recent rotary aircraft additions, it is apparent that the KAF is developing more rapid response capabilities, from troop transport to medical evacuation to attack and reconnaissance missions. The evolution of KAF capabilities requires us to examine more closely Kenya's strategic history. How and why did the KAF go from possessing F-5 combat aircraft to AH-1 attack helicopters, which seems counterintuitive? In other words, why would Kenya's air force apparently devolve from possessing fixed-wing combat aircraft, in the main, to focus on attack helicopters? This is the story of the KAF—past, present, and future.

---

16 These figures, and the figures below, are a cross tabulation from the following sources: Anton Kruger and Guy Martin, "Kenya Defense Forces," *defenceWeb*, Jul. 9, 2013, <http://www.defenceweb.co.za/index.php?option=com_content&view=article&id=31108:kenya-defence-forces&catid=119:african-militaries>; "World Air Forces Directory," *Flight International*, Vol. 190, No. 5566 (Dec. 6, 2016): 39; and "Kenya Air Force," *Wikipedia*, Dec. 2, 2018, <https://en.wikipedia.org/wiki/Kenya_Air_Force>.

17 The Z-9s are also operated by the Kenya Army.

18 Job Weru, "Boost for War against Terror As USA Donates Sh11 Billion Helicopters for Kenya's Military," *Standard Media*, Dec. 2, 2016, <https://www.standardmedia.co.ke/article/2000225559/boost-for-war-against-terror-as-usa-donates-sh11-billion-helicopters-to-kenya-s-military>.

19 Erwan de Cherisey, "Kenya Takes Delivery of AH-1 Cobras," *Jane's Defence Weekly*, June 30, 2017.

20 Erwan de Cherisey, "Kenya Receives Fennec Helicopters," *Jane's Defence Weekly*, Sep. 6, 2018.

Jomo Kenyatta Years (1963–1978)

It is nearly impossible to discuss the air force of Kenya without reaching back to the founding of Kenya. This requires an examination of the role of Kenya's founding father, Mzee Jomo Kenyatta (*mzee* means elder). Kenyatta is credited with having the foresight to establish an air component to Kenya's emerging military in the early 1960s. He dealt firsthand with Kenya's only military campaign prior to 2011, Somali irredentism in the form of the *shifta* (or bandit) conflict. While an in-depth examination of the conflict is not necessary, an overview sets the context for our discussions of the KAF's birth.

Briefly, the *shifta* campaign in northeast Kenya began prior to Kenya's independence in 1963 and continued until roughly 1968. Due to historical and geographic circumstances, the Somali irredentist insurgency claimed large tracts of territory in what would become Kenya's North Eastern Province (now Mandera, Wajir, Garissa, and Lamu counties).[21] The insurgency and subsequent counterinsurgency gave the nascent Kenya military and security forces operational experience under the leadership of British officers and advisers. It also provided Kenyan civilian leadership with command and control experience against a diverse, disjointed Somali enemy—the *shifta*—who roamed, ambushed, and raided Somali-claimed Kenyan territory along the Kenya-Somalia border.[22]

To counter the *shifta*, British and Kenyan ground and, notably, air forces were deployed to the northeastern portions of Kenya. This was required due to the geographic distances and lack of transportation infrastructure that needed to be overcome, from Kenya's main population centers in the lower third of its territory to the conflict zones in the upper third of its territory. By December 1964, the counterinsurgency force, on the Kenyan side, comprised over half of Kenya's only light infantry brigade, plus a police air wing.[23] This is significant, considering the KAF was officially established

---

21 For a detailed look at the Kenyan-Somali strategic relationship, see Donovan C. Chau, "At the Cross-Roads of Cultures? An Historic and Strategic Examination of Kenya-Somalia Relations," *Journal of the Middle East and Africa*, Vol. 1, No. 1 (Jan.–June 2010), 67–83.

22 Ethnic Somalis play unique and distinctive roles across the Kenya-Somalia borderlands. For one analysis, see Donovan C. Chau, "The Fourth Point: An Examination of the Influence of Kenyan Somalis in Somalia," *Journal of Contemporary African Studies*, Vol. 28, No. 3 (Jul. 2010), 297–312.

23 Nene Mburu, *Bandits on the Border: The Last Frontier in Search for Somali Unity* (Trenton, NJ: Red Sea Press, 2005), 142.

six months prior. Clearly, Mzee Kenyatta knew the Kenyan military faced a shortcoming in, through, and from the air. And he needed external support to compensate and establish the KAF.

The KAF began from a squadron of colonial-era light aircraft left behind by the British Royal Air Force (RAF) in East Africa. Equipped with approximately nine trainer aircraft in 1964, the nascent KAF was led by British officers and technicians to start. About the same time, Kenya and Israel concluded a cooperation agreement that included various types of technical assistance and military training. Significant to our discussions here, Israel trained approximately five Kenyans as pilots in Haifa from 1962 to 1963; notably, this occurred in a period before Kenyan independence and prior to the official establishment of the KAF.[24] These pilots later received additional training in Kenya and the United Kingdom.[25] So, from the very beginning of its existence, the KAF relied heavily on external support for equipment, technical assistance, and training.

None of this is particularly surprising, given that Kenya had just achieved independence. What is unique, however, was the KAF's original mission priorities. In addition to being formed to defend Kenyan airspace against any and all external aggression, the KAF was tasked early on with assisting related services as well as civilian authorities of the non-military portions of the Kenyan government. This was a clear recognition of the geographic and infrastructural challenges that faced Kenya and also the limitations of a newly established nation with very limited industrial capabilities. After the *shifta* campaign, as an example, the KAF performed aerial photography in support of national development, fisheries, and agriculture. Other examples of the KAF's early duties included spotting forest fires, spraying locusts, and supplying relief food in drought-stricken areas.[26] These were instances of Kenya attempting to promote a positive image of the KAF among the Kenyan populace, thereby enhancing civil-military relations, as well as the military supporting Kenya's national development.

---

24 Chau, *Global Security Watch – Kenya*, 57.
25 Only four of the five pilots successfully completed all the training. "Interview: Maj Gen (Rtd) Dedan Gichuru," interviewed by Zipporah Kioko, *Majeshi Yetu*, Kenya Defence Forces, 10 (2017), 23.
26 Ibid., 24.

As the Kenyan nation approached the first decade of its existence, Mzee Kenyatta would, again, play a decisive role in the KAF's evolution. In the early 1970s, the KAF comprised an assortment of light and utility transport aircraft along with trainer aircraft—but no combat aircraft. This all changed, beginning in 1974. Around this time, and in the context of the Cold War between the United States and the Soviet Union, geopolitical shifts occurred in the Horn of Africa. Specifically, Ethiopia experienced a communist revolution and the Soviet Union signed a friendship and cooperation treaty with Somalia. In addition, Uganda was under the leadership of the anti-western Idi Amin. Given this threat environment, Kenya began a series of strategic discussions with a newfound ally, the United States.[27] With influential roles by U.S. secretary of state Henry Kissinger and U.S. secretary of defense Donald Rumsfeld, between 1974 and 1976, Kenya under Mzee Kenyatta agreed to the military procurement of a dozen F-5 combat aircraft, plus spare parts and training. As a part of the agreement, Kenyan pilots were trained in the United States and U.S. military advisers and contractors were sent to Kenya to build ground infrastructure and a local training program. The F-5s were delivered to Kenya in the spring of 1978, which was also around the finale of a war between Ethiopia and Somalia.

Both internal and external factors molded the KAF's first few decades of existence. British influence, from equipment to training, remained significant throughout. The RAF had knowledge of local and regional air environments; this, no doubt, served as the foundation of the KAF. An interesting, and often overlooked, external factor in the KAF's growth, and Kenya's development in general, was the role played by Israel. By the time of the KAF's establishment, Israel had nearly two decades of combat experience, including from the air and in counterinsurgencies. This influence on the KAF, as well as the KDF, should not be discounted. Later, the United States offered what became the major showpiece of the KAF, namely the F-5s. These U.S.-supplied combat aircraft to Kenya immediately tipped the military balance in a region, where Soviet-supplied combat aircraft were the norm, specifically in Ethiopia and Uganda. At the same time, one might question the actual strategic necessity of F-5s to the KAF, given

---

27 This brief account is derived from a lengthier explication in Chau, *Global Security Watch – Kenya*, 127–28.

Kenya's strategic history up to that point in time. Yet, there is no question whatsoever that Mzee Kenyatta contributed tremendously to the growth and professional reputation of the KAF during his tenure as the president of Kenya. These early years of the KAF set the tone and foundation for its later development. Now, let us fast-forward several decades in the history of the KAF, as Kenyatta has continued to play a central role. But this time, it is Mzee's son, Uhuru Kenyatta, known affectionately as Kamwana (*kamwana* means child).

Uhuru Kenyatta Years (2013 to present)

Uhuru Kenyatta inherited not only his father's legacy when he entered office in the spring of 2013 but also Kenya's first and, to date, only active foreign military operation. His predecessor, Mwai Kibaki, ordered the KDF to invade southern Somalia in October 2011. Ostensibly, this was done in response to attacks and kidnappings performed by the Somali terrorist group al-Shabaab, into and on Kenyan territory. But the real motives behind Kenya's first foreign military foray remain in question. Nevertheless, since 2011, the KAF has gained operational and tactical experiences in the battlefields of southern Somalia and the borderlands between Kenya and Somalia. On balance, the KAF performed its role of air counterinsurgency well enough, particularly in the absence of a clear strategic purpose from Kenyan's political leadership.[28] While they are now a part of the African Union Mission in Somalia (AMISOM), KDF forces remain entrenched in southern Somalia. Meanwhile, Kenyatta has capitalized on this strategic circumstance, not unlike his father, to bolster both the reputation and capabilities of the KDF, in general, and the KAF, in particular.

Before we delve further into the KAF's evolution, it is useful to gain a clear sense of Kamwana's foreign and defense policies. His 2018 state of the nation address stated these plainly, although somewhat hyperbolically: "We defend democracy abroad as we do at home; we want for others the peace that we enjoy here; we are grateful for the solidarity extended us by our brothers and sisters on the continent, so we offer it to others in

---

28 For a detailed case study of KDF operations, see Donovan C. Chau, "*Linda Nchi* from the Sky? Kenyan Air Counterinsurgency Operations in Somalia," *Comparative Strategy*, Vol. 37, No. 3 (Nov. 2018), 220–34.

return."[29] Kenyatta continued by applying these policies to Kenya's specific neighborhood: "The region is not at peace. Somalia remains troubled, largely by foreign agents who weaken its government, who divide its peoples, and who threaten to reverse the gains we have so painfully won under AMISOM."[30] Furthermore, he said, "Through it all, we remember that if our brothers and sisters in Somalia prosper, we prosper; if they are safe, so are we. It has been our policy, then, to help them regain the peace and prosperity they once knew."[31] Kenyatta, then, has made it clear that Kenya has a strategic interest in Somalia. Therefore, it is logical to conclude that the KDF and the KAF will continue to play strategic roles for Kenya in Somalia.

These political and strategic contexts are significant to take into account when one examines changes to the KAF force strategic since 2013, all under Uhuru Kenyatta's rule. As aforementioned, in recent years, the KAF has added to its rotary aircraft inventory, specifically multi-role and attack helicopters in the form of UH-1Hs, AH-1Fs, and H125Ms. None of these capabilities existed for the KAF in 2013, and certainly not during the invasion of southern Somalia in 2011. Moreover, between 2015 and 2017, the KDF bolstered its air capabilities by adding unmanned aerial vehicles (UAV) and ordering an aerostat as well as a dozen counterinsurgency aircraft, all from the United States.[32] The Kenya Army may be the primary operator of the UAVs and, as of this writing, the aerostat and counterinsurgency aircraft deals remain in abeyance.[33] Regardless, these moves demonstrate Kenya's resolve to enhance its intelligence, surveillance, and reconnaissance (ISR) as well as its light attack and air counterinsurgency capabilities. This may be contrasted with the role and direction of the legacy the KAF's capabilities, specifically its conventional arsenal of F-5 combat aircraft.

Here, Uhuru Kenyatta has played an interesting role in the KAF's evolution. While his father developed strategic relations between Kenya and

---

29 Kenyatta, "Remarks During the State of Nation Address."

30 Ibid.

31 Ibid.

32 "Kenya-US: Military Purchases," *Africa Research Bulletin*, Vol. 54, No. 2 (Mar. 2017), 21,340A–41C.

33 For an example about the AT-802L Air Tractor deal, see "Sh43bn Bid for US Fighter Jets Hangs in the Balance," *Business Daily*, Feb. 14, 2018, <https://www.businessdailyafrica.com/news/Sh43bn-bid-for-US-fighter-jets-hangs-in-the-balance/539546-4304000-pynirl/>.

the United States, Kamwana further developed strategic ties between the two countries and also with Jordan. In September 2016, the King of Jordan, Abdullah II, visited Nairobi to observe a joint counterterrorism exercise as well as cement a mutual defense cooperation agreement between the two countries. This involved Jordan training nearly 100 KAF officers.[34] In addition, and controversially, Kenya also agreed to purchase 15 F-5s from Jordan as part of their defense cooperation agreement. The military rationale for purchase of these Jordanian F-5s is unclear. But Kenyatta's involvement in the entire affair is not: close Jordan and Kenya relations have continued apace as they finalize agreements to enhance bilateral economic, trade, and investment cooperation, beyond those already realized in the fields of defense, security, and anti-terrorism.[35] It is apparent that Kenyatta has sought to broaden Kenya's strategic alliances, particularly vis-à-vis threats emanating from Somalia. And the KAF has played a role in this strategic broadening.

In addition to ISR enhancements and the strategic gambit with Jordan under Kenyatta, the KAF has also enhanced its transport capabilities. Between late 2017 and early 2018, Kenya made acquisitions of two or three C-27J tactical transport aircraft and four AW 139 multi-mission helicopters. Although the latter may eventually enter the air components of nonmilitary Kenyan government agencies, the former are expected to replace older transport aircraft.[36] Notably, the manufacturer of both platforms, Leonardo, is Italian, which shows a modest diversification of KAF suppliers. The KAF, then, continues to evolve, ever so slowly, from its original force structure. It appears to be becoming more niche and also reliant on strategic allies for both equipment and training.

What has been looming in Kenyatta's tenure as head of the Kenyan government and its military has been the same for both himself and his father—namely, the strategic role of the United States, particularly its foreign and defense policies in the Horn of Africa, as well as the threats emanating from Somalia amid the global insecurity landscape. Both Mzee

---

34 "Kenya." *Africa Research Bulletin*, Vol. 53, No. 10 (Nov. 2016), 21,197A–B.

35 "Joint Jordan-Kenya Committee to Finalise Cooperation Agreements," *Jordan Times*, Dec. 7, 2018, <http://www.jordantimes.com/news/local/joint-jordan-kenya-committee-finalise-cooperation-agreements>.

36 Jon Lake, "Secret Spartans for Kenya," *African Aerospace*, Nov. 28, 2018, <https://www.africanaerospace.aero/secret-spartans-for-kenya.html>.

and Kamwana have demonstrated an understanding of strategy in international politics, which has resulted in the general growth and maturation of the KAF, from the 1960s to the present. In this story of the KAF, there remain some points known and unknown yet to be discussed. Questions about Kenya's strategic past may eventually or unexpectedly cause challenges in the future, which may impinge directly or indirectly on the KAF and its continued evolution.

Unknowns, Past, and Future

We turn our attention now to the gap years in between the reigns of the two Kenyattas. In August 1982, the most ambitious and serious attempt at changing Kenya's political system occurred, and this involved the KAF. At 3:30 am on August 1, 1982, a coup d'état attempt was led by young junior air force officers. Under the self-proclaimed People's Redemption Council, coup forces seized the Voice of Kenya radio station in Nairobi as well as air bases in Nanyuki, Eastleigh, and Embakasi.[37] After an initially delayed response, Army Major General Mahmoud Mohamed led a unit of soldiers that recaptured the Voice of Kenya radio station. He unilaterally assumed command of the forces, authorized deployment of troops, and personally led the counter-assault. The Kenya Army, thus, suppressed the attempted air force–led coup, altering the history of the KAF.

The entire KAF was disbanded after the coup attempt. A total of 900 men were tried and sentenced to up to 25 years in prison. Thirteen coup organizers, all of Luo ethnicity, received death sentences and were executed. The KAF commander at the time, Major General Peter K. Kariuki, was imprisoned for four years on coup attempt–related charges, and the majority of the senior officers arrested or dismissed were from the Kikuyu ethnic group. Given the ethnic makeup of the coup participants, tribal factors may have been a motivating factor, given that the president of Kenya at the time, Daniel Moi, was a Kalenjin. The exact cause of the 1982 coup remains unknown. But the coup attempt, especially the role played by tribe and ethnicity, is an important inflection point in the history of the KAF.

---

37 According to one source, there were three concurrent coup attempts ongoing. Samuel Decalo, *The Stable Minority: Civilian Rule in Africa, 1960–1990* (Gainesville, FL: Florida Academic Press, 1998), 242.

When examining the KAF, specifically, and the KDF, in general, one cannot then overlook ethnic considerations, which influence directly the social and political fabric of Kenya.

Major General Mohamed, interestingly a Kenyan Somali, was given the task of creating a new air force after the coup. Renamed the "82 Air Force," the organization fell under army command with alterations of uniform, flag, and motto. In addition, air bases at Eastleigh and Nanyuki were renamed Moi and Laikipia Air Bases. Operation and maintenance suffered greatly after the coup. As a result, greater assistance was provided by British and American training teams, as well as by representatives of western aircraft manufacturers. By 1989, the 82 Air Force reverted back to its original name, the Kenya Air Force.[38] Although this entire episode is now a part of the strategic history of Kenya, it is a significant example of how distinctly African the KAF is. The ethnic composition of the KAF, and the KDF more broadly, may be influential in the future. It remains an issue in Kenya's political and strategic shadows.

Sadly, another perhaps distinctly African quality of the KAF is the role of corruption. A recent example illustrates the potential danger. As aforementioned, in 2010, Kenya agreed to procure F-5 combat aircraft from Jordan. Soon after the purchase, questions arose as to the airworthiness of the second-hand aircraft. Indeed, by November 2010, Kenya's military chiefs, for the first time ever, appeared before a parliamentary committee to answer questions. Internal communications indicate that the Kenyan military may have been duped into buying the 15 Jordanian aircraft, overlooking the actual age of the aircraft.[39] According to subsequent Kenyan governmental audits, the F-5s have not been operational since the time of their procurement, with parliamentary investigations ongoing.[40] Inspections also reveal that defects identified at the time of delivery of the aircraft have not been rectified. An interesting note in the entire affair is that, in 2010, Uhuru Kenyatta was head of the finance ministry. Six years later, as already mentioned, he welcomed the King of Jordan to Nairobi. Whether there was government malfeasance or not remains unknown. What is clear is that

---

38 This account of the 1982 coup attempt draws from Chau, *Global Security Watch – Kenya*, 38–43.
39 "Kenya: Acquisitions Probe," *African Research Bulletin,* Vol. 47, No. 11 (Dec. 2010), 18,632–3.
40 Moses Nyamori, "Mystery of Sh1.5 Billion Faulty Air Force Jets," *Standard,* Jul. 18, 2018, <https://www.standardmedia.co.ke/article/2001288457/mystery-of-sh1-5-billion-faulty-air-force-jets>.

corruption and its impact on procurements for the KAF may impinge on the future of the force. It cannot be overlooked when examining the KAF.

On a positive note, the KAF has long emphasized the important role of training its forces. For quite some time, the majority of KAF recruits have entered with technical and secondary school backgrounds as well as some with university education and experience. Formal and in-service education was strengthened in the 1980s and 1990s with the establishment of the Defence Staff College and the National Defence College for mid-level and senior officers.[41] On occasion of the 50th anniversary of the KAF in 2014, the KDF Chief of Defence Forces, Julius Karangi, was not overstating the role of training and education when he said that "these achievements [of the KAF] have been realized through a constant and conscious decision to build the required human capacity and to invest in technology relevant to the times."[42] Continuity of the KAF training comes largely from the legacy of British influence. This was reinvigorated when Kenya and the United Kingdom agreed to a new defense cooperation agreement in October 2016. The agreement included specialized training of Kenyan forces overseen by the British Army Training Unit in Kenya as well as continued support to security forces deployed to Somalia, specifically intelligence sharing and cooperation.[43]

As demonstrated by the initial training of Kenyan pilots, KAF training is internationally diverse, with influences from the United Kingdom, Israel, the United States, and, now, Jordan. This multitude of influences no doubt affects KAF doctrine, about which very little is publicly available. We may glean some pertinent information from the 2017 *White Paper*: "In discharging their mission and other assigned duties, the Defence forces shall be a balanced, modern and technologically advanced military Force capable of executing its tasks effectively and efficiently."[44] In addition, there is specific and repeated reference to the KDF adherence to international obligations as well as the law of armed conflict. Therefore, in all likelihood, the KAF doctrine is not so dissimilar from western, particularly British and American, air power doctrine. Doctrinally, the KAF places emphasis on balance,

---

41 Chau, *Global Security Watch – Kenya*, 53.
42 Julius W. Karangi, "Speech on Occasion of the KAF 50th Anniversary Celebrations," Moi Air Base, June 4, 2014.
43 "Kenya," *Africa Research Bulletin*.
44 *Defence White Paper*, Ministry of Defence, 19.

efficiency, and effectiveness. In a sense, this is demonstrated in the KAF force structure, with its mix of rotary and fixed-wing aircraft designed for combat, transport, or multi-mission roles.

The precise nature of the KAF force of the future remains unknown. But the 2017 *White Paper* revealed some characteristics that the KAF may be taking into consideration. For example, "it must be understood that Force capabilities and preparedness are tied up to the availability of resources."[45] Clearly, the budget and procurement of KAF capabilities does not occur in a vacuum. Furthermore, according to the *White Paper*, "emphasis will be on three areas; cost-effectiveness, prioritization and best practices in resource management."[46] An example of these characteristics at work was the formation of the KAF Rapid Deployment Squadron (RDS). According to official Kenyan military publications, the need to establish the RDS arose due to three keystone events: the 1998 terrorist attack on the U.S. embassy in Nairobi, anti-piracy operations in the Indian Ocean that began in 2005, and the aforementioned 2011 KDF invasion of southern Somalia.[47] From these strategic events, the mission set of RDS emerged: disaster response, personnel recovery, as well as combat search and rescue.

The RDS was officially established in 2014, with headquarters at Moi Air Base. In terms of capabilities, the unit will comprise a mix of medium transport aircraft, combat support helicopters, along with armored personnel carriers and Zodiac fast boats. Recent purchases of C-27Js and A 139s are, logically, a part of the emerging RDS capabilities. The force composition of RDS demonstrates an evolution of how the KAF has been organized, from single mission air units to ones that are intentionally designed to be multi-domain, multi-mission, and interoperable. In terms of training, the RDS is already learning from partners abroad. In 2016, for example, the RDS conducted a combat search and rescue training exercise with its U.S. and Ugandan counterparts. This was 10 years after initial training of the unit began, demonstrating a long-term, deliberative approach to the RDS's force development. As the RDS, specifically, and the KAF, in general,

---

45 Ibid., 16.
46 Ibid.
47 Samuel N. Thuita, "RDS: Rapid Deployment Squadron," *Majeshi Yetu*, Kenya Defence Forces, 8 (2016), 18–19.

mature further, procurement of new and different platforms may continue to evolve to address emerging threats.

The clearest strategic challenge facing the KAF is the question of Somalia. Granted, this is a question not only for Kenya but also the region and the world. But Kenya feels the strategic weight the most. There are numerous variables, internal and external, to the situation in Somalia. The KDF will remain a part of the AMISOM force contingent in Somalia, at least until 2020. At the same time, Kenya continues to experience attacks on its territory by al-Shabaab while the KDF and Kenyan security forces attempt to buttress the border areas with Somalia.[48] The role and potential impact of the KAF on the security situation in the Kenya-Somalia borderlands are difficult questions facing Kenyan military leadership. "Irrespective of the nature of warfare and the type of adversary," Kainikara informs us wisely, "air power will be required to achieve a range of effects with discrimination, proportionality and accuracy."[49] The strategic challenge of Somalia has been and remains at Kenya's doorstep. Kenyan air power will play a role; whether it can do so with discrimination, proportionality, and accuracy remains the question.

Conclusion

Oddly, perhaps, the story of the KAF ends how it began: with a Kenyatta in charge and the Kenyan military battling threats of a Somali nature. The KAF has grown out of the remnants of British colonial air capabilities with international assistance from Israel, the United States, and, more recently, Jordan. The KAF played a unique and disruptive role in Kenya's political history with the 1982 coup attempt. It also played a substantive role in Kenya's only foreign military operation, the 2011 invasion of southern Somalia. And the KAF may still be described as a niche air force.

With actual combat experiences now, the KAF was and remains known as one of the most professional in Africa. As a sign of this professionalism, the 2017 *White Paper* stated prudently: "The need to direct and control application of force cannot be over emphasized."[50] The Kenyan military

---

48 "Kenya," *Africa Research Bulletin*, Vol. 55, No. 9 (Oct. 2018): 22,016A–B.
49 Kainikara, "The Future Relevance of Smaller Air Forces," 12.
50 *Defence White Paper*, Ministry of Defence, 19.

deserves credit for this mature doctrinal statement, as do the influences coming from the United Kingdom and the United States. The KAF will play a key role in how Kenya directs and controls its application of force. Recent KAF procurements are indicative of the different strategic landscape facing Kenya, from Cold War geopolitics to twenty-first century terrorist threats and constant border insecurity.

At the end of the day, the story of the KAF reflects "the great stream of time" in Kenya's strategic history.[51] The West—primarily the United Kingdom but also the United States and Israel—has bolstered the air force of an African nation-state in a historically and exceedingly tough geopolitical region. Thus, history, geography, and politics have all influenced greatly the development of the KAF, which, as this chapter has illustrated, was niche from the start. Only time will tell what will become of the KAF. But the future may not look so different from the past.

References

Chau, D.C. *Global Security Watch – Kenya*. Santa Barbara, CA: Praeger, 2010.

Chau, D.C. "At the Cross-Roads of Cultures? An Historic and Strategic Examination of Kenya-Somalia Relations." *Journal of the Middle East and Africa*, Vol. 1, No. 1 (Jan.–June 2010), 67–83.

Chau, D.C. "The Fourth Point: An Examination of the Influence of Kenyan Somalis in Somalia." *Journal of Contemporary African Studies*, Vol. 28, No. 3 (Jul. 2010), 297–312.

Chau, D.C. "*Linda Nchi* from the Sky? Kenyan Air Counterinsurgency Operations in Somalia." *Comparative Strategy*, Vol. 37, No. 3 (Nov. 2018), 220–34.

Churchill, W.S. *Never Give In! The Best of Winston Churchill's Speeches*. New York: Hachette Books, 2015.

De Cherisey, Erwan. "Kenya Takes Delivery of AH-1 Cobras." *Jane's Defence Weekly*. June 30, 2017.

De Cherisey, Erwan. "Kenya Receives Fennec Helicopters." *Jane's Defence Weekly*. Sep. 6, 2018.

Decalo, S. *The Stable Minority: Civilian Rule in Africa, 1960–1990*. Gainesville, FL: Florida Academic Press, 1998.

---

51 Reference to the great stream of time concept in Colin S. Gray, *What Should the U.S. Army Learn from History? Recovery from a Strategy Deficit* (Carlisle, PA: Strategic Studies Institute and U.S. Army War College Press, 2017), 46. For a much deeper look, see Richard E. Neustadt and Ernest R. May, *Thinking in Time: The Uses of History for Decision Makers* (New York: Free Press, 1986).

Defence White Paper. Nairobi: Ministry of Defence, 2017. <http://www.mod.go.ke/wp-content/uploads/2017/05/White-Paper.pdf>.

Gray, C.S. *What Should the U.S. Army Learn from History? Recovery from a Strategy Deficit*. Carlisle, PA: Strategic Studies Institute and U.S. Army War College Press, 2017.

"Joint Jordan-Kenya Committee to Finalise Cooperation Agreements." *Jordan Times*, Dec. 7, 2018. <http://www.jordantimes.com/news/local/joint-jordan-kenya-committee-finalise-cooperation-agreements>.

Kainikara, S. "The Future Relevance of Smaller Air Forces," Air Power Development Centre, Royal Australian Air Force, Working Paper 29, Apr. 2009.

Kioko, Z. "Interview: Maj Gen (Rtd) Dedan Gichuru." *Majeshi Yetu*, Kenya Defence Forces, 10 (2017), 20–26.

Karangi, Julius W. "Speech on Occasion of the KAF 50th Anniversary Celebrations," Moi Air Base, June 4, 2014.

"Kenya: Acquisitions Probe." *Africa Research Bulletin*, Vol. 47, No. 11 (Dec. 2010), 18, 632–3.

"Kenya Air Force." Last modified Dec. 2, 2018. <https://en.wikipedia.org/wiki/Kenya_Air_Force>.

"Kenya." *Africa Research Bulletin*, Vol. 55, No. 9 (Oct. 2018): 22,016A–B.

Kenyatta, Uhuru. "Remarks During the State of Nation Address," Parliament Buildings, Nairobi, Kenya, May 2, 2018. <http://www.president.go.ke/2018/05/02/remarks-by-his-excellency-hon-uhuru-kenyatta-c-g-h-president-and-commander-in-chief-of-the-defence-forces-of-the-republic-of-kenya-during-the-state-of-the-nation-address-at-parliament-buildings-n/>.

"Kenya." *Africa Research Bulletin*, Vol. 53, No. 10 (Nov. 2016): 21,197A–B.

"Kenya-US: Military Purchases," *Africa Research Bulletin*, Vol. 54, No. 2. (Mar. 2017): 21,340A–1C.

Kruger, A., and Martin, G. "Kenya Defense Forces." *defenceWeb*. Jul. 9, 2013. <http://www.defenceweb.co.za/index.php?option=com_content&view=article&id=31108:kenya-defence-forces&catid=119:african-militaries>.

Lake, Jon. "Secret Spartans for Kenya." *African Aerospace*, Nov. 28, 2018. <https://www.africanaerospace.aero/secret-spartans-for-kenya.html>.

Mburu, N. *Bandits on the Border: The Last Frontier in Search for Somali Unity*. Trenton, NJ: Red Sea Press, 2005.

"Kenya Air Force." Ministry of Defence. <http://kaf.mod.go.ke>.

Neustadt R.E., and E.R. May. *Thinking in Time: The Uses of History for Decision Makers*. New York: Free Press, 1986.

Nyamori, Moses. "Mystery of Sh1.5 Billion Faulty Air Force Jets." *Standard*, Jul. 18, 2018. <https://www.standardmedia.co.ke/article/2001288457/mystery-of-sh1-5-billion-faulty-air-force-jets>.

"Sh43bn Bid for US Fighter Jets Hangs in the Balance." *Business Daily*, Feb. 14, 2018. <https://www.businessdailyafrica.com/news/Sh43bn-bid-for-US-fighter-jets-hangs-in-the-balance/539546-4304000-pynirl/>.

Thuita, S.N. "RDS: Rapid Deployment Squadron." *Majeshi Yetu*, Kenya Defence Forces (2016), 18–20.

Weru, Job. "Boost for War Against Terror As USA Donates Sh11 Billion Helicopters for Kenya's Military." *Standard Media*, Dec. 2, 2016. <https://www.standardmedia.co.ke/article/2000225559/boost-for-war-against-terror-as-usa-donates-sh11-billion-helicopters-to-kenya-s-military>.

"World Air Forces Directory," *Flight International*. Vol. 190, No. 5566 (Dec. 6, 2016): 29–49.

14 BRAZIL
Control, Defend, Integrate
Luís E.P. Celles Cordeiro

Introduction

Within public policy is defense policy—decisions made by politicians to define priorities and actions. Brazil's first public defense policy, the National Policy of Defense (PDN), was published in 1996 (with the second published in 2005), and created the Ministério da Defesa (Brazilian Ministry of Defense (BMD)). The PDN's successor, the Política Nacional de Defesa (National Defense Policy (PND)) was made public in 2008, brought in line in 2012, and approved by Congress in 2013. As at mid-2019, the third edition is awaiting National Congress approval.[1]

The 2008 PND was a milestone in Brazil's security policies because for the first time in recent democratic history the words "national security" were the focus of national debate.[2] The PND made explicit "what to do" regarding national security, while the Estratégia Nacional de Defesa (National Defense Strategy (END)), released at the same time, made clear the definition of "how to do it."[3]

Until 2007, national security was not a main focus in Brazil's national policy, but this changed following events that year. In March, the Air Force faced a serious institutional crisis when a union-backed strike by the

---

1 Carlos Wellington Almeida, "Política de Defesa no Brasil: considerações do ponto de vista das políticas públicas" *Opinião Pública*, Vol. 17, No. 1 (2010) 220–230.

2 Eliézer Rizzo de Oliveira, "A Estratégia Nacional de Defesa e a Reorganização e Transformação das Forças Armadas," *Interesse Nacional*, Vol. 2, No. 5 (2009) 71–74.

3 Ministério da Defesa, *Política Nacional de Defesa – Estratégia Nacional de Defesa* (Brasília, 2012) 15.

military air traffic controllers paralyzed air traffic service over some regions in Brazil. As the military services are forbidden to strike, the secretaries of the Navy, Army, and Air Force deemed this a mutiny, while Minister of Defense Waldir Pires advised President Luiz Inácio Lula da Silva to accept the controllers' terms. Ultimately, those leading the strike were arrested and Waldir Pires fired, a clear manifestation of the military leaders' power.[4] The same year, in a congressional hearing military commanders pointed out the obsolescence of Brazilian equipment in comparison with Venezuela where President Hugo Chaves announced not only the purchase of aircraft and warships from Russia, but also an alliance with Cuba, Bolivia, Nicaragua, and Ecuador. Faced with this picture, the new minister in command of the Ministry of Defense, Nelson Jobim, promoted a new approach to national security in the national policy.[5]

While the PDN was a presidential decree, which means that its decision-making process was restricted to the executive branch, the PND was a bill of law. The National Congress promoted a public debate, including private and nongovernmental entities, during the PND's approval process.[6]

In light of the input of various sectors of society, another change was an understanding that it would be necessary to define strategic axes for defense, and three areas were considered strategic priorities for Brazil's national security: nuclear (the responsibility of the Navy), cyber (the responsibility of the Army), and airspace (the responsibility of the Air Force).[7]

As a result, the command of the Força Aérea Brasileira (Brazilian Air Force (BAF)), also known by its Portuguese acronym FAB) developed the Strategic Conception–Air Force 100 (DCA 11-45), with the objective of explaining, both internally and externally, the strategic guidelines of the BAF, which were intended to take it to its 100 year anniversary in 2041, in accordance with the objectives described in the PND. Therefore, the DCA 11-45 became a dynamic document as every change in the PND's political

---

4 Oliveira, "A Estratégia Nacional de Defesa e a Reorganização e Transformação das Forças Armadas," 72.

5 Almeida, "Política de Defesa no Brasil: considerações do ponto de vista das políticas públicas," 222.

6 Ibid., 223.

7 Ministério da Defesa, *Política Nacional de Defesa – Estratégia Nacional de Defesa*, 03.

objectives (which is expected every four years, immediately after the presidential and parliamentary elections) means a change in the BAF strategic guidelines.[8]

The most recent version of the DCA 11-45 was released on October 15, 2018. This chapter will refer to this version, not only because it indicates where the BAF should go, but it also directs this chapter in that it is the source for information on the current BAF structure, future budget pressures, the BAF's role in coalition operations, future threats, doctrine, and potential changes to the current systems, as well as new systems, that would enable the BAF to celebrate its 100th anniversary as a modern and effective military force.

History and Structure

Like many other air forces, the BAF was formed, in 1941, by combining the Navy and Army air branches into a single military force. Influenced by the theories of Giulio Douhet, the Aeronautical Ministry was created as an institution responsible not only for protecting airspace, but also promoting aviation in Brazil (almost nonexistent at the time), managing civil aviation regulation, building the airport infrastructure, encouraging the aircraft industry, and managing air traffic control.[9]

Over the years, as Brazil developed, many of these responsibilities were transferred either to the private sector or to other government agencies. The last major administrative reform, in 1999, transformed the Aeronautical Ministry into the current Comando da Aeronáutica (Aeronautical Command (COMAER)), one of the three commands subordinate to the Ministério de Defesa (Ministry of Defense), created in the same reform.[10]

To understand COMAER's mission, one must look to Brazil´s Constitution, which states that the military forces are "permanent and regular national institutions, organized on the basis of hierarchy and discipline, under the supreme authority of the President of the Republic, and are intended for the defense of the Homeland, the guarantee of constitutional

---

8   Comando da Aeronáutica, *Concepção Estratégica Força Aérea 100* (Brasília, 2018), 9–10.

9   Instituto Cultural da Aeronáutica, *História Geral Da Aeronáutica Brasileira*, Vol. 3 (Belo Horizonte: Itatiaia, 1984) 54.

10 Comando da Aeronáutica, Concepção Estratégica Força Aérea 100, 12.

powers and, on the initiative of any of them, of law and order."[11] In addition, two other laws (nº 97/1999 and nº 136/2010) gave the military subsidiary assignments: stimulating national development, promoting civil defense, assisting in public campaigns (vaccination, elections, epidemics), and combating border and environmental crime.[12]

Article 4 of the Constitution states that Brazil's foreign policy is based on the promotion of peace and noninterference in external affairs,[13] and the PND states that the first National Defense Objective is to "guarantee the sovereignty, the national patrimony and the territorial integrity," indicating that external operations must be have the goal of maintaining peace, regional stability, or the protection of national interests and assets.[14] Hence, it can be assumed that the BAF sees environmental events within the national territory or in areas of responsibility as a result of international agreements as a major threat, although, this threat it is not limited Brazil and surrounding nations.

The word "national" appears in BAF´s motto—"Maintain the sovereignty of national airspace and integrate the national territory with an objective to defend the homeland"—and in its vision to "become an Air Force with great deterrent capacity, operationally modern and acting in an integrated way for the Defense of the national interests,"[15] which seems to corroborate the assumption about environmental threats.

To operate in this scenario, the BAF structure comprises seven commands/departments/secretariats and a few other institutions responsible for public affairs, intelligence, flight safety, etc. A simplified structure of command is seen in Figure 14.1.

In 2018 there were around 70,800 personnel on active duty, in 12 wings across the country. From this total, around 70 percent are "career military," which means that they can remain in active service until retirement after 30 years of service, when they are able to join the remunerated reserve. In this position, they receive the same remuneration as those in active service who hold the same rank, and after their death their husband/wife continues to receive the money. The other 30 percent are "temporary military," who

11 Brasil, Constituição da República Federativa do Brasil (Brasília, 1988).
12 Instituto Cultural da Aeronáutica, *História Geral Da Aeronáutica Brasileira*, 61.
13 Brasil, Constituição da República Federativa do Brasil, 1988.
14 Ministério da Defesa, *Política Nacional de Defesa – Estratégia Nacional de Defesa*, 17.
15 Comando da Aeronáutica, Concepção Estratégica Força Aérea 100, 16.

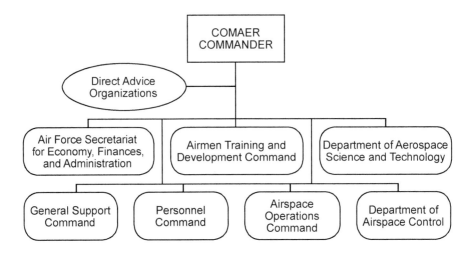

Figure 14.1   COMAER structure
Source: The author, based on Comando da Aeronáutica, "Organograma," (Brasília, 2018).

are limited to eight years in active service and then go to the unpaid reserve and maintain their rank (for example, 1st Lieutenant or 3rd Sergeant).[16]

The aging population of Brazil, caused by the decrease in fertility and increase in longevity,[17] is a social problem that is reflected in the BAF's budget because about 65,300 people receive a full salary without being on active duty (remunerated reserve and spouses of deceased military).[18] This directly affects the budgetary expenditure because, by law, the BAF´s commander cannot change the salary of personnel (active and reserve); therefore, these payments are considered an obligatory expenditure. With an increase in the life expectancy, it is not absurd to think that with the actual retirement plan many personnel will spend more time in the remunerated reserve than on active duty.[19]

---

16 Comando da Aeronáutica, "Comandante da FAB fala sobre resultado da Reestruturação em Anápolis (GO)." (Brasília, 2018), <http://www.fab.mil.br/institucional>.
17 Leandro dos Santos Bernardo, Jacir Leonir Casagrande and Adriana Bainha, "Envelhecimento da População e a Previdência Social: as possibilidades existentes de aposentadorias no Brasil e suas regras." *Ciências Sociais Aplicadas em Revista*, Vol. 17, No. 33 (2017) 145–69.
18 Comando da Aeronáutica, "*Servidores.*" (Brasília, 2018), <http://www.fab.mil.br/servidores>.
19 Comando da Aeronáutica, "Comandante da FAB fala sobre resultado da Reestruturação em Anápolis (GO)." (Brasília, 2018), <http://www.fab.mil.br/institucional>.

This can be seen in Figure 14.2, which shows that the expenditure on real estate maintenance increased, that on maintaining the operational assets and executing the missions decreased, and that on personnel costs increased from 2014 to 2016 (based on the available data).

The personnel costs are the main pressure on the BAF budget, and to solve it the DCA 11-45 proposes a solution based on changing the profile of the workforce. It is expected by 2041 that 70 percent of personnel on active duty will be "temporary military," decreasing personnel costs by the use of reserve personnel and reducing the number of "career military" personnel and thereby reducing the number of people paid as reserves. In addition, the number of higher-rank personnel will be reduced, meaning that more people will be reserves with consequent lower earnings.[20]

This will enable the BAF to allocate a larger share of its expenditure to new operational assets that will start operating in 2041. In addition, more personnel in the lower ranks will mean more specialties than are available today to the extent required (e.g., cyber, public affairs, journalism, law, logistics, psychology), which is a prerequisite for the battlefield where the BAF is expected to operate in the future.[21]

Threats and Doctrine

Military forces are expensive resources—both the price of the equipment and the hours needed training personnel to operate the weaponry available effectively. Lack of funds is a problem for every nation. A rational planning process for military forces should first answer three questions: "what?" are the threats, "where?" are the forces to be used, and "how?" are the forces to be used.

The answer to "what?" is that although the national territory is the main strategic environment, the BAF's responsibilities are not restricted to it. In Brazil, the preparation of the military forces is based on an employment hypothesis (EH) that anticipates where a military contingent could be needed in regions of strategic interest for national security reasons. In other words, it involves anticipating the major threats the nation's military

---

20 Comando da Aeronáutica, "Comandante da FAB fala sobre resultado da Reestruturação em Anápolis (GO)." (Brasília, 2018), <http://www.fab.mil.br/institucional>.
21 Ibid.

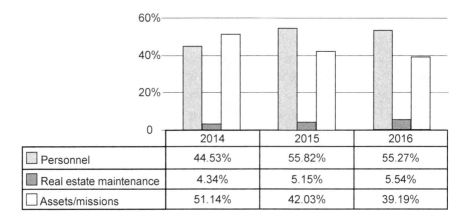

	2014	2015	2016
Personnel	44.53%	55.82%	55.27%
Real estate maintenance	4.34%	5.15%	5.54%
Assets/missions	51.14%	42.03%	39.19%

Figure 14.2   COMAER percentage of expenditure by area
Source: The author, based on Prestação de Contas Ordinária Anual – Relatório de Gestão do Exercício de 2016 [2017 BAF Staff Financial Report], Comando da Aeronáutica (Brasília, 2018), 221.

forces should expect to encounter in the future. There are four types of EH: peacetime, times of crisis, conflict/wartime, and post-conflict/wartime. EHs should take into account domestic and foreign political situations, in accordance with Brazilian national policies and legislation.[22]

EHs are confidential documents. But a generic unclassified guideline gives a good idea of the priorities of Brazilian decision makers. The following scenarios, at least, should be considered when determining the EHs:

- Threats to the sovereignty of Brazilian national airspace, borders, and jurisdictional waters
- Presence of superior military forces in the Amazon region
- Influence of conflicts abroad on domestic interests, especially in relation to chemical, biological, radiological, and nuclear threats
- Operations beyond Brazilian national borders that support the national foreign policy
- Domestic operations where the armed forces act as a police force or support the police force
- Threat of war in the South Atlantic.[23]

---

22 Ministério da Defesa, *Política Nacional de Defesa – Estratégia Nacional de Defesa*, 121–23.
23 Ibid., 122.

The answer to the "where?" question is Brazil's strategic environment, that is the South Atlantic (Antarctica and West coast of Africa included) to the south and east, the borders of the Caribbean Sea to the north, and the Amazon and Central South America to the west.[24]

In light of the "what?" and the "where," the BAF prepared a list of potential actions, to prepare troops and consider equipment acquisition, and delineate the basic doctrine of operation. The following are possible situations for the employment of BAF's assets:

- Guaranteeing the sovereignty, territorial integrity, and patrimonial defense of the country
- Providing humanitarian aid and mitigating the effects of disasters
- Preventing transnational crimes
- Acting as a police force or in support of the police forces to guarantee law and order
- Contributing to world order and peace through international commitments
- Safeguarding Brazilian assets and citizens abroad
- Maintaining neutrality in conflicts external to Brazil's strategic environment
- Operating in space missions for the benefit of the development and defense of the Brazilian state
- Dealing with cyberspace issues.[25]

It is expected that for each EH a single, joint, or combined operational command would be created, based on the forces allocated. A single operational command needs only resources from one of the three branches of the military, a joint operational command needs resources from more than one, and a combined operational command is an international command, which requires resources from more than one country (a coalition).[26]

Although potentially possible, the use of just one branch of the military should be avoided because such a type of isolated operation isn't compatible contemporary global conflicts.[27] Almost every scenario where

---

24 Ibid., 86–88.
25 Comando da Aeronáutica, Concepção Estratégica Força Aérea 100, 23–25.
26 Ministério da Defesa, *MD 30 M 01 – Doutrina de Operações Conjuntas* (Brasília, 2011), 37.
27 Ministério da Defesa, Política Nacional de Defesa – Estratégia Nacional de Defesa, 123.

the use of military forces is needed requires synergistic and joint operations to fill the capabilities gaps; the notion of a "silver bullet" weapon or ability that could alone solve an operational problem is not realistic. The real world demands personnel, weapons, and abilities working together under the same strategic objectives, subordinated to a single command.[28]

So, interoperability is not only a necessity but also a basic characteristic of the BAF since it must be ready to operate in more than one area of interest, but also prevent the enemy from doing so, as well as being able to support naval and terrestrial operations through air, space, and cyberspace operations. For interoperability to be realistic, four elements must be present in the BAF's operations: coordination and integration of command and control at different levels; complementarity between the BAF and other branch assets when operating together; real-time (or least time possible) flow of information improving situational awareness; and an updated joint operation doctrine.[29]

For the BAF in particular, answering the "how?" question, there are four strategic objectives: prioritize the airspace defense, build capabilities to assure the level of air control desired, build mobility capabilities to support ongoing operational demands, and prepare troops focusing on capabilities and not a specific enemy.

Therefore, there are six capabilities that should be developed to achieve the strategic objectives mentioned above:

1. Strategic Projection of Power: the capability to employ air power wherever it is needed, at the desired moment, and in a suitable dimension. The air power ability to deliver weaponry and assets in any area of interest must determine the type of equipment used as well as the training of those who will use it, taking into account Brazil's strategic environment.

2. Airspace Domain Superiority: the capability to control specific portions of air and space domains for a limited period and/or prevent the enemy from doing so. This is related to the basic assumption that dominance of the air and/or space over an area of interest is a

---

28 Ministério da Defesa, *MD 30 M 01 – Doutrina de Operações Conjuntas,* 17
29 Comando da Aeronáutica, Concepção Estratégica Força Aérea 100, 28–29.

crucial element in military operations, enabling effective ongoing operations.

3. Command and Control: the capability to allow the commanders, at all levels, to exercise authority over the subordinate forces and verify the results of the actions performed. Indispensable in all domains of operation, command and control is vital and therefore requires a systematic flow of information.

4. Information Superiority: the capability to collect, process, store, disseminate, produce, and protect data in the appropriate form and in the actual operational environment of interest, and at the same time prevent the enemy from doing so. The interoperability between air, land, and sea assets should first focus first on ensuring communication among the forces in the battlefield; without communication nothing effective can be done.

5. Logistic Support: the capability to provide the means to keep assets operating, by the prevision, provision, and maintenance of the resources necessary for the continuity of the campaign. Aircraft and related matériel are, by nature, fragile and depend on high technology, so the rhythm, duration, and intensity of any air campaign is directly related to efficient logistic support.

6. Force Protection: the capability to protect the personnel, matériel, installations, information, and communications against threats from nature and enemies. Aircraft are easy targets when not flying, therefore they need protection not only while flying but also on the ground. Related resources (e.g., personnel, hangars, barracks, air traffic service installations, etc.) also must remain safe from natural and man-made hazards.[30]

These skills will enable the BAF to be a more prepared air force, ready to operate in coalition either in national or multinational operations. Nevertheless, ideas without equipment are no more than an idea; therefore the materialization of those ideas is based on three programs that are under way today, which are expected to provide the technical skills needed to transform theory into practice.

---

30 Ministério da Defesa, Política Nacional de Defesa – Estratégia Nacional de Defesa, 29–32.

Control, Defend, and Integrate

In essence, what the BAF command did was to create a concept, described by the three objectives, control, defend, and integrate, to justify the acquisition of the systems necessary to acquire the skills described above. Launched in 2017, the Dimension 22 concept gives the general public the main ideas contained in the DCA 11-45. Aimed at those not familiar with issues related to air power, the form and content of the message was prepared to equate what the BAF does with the daily routine of citizens. These three words were chosen because they represent the BAF's constitutional and subsidiary assignments, and the number 22 is a reminder of the total airspace under the BAF control: 22 million square kilometers. Connected to each of these objectives is a main program, selected by the BAF to provide their desired abilities by 2041.[31]

Before discussing each of the three programs, it is necessary to understand the ideas that support them. The first objective, "control," is a reminder that the Air Traffic Service (ATS) in Brazil is under the responsibility of the BAF, including the search and rescue (SAR) missions (undertaken exclusively by the military). Because of international agreements, a huge part of the south Atlantic (around 10 million square kilometers) outside Brazil's exclusive economic zone (EEZ) is the responsibility of the BAF ATS/SAR.[32]

A slightly larger area, around 12 million square kilometers, is the airspace over continental land and the EEZ. The second objective, "defend," is connected to this area and BAF personnel should be prepared to undertake all of the missions in this zone. Its aggregate is a fighting idea because, under the Brazilian legal framework, in this region the Air Force, in addition to the ATS and SAR services, has the authority to act (and if necessary, use force) against land, naval, and air threats.[33]

Inside this larger area is the Brazilian continental zone, around 8.5 million square kilometers (half of South America's total area) with 23,000 kilometers of land border connecting Brazil to 10 of 12 South American continental countries. Since its foundation, the BAF has played a crucial role integrating the country by providing logistic support to remote cities, building and maintaining airport infrastructure in places where such

---

31 Comando da Aeronáutica. "Dimensão 22." <http:www.fab.mil.br/dimensao22/>.
32 Ibid.
33 Ibid.

activity is not profitable, and realizing all sorties of humanitarian missions (e.g., medical evacuation, transportation of organs, supporting nongovernmental organizations, evacuation of refugees, etc.). These activities justify the "integrate" objective.[34]

Space Systems Strategic Program—Control

Brazil's space program was developed to have a dual function, attending to either military or civilian needs. Created in 1961, it has experienced many changes and delays over the decades, but in 2008, with the PND, the BAF was able to assume program management and started to create a long-plan program with the intention of providing Brazil with space equipment to meet its needs. In 2012 the Space Systems Strategic Program (SSSP) was created along with an Implementation of Space Systems Coordination Commission, chaired by an air force officer, in charge of ensuring the dual use of space systems, the promotion of technological progress, and the expansion of the country's industrial base.[35]

The core of SSSP is three integrated projects designed to provide Brazil with integrated (civil and military) access to space, thus meeting the needs of the Brazilian state. The first project encompasses the two Space Operation Centers (SOCs), where the satellite and its payload can be operated according to need. Equal in control capacity in order to generate redundancy to the system, the SOC-P (P standing for primary) is in the federal capital (Brasilia) just a few meters away from the Airspace Operations Command,[36] which is responsible for leading and integrating the country's aerospace monitoring facilities.[37] The other SOC, called SOC-S (S standing for secondary) is located in a naval base in Rio de Janeiro and has the same command and control capability of the SOC located in Brasilia, but

---

34 Ibid.

35 Ministério da Defesa, MD20-S-01 – Programa Estratégico de Sistemas Espaciais (Brasília, 2018), 17.

36 José Vagner Vital, PESE/CCISE. <https://www.ontl.epl.gov.br/html/objects/_download-blob.php?cod_blob=5177>.

37 Ministério da Defesa, MD20-S-01 – Programa Estratégico de Sistemas Espaciais, 18.

the 1,000 kilometer distance between the cities is a safety factor in case the SOC-P suffers any damage, whether due to nature or man-made causes.[38]

The second project is focused on space access, and aims to deliver the capacity to put payloads into orbit through the development of domestic microsatellite launch vehicles (MLVs) and a launch center in Brazil. The first test launch of an MLV was scheduled to take place in 2019. Developed in partnership with Germany, when operational it is expected that the MLV will be able to ascend to 300 kilometers with a maximum payload of 50 kilograms. If everything is on schedule, the MLV should be declared operational in 2021 and it will not only be used by Brazil and Germany but will also be placed on the market as a commercial product. Building on the knowledge from the MLV development process, the next step is to start a family of vehicles named "Aquila," and Aquila 1 is expected to be operational by 2023 and able to ascend up to 500 kilometers with a payload of 300 kilograms, and by 2026 Aquila 2 will be launched and able to ascend to 700 kilometers with a payload of 500 kilograms.[39]

The other part of the second project is the operation of the Alcántara launch center. Because of its location near the Equator, it provides a significant cost reduction for launching into equatorial orbit, thus making it one of the best locations in the world for launch operations. In addition, it has a wide range of launch azimuth and an unobstructed flight path over the sea, which makes Alcántara commercially viable, stimulating the creation of partnerships with countries and private companies that are interested in using the launch center. The main obstacle to its exploitation lies in solving the environmental problems related to local fishing communities. It is estimated that in 2019, with an adjustment to the launch center project made by the BAF to meet the required sustainability criteria, these problems will be solved and an area capable of holding up to six launch stations will be ready to exploration.[40]

The third project of the SSSP is the creation of a fleet of satellites for civil and military use, divided into four groups, each named after species of Brazilian birds: Carponis, Lessonia, Atticora, and Calidris. When all four

38 Vital, PESE/CCISE. <https://www.ontl.epl.gov.br/html/objects/_downloadblob.php?cod_blob=5177>.
39 Ibid.
40 Ibid.

groups are operational they will be able to undertake communications, earth observation, information mapping (data gathering and relay), space monitoring, and positioning.[41]

The Carponis group is one non-geostationary satellite (NGS) aimed at providing earth observation images using optical sensors.[42] It is expected to be launched in 2021 and will be used by civil agencies to aid landuse planning and crop monitoring, by government agencies for onsite evaluation of damage or loss caused by an accident or natural event, and by the military for target acquisition and reconnaissance.[43]

The Lessonia group is another NGS satellite with similar use to the Carponis, but instead of being optical, it will be loaded with radar sensors, providing the ability to gather information in a different frequency.[44] This will be helpful in huge areas where the meteorological or surface conditions can be a problematic for optical imaging, for example in the rain forest. The Lessonia is expected to be in orbit by 2026.[45]

The Atticora group has four NGSs focusing on tactical communications (Atticora I, II, and III) and navigation by a native global position system (Atticora IV). Its first NGS kind is expected to enter into operation after 2025, and the main objective will be to provide link (communication/ position) capabilities to people on the ground by means of portable receivers (the antenna of the receptor is less than one meter in diameter).[46]

The last of the groups, the Calidris, is composed of three geostationary satellites (GES) aimed at providing strategic and operational communications.[47] The Calidris I was launched in 2017 and is fully operational and provides coverage to all of Brazil's strategic environments. The main differences between the Calidris and the Atticora are the size of the receptor antenna (it must be larger than one meter in diameter to receive the Calidris signal) and the GES/NGS characteristics. Calidris II and III are expected to be in orbit in the next decade, making Brazilian strategic

41 Ibid.
42 Ministério da Defesa, MD20-S-01 – Programa Estratégico de Sistemas Espaciais, 29.
43 Vital, PESE/CCISE. <https://www.ontl.epl.gov.br/html/objects/_downloadblob.php? cod_blob=5177>.
44 Ministério da Defesa, MD20-S-01 – Programa Estratégico de Sistemas Espaciais, 31.
45 Vital, PESE/CCISE. <https://www.ontl.epl.gov.br/html/objects/_downloadblob.php? cod_blob=5177>.
46 Ibid.
47 Ministério da Defesa, MD20-S-01 – Programa Estratégico de Sistemas Espaciais, 29–31.

satellite communication systems redundant.[48] When fully developed, the SSSP program will enable the BAF to conduct operations beyond the line of sight within its strategic environment, thereby improving its operational capabilities.

The Gripen-NG/F-39 Program—Defend

The F-X program was initiated at the turn of the century, with the aim of modernizing the BAF's combat capabilities by replacing the aged Mirage III EBR/DBR and F-5 E/F fighters, which had then been operating for about 20 years without any modernization in avionics. In 1995, the BAF's Air Staff released an internal guideline, the Preliminary Operational Demand (POD), which in turn generated the Preliminary Technical, Logistic and Industrial Requirements (PTLIR), a report in which the BAF's commands/departments/secretariat described their requirements of a new combat aircraft: its operational performance, logistical support, cost, weaponry, etc.[49]

But at the same time some other programs were occurring in parallel with the F-X: the ALX (light attack aircraft), the P-X (maritime patrol aircraft), the CL-X (medium transport aircraft), and the implementation of the Amazon Surveillance System (SIVAM). The lack of funds, the existence of multiple projects, and political problems delayed the F-X program and in 1998 the BAF decided to modernize part of its fleet of fighters, contracting EMBRAER and Elbit to upgrade the F-5 E/F in a version called F-5BR, while the Mirage EBR/DBR would remain as they were. The idea was to use the F-BR to fulfill the gap until the F-X became operation, replacing first the Mirage and later the F-5.[50]

A request of proposal (ROP) was released by the BAF in 2001, and in 2002 the rival companies in the process submitted a best and final offer (BAFO). All of the offers were presented to the policy makers, a short list created, and the winner was expected, but there were presidential elections in 2002 and the cabinet decided to leave the decision with the next president. In 2003 the new government decided to halt the program, and in 2004

---

48 Vital, PESE/CCISE, <https://www.ontl.epl.gov.br/html/objects/_downloadblob.php?cod_blob=5177>.
49 Peron, *O programa FX-2 da FAB*, 24.
50 Peron, *O programa FX-2 da FAB*, 20.

it was closed. The Mirage was retired from service in 2005 and replaced with used Mirage 2000s leased from France.[51]

In 2007 the program started again from scratch and named F-X2. At first it seemed that the previous delay would not reoccur, since the POD, PTLIR, ROP, and the BAFO occurred rapidly (compared to F-X) and in October 2008, the short list was announced. The three finalists were the Rafale, the Gripen, and the Hornet.[52] But the process crawled for five years, and in 2013 the Mirage 2000s were retired from service and replaced with F-5BRs. At the beginning of the modernization process the F-5s were operated by two squadrons, but because of the delay in choosing the F-X/X2 they were now operated by four units, highlighting the urgency for a decision on the program. Fortunately, in December of 2013, it was announced that Brazil would buy 36 Gripen-NG models from Sweden—28 single seaters (E model) and 8 twin seaters (F model)—to replace the entire fleet of jet fighters then in operation.[53]

The model chosen received the designation F-39 E/F and will have some specifics that were Brazilian demands. It will use a wide-area display developed by the Brazilian AEL Systems (subsidiary of the Israeli company Elbit) in accordance with BAF requirements. In 2018, Brazil decided to adopt the same equipment on its Gripen-NG fighters as Sweden, although Brazil will not buy the model used by Sweden but rather an aircraft developed by Brazilian engineers based on the Swedish model. For example, the twin seater is, so far, a specific Brazilian order. The manufacturing process will begin at SAAB but will gradually transfer to EMBRAER, giving Brazil the independence to change the project design if it wishes. From a strategic point of view, this is a game changer because for the first time Brazil will produce a supersonic fighter, expanding the possibilities of future projects in the aeronautical area.[54]

It is expected that the program will enhance the BAF capabilities. Being a multi-role plane, it means that the BAF will develop skills that will close some operational gaps. Today the only asset in the BAF's inventory capable of using an air-to-ground missile is its MI-35, and an air-to-sea

---

51 Ibid., 46.
52 Ibid.
53 Comando da Aeronáutica, "Tags: FX-2," <http://www.fab.mil.br/noticias/tag/FX-2>.
54 Ibid.

missile, the P-3 AM. Of the BAF's two fighter jets, only the F-5 BR can carry beyond visual range (BVR) missiles, but only the A-1M can operate designation, reconnaissance, or jammer system pods. According to the program's plan, the F-39 will be able to carry out all of these missions and new technologies, such as the possibility of operating in a network or using weapons through passive sensors such as the infrared search and tracker, will be introduced.[55]

Regarding the weaponry, the BAF official statement is that information about type and amount is classified. However, it is known that the development and integration of new weapons demands time and money, two things that the BAF (and almost any military program) does not have in abundance. So, thinking of fast and less costly solutions, it is likely that the F-39 dumb ammunition (bombs, rockets, and gun shells), which are produced in Brazil, will be used.[56]

As to sophisticated weapons, in 2015 it was revealed that Brazil intended to buy from Israel the Reccelite reconnaissance pod, the Litening targeting pod along with Spice 1000 and 250 smart bomb kits (all from Rafael Advanced Defense Industries), and the short-range infrared missiles Iris-T (from the German Diehl BGT Defense), and the A-Darter (a South African/Brazilian project).[57] The BAF already uses the Sky Shield electronic warfare pod, the Python short-range infrared missile, and the Derby BVR missile, all of which are Rafael products.[58]

So, on the basis that time and money are important and the certification flights of the F-39 monoplace are scheduled to start in 2019 and its operational debut in 2021,[59] it is plausible to say that the BAF will certify its Rafael pods, therefore their smart bombs kits, to operate in the F-39. As the A-Darter is a binational development and its integration process in the South African JAS-39 C/D is already in process, it is expected that it will be

---

55 Ibid.
56 Comando da Aeronáutica, "Munição usada pela FAB é tema da estreia do FAB & Indústria de Defesa," (Brasília, 2014), <http://www.fab.mil.br/noticias/mostra/17983/FAB-TV---Muni%C3%A7%C3%A3o-usada-pela-FAB-%C3%A9-tema-da-estreia-do-FAB-&-Ind%C3%BAstria-de-Defesa>.
57 G1, "Brasil compra 70 mísseis e bombas israelenses para armar o Gripen," (São Paulo, 2015) <http://g1.globo.com/politica/noticia/2015/08/brasil-compra-70-misseis-e-bombas-israe-lenses-para-armar-o-gripen.html>.
58 Comando da Aeronáutica, "Tag: Armas," <http://www.fab.mil.br/noticias/tag/ARMAS>
59 Comando da Aeronáutica. "Dimensão 22," <http://www.fab.mil.br/dimensao22/>.

the first choice of short-range weapon. Although the interest in the Iris-T suggests that maybe more than one type of infrared missile could be used, this is very unlikely due to the BAF's historical problem of limited funds.[60]

Concerning BVR, air-to-ground, and air-to-sea missiles, the BAF should probably choose from five possibilities: 1. develop its own (costly and prolonged); 2. enter a partnership with another country (e.g., Sweden, from which it is buying 60 Gripen E/F,[61] or a JAS-39 C/D operator); 3. buy from or join a consortium of countries; 4. deepen its relationship with Rafael (which produces all of the types of weaponry mentioned);[62] or 5. one of the above options for every specific case.

It is known that the weaponry the BAF buys is more a political than an operational choice, so the last option seems to be the most plausible as it provides financial and strategic flexibility. It also establishes solutions with different partners, avoiding all the chips being placed in a single solution. For example, Brazil could choose to retake its indigenous anti-radar missile program, retain the Derby on its arsenal, and develop a new version of the RBS-15 anti-ship missile with Sweden.

The KC-390 Program—Integrate

During the 2007 Latin America Aero and Defense (LAAD) conference, hosted in Rio de Janeiro, EMBRAER revealed its plans to develop a heavy military transport plane. At the time, the program was called C-390 and the Brazilian aerospace company understood that there would be a niche in the market for this type of airplane in the coming decades. The initial idea was to take advantage of the technology used during the development of the E-190 regional jet to build a military version of it. Initial conversations began and in the 2009 LAAD, BAF and EMBRAER announced a partnership to develop a military transport plane, now called the KC-390, with the objective of replacing the Brazilian Lockheed C-130 Hercules fleet. Later, EMBRAER became a contractor responsible for building planes as

---

60 UOL, "Forças Armadas sofrem corte de 44% dos recursos," (Brasília, 2014), <https://noticias.uol.com.br/ultimas-noticias/agencia-estado/2017/08/14/forcas-armadas-sofrem-corte-de-44-dos-recursos.htm>.
61 UOL, "FAB vai receber primeiros caças Gripen em 2021," (Brasília, 2017), <https://airway.uol.com.br/fab-vai-receber-primeiros-cacas-gripen-em-2021/>.
62 Rafael Advanced Defense Systems, "Capabilities,"<www.rafael.co.il>.

requested by the BAF, and the BAF agreed to buy at least 28 units of the plane.[63]

According to EMBRAER's own research, there would be a market demand for 700 aircraft of this class in the future and the best way to involve potential buyers would be through the creation of partnerships, as once a government engages in production, it becomes politically easier to justify the purchase to either a domestic or foreign audience. EMBRAER thought that a partnership would be necessary as the KC-390 would directly compete with the aviation legend, the C-130 Hercules, which has the support of the world's largest military power, the United States.[64]

The approach was correct because although still on the drawing board, in addition to the KC-390 serial units ordered by the BAF, EMBRAER received 32 letters of intent, 37 with other countries such as Chile (six), Colombia (twelve), Portugal (six), Argentina (four) and the Czech Republic (two). With the prototype already in the certification phase, other European and Middle East countries showed interest and requested visits and demonstration flights at their bases. In the development phase, Portugal, Argentina, and the Czech Republic became industrial partners and contributed financially to the program, in return receiving the right to produce parts of the aircraft in their respective countries.[65]

As a transport task aircraft, the KC-390 was initially built to replace the BAF's C-130 and should be able to execute its missions: air logistic transport, mid-air refueling, airbsorne drop of personnel and matériel, fire-fighting and SAR; operating from a 12,000 meter long/30 meters wide runway, from a wet/hot Amazon to a frozen ice runway in the Antarctica—all of this with a higher cruising speed, less maintenance time, longer range, and state of the art avionics and navigation systems, similar to those used in the EMBRAER E-190 E2 commercial jet series.[66] But there are some

---

63 Comando da Aeronáutica, "Tags: KC-390," <http://www.fab.mil.br/noticias/tag/KC-390>.

64 Ibid.

65 Cassio Garcia Ribeiro "Desenvolvimento Tecnológico Nacional: o caso KC-390" in *Políticas de inovação pelo lado da demanda no Brasil*, edited by André Tortato Rauen (Brasília: Ipea, 2017) 263.

66 Comando da Aeronáutica, "Tags: KC-390," <http://www.fab.mil.br/noticias/tag/KC-390>.

signs that make one wonder whether the KC-390 will be more of a multi-task platform than it seems.

The first sign was the report that the R-99, an intelligence, surveillance, and reconnaissance (ISR) platform built in 1999 on an EMBRAER E-145 regional jet and operating at the Anápolis Air Force Base (the base that will receive the KC-390), will not have a midlife upgrade.[67]

Second, in Brazil the maritime patrol mission is in the hands of the Air Force, which currently uses a modernized version of aircraft from the 1950s and 1970s: the P-95M Bandeirulha (a maritime patrol version of the EMBRAER first project, the Bandeirante) and the P-3AM (a modernized version of the P-3A Orion).[68]

Third, is the fact that the Group Kilo, the unit selected by the BAF to receive the KC-390, is composed of personnel from all fixed-wing aircraft types (fighters, reconnaissance, transport, and maritime patrol) who will oversee the verification of the existing BAF mission demands to ascertain what the KC-390 is capable of fulfilling.[69]

Further, the KC-390 will arrive with a tactical synthetic-aperture radar capable of gathering ISR information from sea and ground targets, acquiring and identifying ships, and providing terrain and weather mapping.[70] In addition, the Litening pod will also be integrated to operate from the KC-390, which is loaded with a refueling probe and will be capable of providing mid-air refueling not only to fixed-wing aircraft but also to helicopters.[71]

67 Comando da Aeronáutica, "Comandante da FAB fala sobre os resultados da Reestruturação em Anápolis (GO)," <http://www.fab.mil.br/institucional>.

68 Comando da Aeronáutica, "Tags: Patrulha," <http://www.fab.mil.br/noticias/tag/patrulha>.

69 Comando da Aeronáutica, "Novos Ares," <http://www.fab.mil.br/noticias/mostra/31692/AEROVIS%C3%83O%20-%20Novos%20Ares>.

70 UPI, "Selex ES radar chosen for Brazil's KC-390 transports," <https://www.upi.com/Defense-News/2015/06/18/Selex-ES-radar-chosen-for-Brazils-KC-390-transports/5291434655370/>.

71 Comando da Aeronáutica, "Novos Ares." <http://www.fab.mil.br/noticias/mostra/31692/AEROVIS%C3%83O%20-%20Novos%20Ares>.

Conclusion: Doctrine Changes and Challenges

The BAF expects to totally change its way of fighting by 2041. It is expected to go beyond the line of sight in communication operations, and to move from assent-centered warfare to network-centered warfare with a huge amount of information to be processed and secured.

The first doctrinal change will be related to the training process. By changing the characteristics of its personnel (from career to temporary), the military in the higher positions must be prepared to manage more than execute, and those in the lower ranks to take more initiative beyond only executing orders. Developing, integrating, and operating space systems with dual goals (civil and military) will create the need for a more flexible doctrine and where the priorities are well defined. Although that already exists in relation to ATS/SAR operations, it must be evaluate for adapted to satellite operations.

Introducing a multi-role fighter demands the creation of a multi-role doctrine, encompassing some missions not actually executed (like air to sea) and adapting the existing ones. One must also keep in mind that the F-39 is a network system, designed to operate by exchanging information with other platforms in real time. Not using this function would decrease its operational benefits, and therefore the doctrine of operation must be changed to not only use these characteristics but also to extract from them as much advantage as possible.

Finally, the KC-390 is not only a BAF project—it is a Brazilian project. Its success or failure could influence the future development of Brazil's domestic airspace industry. Air Force commanders are preparing airmen to not only operate in the known terrain but also to explore the what-if options of the project that, if not used by the BAF, could be tested as possibilities for newer versions of the plane.

References

Bernardo, L. dos S., Casagrande, J.L., and Bainha, A. "Envelhecimento da População e a Previdência Social: as possibilidades existentes de aposentadorias no Brasil e suas regras." *Ciências Sociais Aplicadas em Revista* (2017) 145–169.

Brasil Constituição da República Federativa do Brasil. Brasília, 1988.

Ribeiro, C.G. "Desenvolvimento Tecnológico Nacional: o caso KC-390." *Políticas de inovação pelo lado da demanda no Brasil*, edited by André Tortato Rauen, Brasília: Ipea, 2017.

Comando da Aeronáutica (2018). "Dimensão 22." <http://www.fab.mil.br/dimensao22/>.

Comando da Aeronáutica Concepção Estratégica Força Aérea 100. Brasília, 2018.

Comando da Aeronáutica Prestação de Contas Ordinária Anual – Relatório de Gestão do Exercício de 2016. Brasília, 2018.

Comando da Aeronáutica, "Munição usada pela FAB é tema da estreia do FAB & Indústria de Defesa," Brasília. <http://www.fab.mil.br/noticias/mostra/17983/FAB-TV---Muni%C3%A7%C3%A3o-usada-pela-FAB-%C3%A9-tema-da-estreia-do-FAB-&-Ind%C3%BAstria-de-Defesa>.

Comando da Aeronáutica, "Comandante da FAB fala sobre os resultados da Reestruturação em Anápolis (GO)." Brasília. <http://www.fab.mil.br/institucional>.

Comando da Aeronáutica, "Novos Ares." Brasília. <http://www.fab.mil.br/noticias/mostra/31692/AEROVIS%C3%83O%20-%20Novos%20Ares>.

Comando da Aeronáutica, "Organograma." Brasília. <http://www.fab.mil.br/organograma>.

Comando da Aeronáutica, "Tag: Armas." Brasília. <http://www.fab.mil.br/noticias/tag/ARMAS>.

Comando da Aeronáutica, "Tags: FX-2." Brasília. <http://www.fab.mil.br/noticias/tag/FX-2>.

Comando da Aeronáutica, "Tags: KC-390." Brasília. <http://www.fab.mil.br/noticias/tag/KC->.

Comando da Aeronáutica, "Tags: Patrulha." <http://www.fab.mil.br/noticias/tag/patrulha>.

Comando da Aeronáutica. "Dimensão 22." <http://www.fab.mil.br/dimensao22/>.

Comando da Aeronáutica. "Servidores." Brasília. <http://www.fab.mil.br/servidores>.

G1, "Brasil compra 70 mísseis e bombas israelenses para armar o Gripen." <http://g1.globo.com/politica/noticia/2015/08/brasil-compra-70-misseis-e-bombas-israelenses-para-armar-o-gripen.html>.

Instituto Cultural da Aeronáutica. *História Geral Da Aeronáutica Brasileira.* Belo Horizonte: Itatiaia, 1984.

Vital, J.V. (PESE) – CCISE. <https://www.ontl.epl.gov.br/html/objects/_download-blob.php?cod_blob=5177>.

Ministério da Defesa MD 30 M 01 – Doutrina de Operações Conjuntas. Brasília, 2011.

Ministério da Defesa MD20-S-01 – Programa Estratégico de Sistemas Espaciais. Brasília, 2018.

Ministério da Defesa Política Nacional de Defesa – Estratégia Nacional de Defesa. Brasília, 2012.

Oliveira E.R. "A Estratégia Nacional de Defesa e a Reorganização e Transformação das Forças Armadas." *Interesse Nacional* (2009) 71–74.

Peron, A.E.R. "O programa FX-2 da FAB: um estudo acerca da possibilidade de ocorrência dos eventos visados." Master´s thesis, University of Campinas, 2011. <http://repositorio.unicamp.br/jspui/handle/REPOSIP/286702>.

Rafael Advanced Defense Systems. "Capabilities." <http://www.rafael.co.il>.

UOL. "FAB vai receber primeiros caças Gripen em 2021." <https://airway.uol.com.br/fab-vai-receber-primeiros-cacas-gripen-em-2021/>.

UOL. "Forças Armadas sofrem corte de 44% dos recursos." <https://noticias.uol.com.br/ultimas-noticias/agencia-estado/2017/08/14/forcas-armadas-sofrem-corte-de-44-dos-recursos.htm>.

UPI. "Selex ES radar chosen for Brazil's KC-390 transports." <https://www.upi.com/Defense-News/2015/06/18/Selex-ES-radar-chosen-for-Brazils-KC-390-transports/5291434655370/>.

Wellington A.C. "Política de Defesa no Brasil: considerações do ponto de vista das políticas públicas." *Opinião Pública* (2010) 220–30.

15 CONCLUSION
James R. Beldon

Having collected such an informed set of insights into the political, budgetary, strategic, and doctrinal considerations of so many states' air arms, the editor of this compendium is to be congratulated. My ordering of considerations is deliberate: in an era of monumental political change and budgetary pressure, military professionals who unerringly trumpet "strategy" as being the *primus inter pares* of all factors need to take a reality check.

This book is replete with examples of where the predominance of nonmilitary (and even nonstrategic) factors determines a country's defense arrangements. However, some nations, including the United States, Russia, China, the United Kingdom, and France, do at least proclaim to have a strategy-based foundation for their capabilities—even if those capabilities don't always match their national ambition. "Doing more with less" has been a mantra for the British armed forces for a generation, and as the editor highlighted in his introduction, is a factor that has not been limited to the United Kingdom in the post–Cold War era. While the reemergence of Russia as an expeditionary player in the Middle East and as a menace on the North Atlantic Treaty Organization's (NATO) eastern flank has caused western nations to awake from their 25-year strategic slumber, only slowly are NATO nations beginning to work toward fulfilling the spending requirement of 2 percent of gross domestic product (GDP) on defense. And what does that headmark figure really mean anyway? As the Brazilian case study shows, if personnel costs (including pensioners or fully salaried "remunerated reserves") continue to spiral and absorb ever more of the percent GDP allocated to defense, how does an arbitrary percent GDP figure actually translate into military capability? In countries where salaries and pensions are lower, does that mean that more investment can be made into technology, mass, and training? Such political and budgetary shorthand for determining whether a nation's investment in defense is healthy or not is misleading and strategically illiterate. The disappointing reality for strategists is that it is in the political and budgetary domains (rather than a strategic one) that decisions on how a state allocates its resources are mostly made. The challenge for military leaders and strategists is to ensure that strategy is translated in such a way that it can be interpreted meaningfully by those who can only speak fluently the languages of politics and budgets.

Happily, the decline of defense budgets in many countries seems to have abated (whether a temporary respite or otherwise remains to be seen). So perhaps the challenge now for all air forces—which are inherently expensive owing to the technology involved—is not so much to "do more with less," but to "do better with what we have." As discussed above, personnel costs impact on the resources available to invest in the development and procurement of more advanced technologies, a factor that contributes to the growing desirability of unmanned systems, especially as their capabilities are expanding through digitization, networked systems, and artificial intelligence. As the U.K.'s former Commander of Joint Forces Command, General Sir Richard Barrons, commented, in an unmanned system "[y]ou have a machine that doesn't get bored, that doesn't have to be replaced when it resigns early, that never has a pension, or [needs] a hospital."[1]

Embracing the potential of unmanned systems will undoubtedly be a major feature—if not the predominant feature—of the intelligence and combat capabilities of the more advanced air forces in the future, but the human operator, maintainer, planner, and strategist is far from near extinction. Take, for example, the Franco-German and British decisions to develop sixth-generation combat aircraft—with manned and unmanned options currently under development. Through its Tempest project, the United Kingdom is developing a blended "system of systems" including manned elements and "loyal" autonomous wingmen. This approach provides a useful signpost toward an era in which new technology enabled by artificial intelligence and autonomy will be increasingly exploited, while retaining the superior flexibility that humans will continue to have over machines for the foreseeable future to deal with multifarious and unpredictable

---

1 Ben Farmer, "Prepare for Rise of 'Killer Robots' Says Former Defence Chief," *Daily Telegraph*, Aug. 27, 2017, <https://www.telegraph.co.uk/news/2017/08/27/prepare-rise-killer-robots-says-former-defence-chief/>.

situations. As highlighted by several of this collection's authors, there is growing reluctance to put highly trained and valuable personnel in harm's way, so the delegation of the dull, dirty, and dangerous jobs to uninhabited air (and other vehicles) has an obvious attraction too—especially in an era when anti-access/area denial (A2/AD) capabilities are becoming more prevalent and potent. An important factor which must, however, be confronted by politicians and military professionals alike is to resist the temptation to lower the threshold of conflict through the use of uninhabited vehicles—a point articulately made by Ron Gurantz in the opening chapter.

This volume provides a particularly refreshing insight into non-western air force developments and ambitions, informed by the historical factors that have brought the air forces to their current state. It seems to be the case that the air forces of the developing world and China have been regarded as the Cinderella services in their nations—owing principally to the preponderance of political power enjoyed by the army in those countries. For the older and most advanced air forces, this seems not to be such a factor, with more equitable and reasoned allocations of resources to the armed forces of those countries. In some countries, such as Kenya and Indonesia, the involvement of their air forces in coup attempts seems to have exerted a long-term spell in the "dog house," with a skewing of national resources away from them as a consequence. Such factors reinforce the political, rather than strategic factors that dominate force structures and capabilities. Even in the most advanced nations, the effects of politics on procurement decisions is palpable, not least in Europe where the effects of Brexit are already exerting profound consequences, as evidenced by the ejection of Britain from the European Union's Galileo global satellite navigation project. Furthermore, it can be inferred that Brexit played a significant role in Germany and France, on the one hand, and Britain, on the other, forging an alliance in the development of sixth-generation fighters—an enterprise that otherwise should have represented a golden collaborative opportunity for Europe's three most advanced aerospace nations. Italy is currently assessing its options, but is a nation (Australia is another) that has been commendably adept in delivering a highly effective and modern air force based on its strategic imperatives, although political winds continue to menace its future capabilities.

Thankfully, despite the problems being thrown up by Brexit, NATO looks secure as an organization, and provides a unique cohering function for the strategy, doctrine, and capabilities of western European and North American air forces, as well as providing the procedural baselines for other multinational operations, such as the 80-nation Global Coalition against Daesh. NATO remains the world's most powerful incubator of interoperability between nations and components—conferring on the western allies an operational (as well as political) advantage that is often undersold. It should be remembered that however powerful the Chinese and Russian militaries may appear to be, neither has meaningful allies (although there is evidence of growing cooperation between Russia and China themselves), and therein lie some political as well as operational disadvantages when compared with their western competitors.

The changing tides of international politics are well documented throughout this volume. Taiwan's experience provides an extremely apposite example. For those countries reliant on outside sources for their aerospace capabilities, shifting strategic sands have presented many challenges over the years, with alliances of convenience all too often giving way to arms embargoes resulting in a feast or famine situation that has resulted in several nations having to hedge their bets by purchasing a range of different (and sometimes duplicative) capabilities from multiple states. India and Pakistan head the list of such countries, both of which seem encumbered by a mixture of capabilities sourced from opposing countries, with the inherent cost, interoperability, and efficiency penalties that such an approach—however necessary—imposes. None of this helps to promote the emerging doctrine of multi-domain operations that seek to synchronize effects in the land, maritime, air, cyber, and space domains to overwhelm an opponent's ability to respond effectively—in essence, by paralyzing an opponent by presenting insoluble operational dilemmas across multiple domains and axes all at the same time. Air forces that are not in themselves integrated from an intelligence and information perspective, let alone from a joint services perspective, will be at a huge disadvantage compared with foes that are able to orchestrate synchronized operational effects across all domains simultaneously.

Finally, while acknowledging the political and budgetary factors that enable and constrain an air force's structure, capabilities, and effectiveness, it will always be down to the professional airman and woman to deliver the air and space effects demanded by their nations in the preservation of peace or in the execution of combat operations. To that end, it is essential that the conceptual underpinnings are in place to ensure that the right doctrine is employed to ensure success, and this is very much the responsibility of senior air force leaders (and their army and naval counterparts) to get right. As Air Commodore Peter Gray highlighted in his essay "The Thin Blue Line," the Royal Air Force has sensibly continued to invest in the intellectual capital of its personnel, which together with its constant testing in the crucible of war, has delivered to the United Kingdom an air force that is highly effective across all the air power roles, and is set to remain so despite the political influences connected with Britain's relationship with Europe. Australia's Plan Jericho provides an excellent strategy for the delivery of a lean and capable fifth-generation air force that embraces multi-domain operations at its heart. Air forces, both large and small, can learn much from the Royal Australian Air Force's approach—not least the power that intellect, reason, and strategy can still exert in an era dominated by politics and budgetary competition.

Reference

Farmer, B. "Prepare for Rise of 'Killer Robots' Says Former Defence Chief," *Daily Telegraph*, Aug. 27, 2017. <https://www.telegraph.co.uk/news/2017/08/27/prepare-rise-killer-robots-says-former-defence-chief/>.

INDEX